U.S. Marine Corps Wilderness Medicine Survival Course

Plus

U.S. Marine Corps Physical Readiness Training for Combat

Department of Defense

Pentagon
Publishing

UNITED STATES MARINE CORPS
Mountain Warfare Training Center
Bridgeport, California 93517-5001

WILDERNESS MEDICINE COURSE
STUDENT HANDBOOK

TABLE OF CONTENTS

TOC

TABLE OF CONTENTS (CONT.)

UNITED STATES MARINE CORPS
Mountain Warfare Training Center
Bridgeport, California 93517-5001

FMST.07.18
04/02/02

STUDENT HANDOUT

MOUNTAIN SAFETY (SUMMER)

TERMINAL LEARNING OBJECTIVE: Given a unit in a wilderness environment and necessary equipment and supplies, apply the principles of mountain safety to prevent death or injury per the reference. (FMST.07.18)

ENABLING LEARNING OBJECTIVE: Without the aid of references and given the acronym "BE SAFE MARINE", list in writing the 12 principles of mountain safety, in accordance with the references. (FMST.07.18a)

OUTLINE

1. **PLANNING AND PREPARATION:** (FMST.07.18a) As in any military operation, planning and preparation constitute the keys to success. The following principles will help the leader conduct a safe and efficient operation in any type of mountainous environment. We find this principle in the acronym "BE SAFE MARINE". Remember the key, think about what each letter means and apply this in any type of environment.

    B - Be aware of the group's ability.

    E - Evaluate terrain and weather constantly.

    S - Stay as a group.

    A - Appreciate time requirements.

    F - Find shelter before storms if required.

    E - Eat plenty and drink lots of liquids.

    M - Maintain proper clothing and equipment.

1

A - Ask locals about conditions.

R - Remember to keep calm and think.

I - Insist on emergency rations and kits.

N - Never forget accident procedures.

E - Energy is saved when warm and dry.

a. Be Aware of the Group's Ability. It is essential that the leader evaluates the individual abilities of his men and uses this as the basis for his planning. In his evaluation, the leader must include the group's overall physical conditioning, and the consideration of change in climate and how long the unit has had to acclimatize.

(1) Mental attitude of your group. Is morale high? How much tactical training has the group had in a particular type of terrain?

(2) Technical aspect of your group. Have they been on skis, snowshoes, etc.?

(3) Individual skills. At this point, you must choose who is most proficient at the individual skills that will be required for your mission, navigation techniques, security, call for fire, rope installations, track plans, bivouac site selection, skijoring, etc.

b. Evaluate Terrain and Weather Constantly.

(1) Terrain. During the planning stages of your mission, the leader must absorb as much information as possible on the surrounding terrain and key terrain features involved in your area of operation. Considerations to any obstacles must be clearly planned for. Will you need such things as fixed ropes, rope bridges, climbing gear, etc?

(a) Stress careful movement in particularly dangerous areas, such as loose rock and steep terrain.

(b) Always know your position. Knowing where you are on your planned route is important.

(2) Weather. Mountain weather can be severe and variable. Drastic weather changes can occur in the space of a few hours with the onset of violent storms, reduced visibility, and extreme changes. In addition to obtaining current weather data, the leader must plan for the unexpected "worst case". During an operation he must diagnose weather signs continually to be able to foresee possible weather changes.

(a) Constantly evaluate the conditions. Under certain conditions it may be advisable to reevaluate your capabilities. Pushing ahead with a closed mind could spell disaster for the mission and the unit.

c. Stay as a Group. Individuals acting on their own are at a great disadvantage in this environment.

(1) Give the unit adequate rest halts based upon the terrain and elevation, physical abilities of the unit, combat load and mission requirements.

(2) Remember to use the buddy system in your group.

(3) Maintain a steady pace so that it will allow accomplishment of the mission when all members of the unit reach the objective area.

d. Appreciate Time Requirements. Efficient use of available time is vital. The leader must make an accurate estimate of the time required for his operation based on terrain, weather, unit size, abilities, and on the enemy situation. This estimate must take into account the possibility of unexpected emergencies and allow sufficient leeway to make unplanned bivouacs in severe conditions.

(1) Time-Distance Formula (TDF). This formula is designed to be a guideline and should not be considered as the exact amount of time required for your movement. Furthermore, this formula is for use in ideal conditions:

3 km/mph + 1 hour for every 300 meters ascent; and/or + 1 hour for every 800 meters descent.

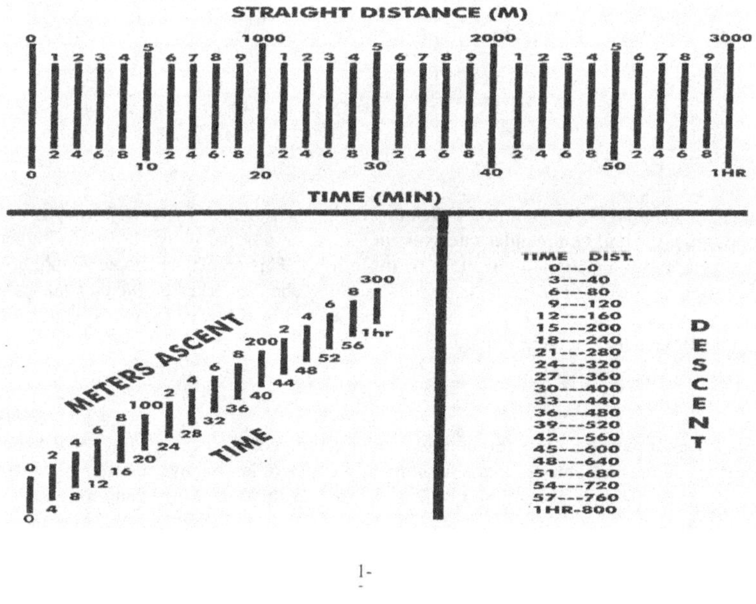

(2) Route Planning. Route cards are not to be used in place of an overlay, but as a tool to be used in route planning. Overlays/Route cards should contain the following information at the minimum:

Unit Designation:
Unit Commander.
Number of personnel.
Inclusive dates and times of movement.
Grid coordinates of each checkpoint and bivouac.
Map references.
Azimuth and distances for each leg.
Elevation gain/loss per leg.
Description of the ground.
ETA and ETD.

ROUTE CARD

UNIT I.D.		UNIT COMMANDER		NUMBER OF PERSONNEL		DATE AND TIME	MAP REFERENCE
LEG	AZM	DIST	GRID	ETA	ETD	ELEVATION	DESCRIPTION

(3) As in any military operation, route planning and execution are of vital importance. Prior to departure, the unit commander must submit a route card or patrol overlay to his higher headquarters and keeps a duplicate copy for himself. This preplanned route should be followed as closely as possible, taking into account changes based on the tactical situation. In non-tactical situations, the preplanned route should be followed to reduce search and rescue time in an emergency situation.

e. Find Shelter before Storms if required. Under certain conditions, inclement weather can provide tactical advantages to the thinking unit commander, but by the same token it can reduce the efficiency of a unit to nil if an incorrect evaluation of the situation is made.

(1) Bivouac. If the group decision is to bivouac, then it's vital that we know the principles for an unplanned bivouac.

(a) Unplanned bivouac. The principles and techniques discussed here apply both to unplanned and tolerated bivouacs. In any survival situation, especially in a mountainous environment, the most immediate danger is from exposure to the elements. Being lost will not directly kill an individual. Starvation takes time, but hypothermia can manifest itself in a matter of hours resulting in death. Adhering to the following principles will give an individual the best chance to spend a relatively safe bivouac with the prospect of continued effort toward mission accomplishment.

1. Make shelter. The requirements for expedient shelters and the building procedures will be covered in another section. The basic requirement for protection from the elements is essential.

2. Keep warm. The retention of body heat is of vital importance; any action in which body heat is lost should be avoided. The following points should be considered:

a. Adequate shelter.

b. Insulation from the ground using branches, a rucksack, etc.

c. Wear extra clothing.

d. Use extra equipment for insulation.

e. Produce external heat while trying to conserve fuel for future use.

3. Keep dry. Being wet causes the loss of body heat 24 times faster than when dry. Adequate protection from the elements is of prime importance to prevent the onset of hypothermia.

f. Eat Properly and Drink Plenty of Fluids

(1) Food. The human body can be compared to a furnace, which runs on food to produce energy (warmth). By planning the consumption of food to suit the specific situation, adequate nutrition and extra warmth can be supplied.

(2) Water. The intake of adequate amounts of water will maintain the body in proper working order. Danger from dehydration is as high in mountain regions as in hot dry areas. Loss of liquids is easily seen and felt in hot climates; whereas in the

mountains, the loss of body fluids is much less noticeable. High water intake, at least 6 quarts per day when in bivouac, 8 quarts per day when active, will help to prevent dehydration.

g. Maintain Proper Clothing and Equipment

(1) Clothing

(a) Our clothing has to perform an important function in our mission; therefore, when choosing our clothing we have to take into account some essential requirements.

1. Protection against wind and rain.

2. Layered and easily adjustable.

3. Lightweight and durable.

(b) To help us remember how to maintain and wear our clothing, we use the acronym, "COLD".

C - Keep clothing Clean.

O - Avoid Overheating.

L - Wear clothing loose and in Layers.

D - Keep clothing Dry.

NOTE: In the mountains a man should never be separated from his gear. Here are some basic and essential items that should be considered during your planning stage.

(2) Required equipment:

(a) A daypack should always be carried per squad and one Marine should carry a combat load.

(b) Map and compass. Every individual in a leadership position and his assistant should carry a map and compass. The maps should be weatherproofed and extra maps should be distributed throughout the unit.

(c) Repair kit. This kit should include those items necessary to do emergency repairs on your equipment.

(d) Survival Kit. Always carried on your person. The contents of a survival kit will be covered in another period of instruction dealing specifically with survival kits.

h. Ask Locals About Conditions. An often-overlooked source of information is the indigenous population of an area. Local weather patterns, rockslide areas, watering points, and normal routes can all be obtained by careful questioning. The leader must obtain current information of the actual conditions along his intended route. Of particular importance are recent precipitation and enemy sightings.

i. Remember to Keep Calm and Think.

(1) Emergency situation. Having recognized that you are lost and that an emergency situation exists, the following principles should be followed:

(a) Keep calm and do not panic. At this point you must make every effort to conserve body heat and energy.

(b) Think. When an individual is cold, tired, hungry or frightened he must force himself to organize his thoughts into a logical sequence.

(c) The group must try to help itself by either finding the way back to safety or by preparing shelters and procuring food.

(d) Above all else, the group must act as a tight-knit unit. In emergency situations, individual dissension can cause a total loss of control and unit strength.

(2) If the decision is reached that the group should seek its way back to safety, several possibilities exist. In most situations, the safest approach will be to retrace the route to the last known point and continue from there. The other course of action is to get a group consensus on the present location and send out a small search party to locate a known point. This party must ensure that they mark their trail adequately to return to the group. If all attempts at finding a way back to known terrain fails, a definite survival situation exists and actions discussed later in this section must be instituted.

j. Insist on Emergency Rations and Kits. Just like what was covered in the SUMMER MOUNTAIN WARFIGHTING LOAD REQUIREMENTS class, survival rations and a survival kit should always be carried.

k. Never Forget Accident/Emergency Procedures.

(1) Causes of accidents. The general procedures used to handle accidents differ little in this environment, but several distinct points should be kept in mind. The most frequent causes of accidents are as follows:

(a) Overestimation of physical and technical abilities.

(b) Carelessness.

(c) General lack of observation of one's surroundings.

(d) Lack of knowledge and experience by leaders.

(e) The failure to act as a group.

(f) Underestimation of time requirements to move through mountainous terrain and underestimation of the terrain itself.

(2) Preventive measures. The only truly effective preventive measures for the above lie in the education and experience of leaders at all levels. Too often, leaders sit by watching during training and as a result have no concept of the requirements involved in the mountainous environment. Only by active involvement can a leader gain the knowledge and experience needed to effectively lead in this environment.

(3) General procedures for handling an accident. These require only a good dose of common sense as outlined below.

(a) Perform basic first aid.

(b) Protect the patient from the elements to include insulation on top and bottom.

(c) Evacuate if necessary.

(d) Send for help if required. If possible, never send a man for help alone.

(e) Send the following information regarding the accident:

1. Time of accident.
2. Nature and location of accident.
3. Number injured.
4. Best approach route to accident scene.

(4) If one man of a two-man team is injured, the injured man must be given all available aid prior to going for help. If the injured man is unconscious, he should be placed in all available clothing and sleeping gear and anchored if on steep terrain. A note explaining the circumstances, and reassuring him, should be left in a conspicuous spot. This note must also contain the following information:

(a) When you expect to return.

(b) Where you went.

(c) What you did before you left (medication, etc.).

(5) International distress signal (whistle):

(a) Six short blasts in 1 minute from person requesting help.

(b) The return signal is three blasts in 1 minute from the respondent.

(6) Other methods if help is required:

(a) Red pyrotechnics.

(b) SOS, (... --- . . .).

(c) "Mayday" by voice communications.

l. Energy is saved when warm and Dry. With the previous 11 principles in mind this one should fall right into place. Save your heat and energy by following these steps:

(1) Dress properly.

(2) Eat properly.

(3) Drink properly.

(4) Ensure shelter meets criteria.

(5) Produce external heat (fires, stove, extra clothing, etc.) to save body heat and energy for future use.

(6) Don't lose body heat by getting wet.

UNITED STATES MARINE CORPS
Mountain Warfare Training Center
Bridgeport, California 93517-5001

FMST.07.01
03/08/02

STUDENT HANDOUT

NUTRITION

TERMINAL LEARNING OBJECTIVE: Given a unit in a mountainous environment, and the necessary equipment and supplies, apply the principles of nutrition in a mountainous environment to prevent death or injury per the reference. (FMST.07.01)

ENABLING LEARNING OBJECTIVES:

(1) Without the aid of references and from a given list, choose the correct caloric intake needed per day in a mountainous environment, in accordance with the references. (FMST.07.01a)

(2) Without the aid of references and from a given list, choose which type of diet increases your tolerance to cold, in accordance with the references. (FMST.07.01b)

(3) Without the aid of references and from a given list, choose one of the preventive measures of excessive water loss in a mountainous terrain, in accordance with the references. (FMST.07.01c)

OUTLINE.

1. **STARVATION FACTORS.** Factors that combine to result in "Near Starvation" during military operations in any mountainous environment:

   a. Difficulty in carrying sufficient food.

   b. Lack of time for preparing meals and beverages.

   c. Loss of appetite and low food palatability resulting from hypoxia/hypothermia.

   d. Fat intolerance resulting from altered metabolism.

2

2. **PHYSIOLOGICAL ASPECT TO COLD STRESS.**

a. Man's adaptation to cold is essentially behavioral. Shelters and clothing are utilized to provide a micro-environment, which will allow survival. Body composition adjustments such as increases in body weight and fat are also examples of the body's adaptation to the cold. The winter months lend themselves to greater periods of inactivity, leading to these changes.

b. In the Falkland War, constipation was a major concern. Dehydration was found to be the culprit in the majority of cases, but diet had a significant role as well. MRE's are low in fiber, containing less than 10 grams. Man needs at least 31 grams of fiber per day to ease passage of stool. Constipation can be a debilitating condition, affecting overall unit readiness.

(1) Low fiber diets can lead to diverticulosis, diverticulitis and Irritable Bowel Syndrome (IBS). IBS flare-ups tend to occur during upsetting situations and emotional stress. Educate your troops on the importance of fiber, especially in a cold weather environment.

(2) Fiber is supplied by foods derived from plants. The best sources are:

(a) Whole grain cereals.

(b) Whole wheat breads.

(c) Fresh and dried fruits.

NOTE: Vegetables also help prevent constipation, but the fiber from whole grain cereals and whole wheat bread is more effective than vegetables.

3. **CALORIC REQUIREMENTS.** (FMST.07.01a) Recommended caloric intake (Naval Medical Command Instruction 10110.1) for troops operating in a cold weather environment is 4,500 calories per day.

a. Energy Expenditure. Military operations in a cold weather environment usually involve strenuous activity. Walking on snow expends roughly twice the energy that walking on hard ground does. If troops are unfamiliar with personal over-snow mobility equipment, i.e. snowshoes and skis, awkward movement and thus wasted energy will result.

(1) The heavy clothing necessary for functioning in the cold increases energy expenditures by 5-15%. You will have the opportunity to experience this effect in the upcoming field evolution.

(2) Properly clothed, experienced individuals should have negligible heat loss from radiation, conduction, and convection. However, losses associated with respiration and evaporation can be significant.

(3) In intense cold, particularly with high winds, heat loss to the environment may be inevitable.

4. **NUTRITION IN A HIGH ALTITUDE ENVIRONMENT.** A diet high in carbohydrates and high in fat is the most beneficial, while a high protein diet is generally the least desirable.

a. The marked weight loss seen in individuals at high altitude is largely due to increased metabolic demands and anorexia secondary to hypoxia. Total caloric intake commonly does not meet energy demands.

b. Altered metabolic processes that have been reported at high altitude include delayed gastric emptying, intestinal malabsorption and decreased gastric enzyme secretion.

5. **BEST DIET FOR COLD WEATHER**. (FMST.07.01b)

a. High carbohydrate and high fat foods should be included rather than high protein foods.

b. Small meals (snacks) spaced at time intervals of two hours or less should be consumed rather than the usual three meals-a-day.

c. During a typical eight-hour day of cold exposure, three high-fat (unsaturated fat) meals providing 60% of the day's calories, coupled with two normally scheduled meals (breakfast and dinner) should provide good cold tolerance.

6. **BASIC CONSTITUENTS OF FOOD**. The three basic constituents of food are carbohydrates, fats and proteins.

a. Carbohydrates. Also known as the quick energy foods that provide energy to produce heat with the byproducts of carbon dioxide and water when oxidized by the body.

(1) Carbohydrates are present in food mainly as sugars and starches and are broken down during digestion into simple sugars that are converted into glucose.

(2) Carbohydrates are stored as glycogen in the liver and muscles and can be broken down quickly into glucose to provide a rapid source of energy. However, these stores are not large and are markedly depleted by fasting for as little as 24 hours.

(3) Eating a high carbohydrate diet for several days will double one's glycogen stores and can increase endurance by up to three times compared to the ordinary balanced diet.

(4) The main carbohydrate food sources are: CARBOHYDRATE SOURCES

(a) Fruits.

(b) Vegetables.

(c) Cereals.

(d) Sugar.

b. Fats: In the body, fat serves as the main storage form of energy. One gram of fat has the equivalent of 9.3 Cal/gm as opposed to 4.1 Cal/gm for carbohydrates and proteins. They also produce energy, heat, CO_2 and H_20.

(1) In cases of starvation, body fat tends to be broken down into acidic compounds. If these compounds accumulate in the blood faster than they can be burned, they cause the body tissues and blood to become excessively acidic (acidosis).

(2) During brief physical activity, energy is derived equally from fat and carbohydrates, but as the duration of the activity lengthens, the percent of energy supplied by fat increases as carbohydrate energy is quickly depleted.

(a) Fat sources:

1. Butter.
2. Lard.
3. Cooking oil.
4. Mayonnaise.
5. Ice cream.

FAT SOURCES

(4) Fat is also found in lesser amounts in:

(a) Dairy products. (d) Nuts.

(b) Meat. (e) Vegetables.

(c) Eggs. (f) Cereals.

c. Proteins: Protein is a reparative food of complicated molecules composed of chains of amino acids.

(1) The protein structure is made up of 25 different amino acids. Eleven of these are important for growth while eight are essential for normal tissue maintenance.

(2) Common protein food sources are:

PROTEIN SOURCES

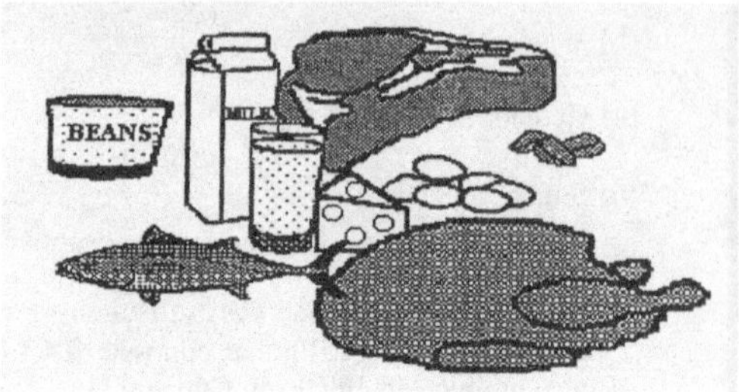

(a) Eggs.

(b) Dairy products.

(c) Meat.

(d) Poultry.

(e) Fish.

(f) Legumes (peas and beans, nuts and cereals with animal products, especially eggs, being better sources of complete protein).

(3) A pure protein diet may cause fatalities in 3-8 weeks from "rabbit starvation". This refers to those who have tried to live on relatively fat-free rabbit meat only. Fat must be a part of a normal diet.

7. **PERCENTAGE OF FATS, PROTEINS AND CARBOHYDRATES IN A SUMMER MOUNTAINOUS ENVIRONMENT**.

a. Requirements:

(1) 20% Protein.

(2) 35% Fats.

(3) 45% Carbohydrates.

b. When troops move to higher altitudes and/or are doing strenuous activity, carbohydrates should definitely be increased to at least 70 - 80% of energy intake. (Experienced mountaineers find carbohydrates more palatable at altitude).

8. **VITAMINS.** Vitamins are essential to the metabolic functioning of the body and are even more important in a cold weather environment because of an increase in metabolism and stress. Because the body cannot make vitamins, they are supplied in the food we consume. A daily multivitamin tablet supplement is recommended in a cold weather environment, but mega doses (more than 10 times the RDA) may be harmful.

a. Water-soluble vitamins are not stored in the body and need to be consumed daily. They are not stable to cooking and oxidation.

(1) B complex, Niacin, Pantothenic acid, Biotin, Choline, Folic Acid and C. (Note: Vitamin C facilitates absorption of iron from food.)

b. Fat-soluble vitamins can be stored in the body and are more stable to cooking and oxidation.

(1) A, D, E and K.

9. **FOODS THAT MAKE YOU FEEL WARM**.

a. Hot Beverages. Increases peripheral vasodilatation.

b. Adequate Dietary Iron. Mild iron deficiency increases chilling, thus decreasing cold tolerance.

c. Fat in the Diet. Fat slows gastric emptying and anecdotal evidence suggests less chilling and more restful sleep at night.

d. Carbohydrates. Keeps energy levels up. "Tired" is associated with "Cold".

e. Spices. Red pepper (Tabasco sauce) and monosodium glutamate increase peripheral vasodilatation, while capsicum and capsaicin, which are found in Cayenne, produces the same result when applied externally to the feet.

f. Water. Dehydration leads to lethargy, which tends to reduce food intake resulting in less heat production.

10. **INCREASING COLD TOLERANCE**.

a. Frequent Feedings. A high carbohydrate diet is superior to a high protein diet in increasing tolerance to cold.

(1) Diets deficient in both calories and protein reduce physical work capacity in the cold. The provision of both components is required to maintain work capacity.

11. **HIGH FAT DIETS AND SUSCEPTIBILITY TO FROSTBITE**.

a. High fat diets consumed at frequent (2 hour) intervals, maintain body temperatures better than high carbohydrate diets, but not if normal (4 hour) intervals are observed between meals.

b. Fat enriched diets lead to a propensity toward increased arterial thrombosis and increased platelet aggregation, which translates into increased potential for thrombosis.

c. Frostbite injuries are associated with disruption of the microcirculation, progressive micro vascular thrombosis, alteration of platelet function, and fibrin deposition in affected areas.

d. Therefore, diets high in (saturated) fat may predispose to frostbite.

CAUTION! A high-fat diet is important in a high-altitude, cold-weather environment. The benefits associated with the high-energy source called Fat far outweigh the risks of arteriosclerosis from the diet itself. A high-fat diet is appropriate in this specific environment during the time frame of cold, high-altitude exposure. In garrison, at sea level, a high fat diet should be avoided in order to maintain a healthy lifestyle.

12. **IRON DEFICIENCY AND THERMOREGULATION**.

a. Iron deficiency results in 9% less heat production.

b. Consuming only 1/3 of the RDA for iron results in a 29% greater heat loss during cold exposure. "The furnace is going, but the windows are open" is a phrase that helps describe this effect.

c. Iron absorption is increased by consuming iron rich foods:

(1) Lean meat.

(2) Poultry.

(3) Beets.

(4) Beans.

(5) Green leafy vegetables.

(6) Fish.

(7) Vitamin C.

d. Bottom line, IRON improves the body's response to cold.

13. **WATER REQUIREMENTS**.

a. On average, water makes up about 60% of the one's entire body weight. The average sedentary person excretes about 2.7 quarts of water each day. Of this total, 1.3 quarts are lost in the urine, 1.1 quarts through the skin and lungs and 10 ounces in the stool.

b. At high altitude, and during strenuous exercise, the amount of water lost through the skin and lungs due to the cold, dry air increases greatly.

c. Cold weather decreases the sense of thirst, which may lead to a state of chronic, mild dehydration. At temperatures below freezing and at elevations above the snow line, the lack of liquid water and the time and effort required to melt snow compound the problem.

d. Prevention of excessive water loss: (FMST.07.01c)

(1) Force fluids. Drink 6-8 quarts of water daily.

(2) Observe for dark urine and other symptoms of dehydration. (Headache, nausea/vomiting, dizziness, constipation, etc.)

(3) Limit caffeine to 1-2 cups per day.

(4) Do not rely on thirst as an indicator of hydration status.

NOTE: Eating snow or ice will cause extreme inflammation of the mucous membranes of the mouth after two days. Warm canteens of water placed in your sleeping bag will give you water to drink in the morning. They will initially act as hot water bottles. (Place them upright, as they tend to leak ...keep bag dry.)

e. Health Problems Associated with Excessive Water Loss:

(1) Dehydration.

(2) Constipation.

(3) Hypothermia.

(4) Frostbite.

(5) Heat Exhaustion.

f. The importance of hydration is essential while in a cold weather environment. If these guidelines are followed then complications arising from poor hydration should be prevented.

14. **SPECIAL PURPOSE MILITARY RATIONS**.

RATION	APPLICATION
Ration, Cold Weather, RCW.	Cold weather operations where weight and freezing are a problem.
MRE Supplement Packet.	Cold weather calorie supplements for the MRE.
Ration, Light Weight, RLW.	Long range patrols, reconnaissance, and assault.
Nutrition Sustainment Module, NSM.	Calorie dense food, bar ration for mission lasting 3-5 days where soldier must carry his own food.

Family of Operational Rations, FOR.	Modular rations that allow an optimum mix of meals for tactical operations under all climatic conditions.

15. **FIELD STRIPPING OF MRE'S.**

a. Tips on how to conserve space and weight:

(1) Remove the entire contents from the plastic wrapper.

(2) Remove the cardboard, and all materials that you will not use.

(3) Be sure to place the toilet paper and other materials from the accessory pack either back in your pocket or into the brown packaging.

(4) Carefully repack the stripped meal back into the wrapper and seal with tape.

(5) Your meal will be half its original size and two-thirds its original weight.

UNITED STATES MARINE CORPS
Mountain Warfare Training Center
Bridgeport, California 93517-5001

FMST.07.09
03/31/02

WILDERNESS PATIENT ASSESSMENT

TERMINAL LEARNING OBJECTIVE: Given a casualty in a cold weather environment and the necessary equipment and supplies, perform a patient assessment in a mountainous environment to determine the nature and extent of injury per the reference. (FMST.07.09)

ENABLING LEARNING OBJECTIVES:

(1) Without the aid of references, from a given list, choose when CPR may be legally discontinued, in accordance with the references. (FMST.07.09a)

(2) Without the aid of references and from a given list, choose which situations CPR may not be started in a remote wilderness environment, in accordance with the references. (FMST.07.09b)

(3) Without the aid of references, in the proper sequence, correctly perform a patient assessment in the proper sequence in accordance with the references. (FMST.07.09c)

OUTLINE

1. **INTRODUCTION.**

   a. An organized prioritized assessment of a wilderness trauma victim is essential to proper care. A modified Advanced Trauma Life Support (ATLS) assessment system will be introduced in this chapter.

2. **TRIAGE.**

   a. Two situations of triage may exist in a wilderness environment.

(1) The number of patients and the severity of their injuries do not exceed the ability of the care-giver and his/her equipment to render care. In this situation, treat life-threatening problems first.

3

(2) The number of patients and the severity of their injuries exceed the capacity of the care-giver and his/her equipment to render care. In this situation, those with the greatest chance of survival are treated first

3. **CPR: URBAN vs. WILDERNESS SPECIFICS.**

a. In a remote wilderness situation where advanced life support assistance may be hours or days away, CPR probably has no chance of success. In addition, administering CPR under wilderness conditions may put group members in serious danger, due to physical hazards and the effects of exhaustion.

b. Based on the above, several suggestions regarding CPR in remote wilderness areas can be made. Even though the following suggestions seem reasonable, they should be viewed as recommendations, without legal force, and not doctrine. At the present time, a rescuer who begins CPR is legally obligated to continue unless or until one of the following conditions is fulfilled:

(1) CPR may be discontinued legally when: (FMST.07.09a)

(a) The patient revives.

(b) ACLS is started by higher authority.

(c) Relieved by another rescuer.

(d) Too exhausted to continue.

(e) Pronounced dead by proper authority.

(2) In a remote wilderness environment, CPR may not be started under one of the following situations: (FMST.07.09b)

(a) The patient is in cardiac arrest caused by trauma.

(b) The patient is a drowning victim who has been immersed for over an hour.

(c) The patient is in cardiac arrest and advanced life support is more than an hour away, especially if the patient must be carried out.

(d) The patient's cardiac arrest was unwitnessed and the time of onset is unknown.

(e) The patient appears to be dead, based on rigor mortis, lethal injuries, or a body temperature below 60°F.

(f) Giving CPR would be hazardous to rescuers.

(3) After 30 minutes of CPR with no signs of life, further CPR is probably useless and may reasonably be discontinued. Administering CPR to a victim who is being evacuated by litter or sled is very difficult, if not impossible. Unless an ambulance or helicopter can be brought in rapidly, chances of survival are slim.

(4) Exceptions include patients in cardiac arrest caused by hypothermia; patients with another illness or injury complicated by hypothermia (avalanche burial and near drowning in cold water), and patients in cardiac or respiratory arrest caused by lightning injury. In these cases, the outlook is probably more favorable and CPR should be aggressively administered.

4. **PATIENT ASSESSMENT** (FMST.07.09c). There are different considerations to think about in a wilderness setting. The environment plays a major role in treatment, care and even the extraction of the patient. Not only do we have to be mindful of the enemy situation, by keeping tactical. We also have to take into consideration the elements. Snow, rain ice, and the terrain can and will pose a great danger to the patient and the rescuers. We must protect the patient from the elements. We can do this by placing an insulating barrier between the patient and the ground. Also exposing one body part at a time and recovering as you move on will help to fight hypothermia.

A) SCENE SIZE-UP.

1) Determine scene safety:

a. Rescuers: Your safety is priority. You are no good to the team as a casualty.

b. Patient: Protect the patient from further injuries, enemy, and the elements.

c. Bystanders: They can cause many accidents. Limit the personnel on scene.

d. Tactical: Have the Marines maintain security and find cover for the rescuer and the patient.

e. Fire: Forest fires can pose a formidable threat.

f. Electrical: Make sure to shut off electricity first. Rushing in injures many rescuers.

g. Special Rescues: Swift water and high angle rescues are very dangerous and only trained professionals should attempt the rescue.

2) Determine the mechanism of injury / nature of illness:

a. Obtain SAMPLE and Mechanism of Injury (MOI) history from patient, witnesses, or other caregivers.

S = Signs and/or Symptoms.
A = Allergies.
M = Medications.
P = Past Illnesses.
L = Last Meal.
E = Events of Injury.

MOI = What, when, where, and how injury occurred.

b. Environment: Survey the area to get clues. (i.e. falling rocks, next to a cliff, broken branches, etc...)

c. Bystanders: Ask them what they saw.

3) Determine number of patients:

a. Survey area for number of patients.

b. Look for the hidden patients. Patients could be buried in rock, sand or snow. They also could be spread out in bushes and trees. One of the most over looked places is in the bystanders. Be sure to have people search and identify all patients.

4) Request additional help / Organize help if needed:

a. Medical assistance: Due to the lack of medical personnel and equipment, you might need to request additional help.

b. Technical assistance: When the scope of care is over your ability or higher than the casevac corpsman can provide, you should request higher trained personnel on the casevac bird.

c. Organize help at the scene: Use the people around you to help out.

5) Consider stabilization of the spine:

a. If indicated from: Falls of 10 feet or greater, high speed collisions, unwitnessed loss of consciousness, etc...

b. Designate an assistant: Use the bystanders / marines. Tell them exactly what you want them to do and show them how to do it.

c. Improvise control: If you are alone you can use the patient's boots, sand bags, rocks, etc as forms of temporary control.

B) INITIAL ASSESSMENT.

1) Make general impressions of the patient: What can you obtain from the patient's condition as you walk up? Is the patient talking, or unconscious?

2) Determine responsiveness / level of consciousness.

a) Use AVPU.

b) Assess pupils for size, equality, and reactivity.

3) Determine chief complaint / Apparent life threats.

4) Assess Airway and breathing.

a) Assessment of airway.

- Perform chin lift or jaw thrust maneuver, with C-Spine precautions.
- Clear the airway of foreign bodies.
- Insert an oropharyngeal or nasopharyngeal airway.
- Use of other airways as needed, and as equipment and training allow.

b) Assessment of breathing.

- Expose the neck and chest, assure immobilization of the head and neck.
- Determine the rate and depth of respirations.
- Auscultate the chest bilaterally. (Ear to chest wall, if necessary)
- Check for symmetry. (Rise and fall of the chest)

c) Management.

- Ventilate. (Face-mask, bag-valve-mask, if available)
- Alleviate tension pneumothorax.
- Seal open pneumothorax.
- Administer high concentration oxygen, if available.

5) Assess Circulation.

a) Assess and control major bleeding.

- Identify and control major bleeding.
- Pulse: Quality, rate, and regularity.
- Pulse Pressure: Check radial, then carotid.
- Skin color: cyanosis, pallor and temperature.

b) Management.

- Insert two large bore IV's with saline locks. IV fluid replacement, if indicated.
- Place in shock position.
- Prevent hypothermia.

6) Identify priority patients and make transportation decisions.

a) Triage patients at this point.

b) Make transportation decisions. In this type of environment it is possible that a casevac vehicle might not gain access to your position. At this point you will have to organize how the patient will be transported out. Also, you might have to set up relay points and rewarming stations.

C) DETAILED PHYSICAL EXAM: This is the head to toe assessment. We will find and treat all secondary injuries at this time. When performing this exam in the wilderness, you will need to expose one body part at a time and cover it back up as you go. When assessing the patient, you are looking for deformities, contusions, abrasions, penetrations/punctures, bruising, tenderness, lacerations, and swelling (DCAP-BTLS).

1) Head.

a. Assessment.

- Inspect and palpate entire head and face for lacerations, contusions, and fractures.
- Re-evaluate pupils.
- Re-evaluate level of consciousness.
- Assess eyes for retinal hemorrhage, optic disc bulging, visual acuity disturbances, and contact lenses.
- Evaluate cranial nerve function.
- Inspect ears and nose for CSF leakage, hemorrhage
- Inspect mouth, tongue, and teeth.

b. Management.

- Maintain airway, continue ventilation, and oxygenate as indicated.
- Control hemorrhage.
- Prevent secondary brain injury.
- Remove Contact Lenses.

2) Assess the neck.

a. Assessment.

- Inspect for signs of blunt trauma, penetrating injuries, tracheal deviation, and use of accessory muscles during respiration, and jugular vein distention.
- Palpate for tenderness, deformity, swelling, or subcutaneous emphysema.

b. Management.

- Maintain open airway.
- Maintain immobilization and protection of cervical spine.
- Mark trachea.

3) Assess the chest.

3

a) Assessment.

- Inspect chest wall for signs of blunt trauma, penetrating injuries, and use of accessory muscles during respiration.

- Auscultate the chest wall and posterior bases for breath sounds.
- Auscultate for heart sounds.
- Palpate the entire chest wall for evidence of blunt trauma, penetrating injuries, flail chest, tenderness, crepitus, or subcutaneous emphysema.
- Percuss for hyperresonance or dullness.

c. Management.

- Needle thoracenthesis, if tension pneumothorax is suspected and proper training and equipment are present.
- Three-sided dressing for open chest wounds.
- Pericardiocentesis, if cardiac tamponade is suspected and proper training and equipment are present.

4) Abdomen.

a. Assessment.

- Inspect abdomen for signs of blunt trauma, penetrating injuries, and internal bleeding.
- Auscultate for the presence / absence of bowel sounds.
- Palpate the abdomen to elicit subtle tenderness, rebound, or guarding.

- Check pelvis for stability.

b b. Management.

c

- Correctly dress wounds.
- Keep patient NPO until definitive care is delivered.

5) Perineum / Rectum / Penis / Vagina.

d a) Assessment.

e

(1) Perineum.

- Contusions / hematomas.
- Lacerations.

(2) Rectum.

- Blood. Sphincter tone.
- High-riding prostate.

(3) Penis.

- Urethral bleeding.

(4) Vagina.

- Bleeding.
- Lacerations.

6) Musculoskeletal.

a) Assessment.

- Inspect extremities for evidence of blunt / penetrating injury. (deformities, altered range of motion).
- Palpate extremities for tenderness, crepitus, and abnormal movements.
- Palpate all pulses.
- Assess pelvis for stability. (Fractures) - Palpate spine for tenderness, crepitus, step-off's, depressions, and abnormal movements.

b) Management.

- Splint all suspected fractures in the position of function or comfort.

- Reduce dislocations as soon as possible, depending on training.
- Maintain immobilization and alignment of cervical, thoracic, and lumbar spine.

7) Reassess the vitals signs.

a) Assessment.

- Assess the vitals, pulse rate, breathing rate, pulse pressure.
- Re-evaluate the pupils and level of consciousness (GCS).
- Evaluate extremities for motor and sensory deficits.

b) Management.

- Maintain adequate immobilization of entire patient, as indicated.

5. **PATIENT PACKAGING AND EVACUATION.**

a. See the Mountain Casualty Evacuation chapter.

UNITED STATES MARINE CORPS
Mountain Warfare Training Center
Bridgeport, California 93517-5001

FMST.07.13
03/30/02

STUDENT HANDOUT

HIGH-ALTITUDE ILLNESSES

TERMINAL LEARNING OBJECTIVE: Given a simulated casualty, in a high altitude wilderness environment, treat high altitude health problems, in accordance with the references. (FMST.07.13)

ENABLING LEARNING OBJECTIVES:

(1) Without the aid of references, select from a given list the definition of acute mountain sickness, in accordance with the references. (FMST.07.13a)

(2) Without the aid of references, select from a given list the four main mechanisms of acute mountain sickness, in accordance with the references. (FMST.07.13b)

(3) Without the aid of references, select from a given list the most common and prominent symptom of acute mountain sickness, in accordance with the references. (FMST.07.13c)

(4) Without the aid of references, select from a given list the field management of acute mountain sickness, in accordance with the references. (FMST.07.13d)

(5) Without the aid of references, select from a given list the seven high altitude health preventive measures, in accordance with the references. (FMST.07.13e)

(6) Without the aid of references, select from a given list the correct definition of high altitude cerebral edema, in accordance with the references. (FMST.07.13f)

(7) Without the aid of references, select from a given list the pathophysiology of high altitude cerebral edema, in accordance with the references. (WMC.6.4g)

4

(8) Without the aid of references, select from a given list five of the symptoms of high altitude cerebral edema, in accordance with the references. (FMST.07.13h)

(9) Without the aid of references, select from a given list the field management of high altitude cerebral edema, in accordance with the references. (FMST.07.13i)

(10) Without the aid of references, select from a given list the correct definition of high altitude pulmonary edema, in accordance with the references. (FMST.07.13j)

(11) Without the aid of references, select from a given list three of the nine symptoms of high altitude pulmonary edema, in accordance with the references. (FMST.07.13k)

(12) Without the aid of references, select from a given list the field management of high altitude pulmonary edema, in accordance with the references. (FMST.07.13l)

(13) Without the aid of references, select from a given list choose the most important treatment for both high altitude cerebral edema and high altitude pulmonary edema, in accordance with the references. (FMST.07.13m)

OUTLINE

1. **ACCLIMATIZATION**.

a. Successful acclimatization depends on three factors:

(1) Degree of hypoxic stress (i.e. altitude)

(2) Rate of onset of hypoxic stress (i.e. ascent rate)

(3) Individual Physiology (i.e. genetic differences between individuals)

b. Physiologic Changes:

(1) Respiratory - Hypoxic Ventilatory Response (HVR)

- Hypobaric hypoxia
- Triggers carotid body
- Central respiratory center stimulated
- Ventilatory rate increases

HVR can be measured:

- "High" leads to successful acclimatization and superior performance.
- "Low" tends to be found in victims of altitude illnesses.
- HVR can be influenced by depressants (ETOH) and stimulants. (caffeine), and progesterone.
- **May** be negatively influenced by physical conditioning.

(2) Respiratory – Fast response: minutes to hours

- HVR leads to increased Ventilatory rate.
- Increased ventilations "blows off' C02."
- Low C02 levels yield a respiratory alkalosis and rising pH.
- Elevation of pH feeds back to depress the central respiratory center.
- No further immediate rise in ventilatory rate (a brake is placed on any further increase in respiration.)

(3) Renal – Slower response: kidneys respond over the next 24 - 48 hours:

- Kidneys respond to alkalosis by inducing a bicarbonate diuresis.
- PH slowly declines - brake on central respiratory center eases.
- Respiratory rate gradually rises over 4 - 7 days to new baseline.
- Final respiratory rate is altitude dependent.

(4) Cardiovascular - changes are a result of increased sympathetic tone (due to hypoxia).

- Increased: Blood Pressure, venous tone and Cardiac output (increased heart rate and contractility.)
- Stroke Volume declines in first 1 - 3 days due to bicarbonate diuresis.
- Left ventricular pressures are normal, usually; therefore, wedge pressures are usually normal.

(5) Pulmonary Circulation - critical in understanding the pathophysiology of HAPE.

- Ascent results in global Hypoxic Pulmonary Vasoconstriction.
- Elevated Pulmonary Artery Pressure (PAP) occurs.
- Exertion at altitude worsens this Pulmonary Hypertension.
- Damage to pulmonary vascular endothelium leading to leaky capillaries.

(6) Cerebral Circulation - critical in understanding pathophysiology of HACE

- Hypoxia -----------Vasodilatation
- Hypocapnia ----------Vasoconstriction
- Hypoxia is the dominant influence and vasodilatation results.
- On average, Cerebral Blood Flow (CBF) increases 20-40% on ascent to 12.000ft.

(7) Hemopoietic Response

- Erythropoietin levels increase within hours of ascent.

- New immature Red Blood Cells (RBC) seen within days.
- New mature RBC's present within 4-10 days
- RBC mass increase seen over weeks to months (altitude dependent).
- Early changes due to hemoconcentration.
- Polycythemia is a potential danger -- this can lead to a condition known as CMS, or Chronic Mountain sickness.

(8) Muscle Tissue

- Intramuscular capillary and mitochondrial density increases after three weeks exposure to altitude.

(9) Sleep Pattern

- Stage 1 increase, Stage 2 unaffected
- Stage 3 & 4 decreases
- Rapid Eye Movement (REM) decreases
- Arousals increase
- Periodic Breathing (Cheyne-Stokes).
    - Increased respiratory rate leads to respiratory alkalosis
    - Elevated pH depresses central respiratory center
    - Apnea
    - 02 saturation declines, C02 levels rise
    - Carotid body stimulated
    - Ventilatory rate rises

(10)Summary: Physiologic changes seen at altitude result from the body's responses to hypo baric hypoxia. It has been speculated that man would be limited to an 8000-foot elevation limit without supplemental oxygen if not for the acclimatization process. As Messner and Haebler demonstrated on Mt. Everest, the process of acclimatization enables man to explore the entire planet.

2. **ACUTE MOUNTAIN SICKNESS (AMS)**.

a. Definition - AMS is an acute, self-limiting illness, which results when an unacclimatized individual ascends rapidly to high altitude. (FMST.07.13a)

(1) It is rare below 8,000 feet but will occur in approximately 42% of persons rapidly exposed to altitudes greater than 10,000 feet in elevation.

(2) Aerobic fitness is no predictor for risk of AMS with exposure to altitude. There is some evidence that prior aerobic fitness leads to increased AMS incidence, most likely due to excessively rapid ascent or over-exertion at altitude. Elite athletes may also be at increased risk due to depressed cardiovascular and Ventilatory responses to exertion.

(3) Prior ascents to altitude without symptoms of AMS are no guarantee against having symptoms of AMS with future ascents. However, those with past AMS are at an increased risk of AMS with repeat exposure to high altitude. Generally, there is no correlation between the severity of the illness and increasing altitudes. The incidence is slightly increased in younger persons.

(4) There is still much we do not know about altitude-induced physiologic changes. Current explanations in the literature center around genetic differences in hypoxic ventilatory response and vascular endothelium release of inflammatory mediators and Nitric Oxide.

b. Pathophysiology of AMS: The physiologic changes associated with the symptom complex known as AMS are linked to: (FMST.07.13b).

(1) Poor ventilatory response: An individual's Hypoxic Ventilatory Response (HVR) dictates how well he/she will respond to the hypobaric hypoxia of altitude. Those with a brisk HVR will do well, whereas those with a blunted response tend to suffer the symptoms of AMS. A poor ventilatory response leads to hypoxia and hypercapnia (elevated C02 levels). These two combine to augment cerebral blood flow by inducing cerebral vasodilatation.

(2) Fluid retention: The increased sympathetic tone associated with hypobaric hypoxia explains many of the changes seen in the renal system at altitude. Low Renal Blood Flow, increased anti-diuretic hormones, low Glomerular Filtration Rates, and decreased Urine Output all contribute to a Net Fluid Gain. This retained fluid redistributes within the body's fluid spaces. As a result, intravacular and extravascular fluid shifts occur, especially in the brain, lung, and peripheral interstitial tissues.

(3) Increased Intra-cranial Pressure: Increased cerebral blood flow coupled with CNS inflammation resulting in increased vascular permeability leads to increased intra-cranial pressure and associated symptoms.

(4) Altered Pulmonary Mechanics. On ascent to altitude, there is an increase in extravascular lung water. This is thought to be mediated, in part, by a phenomenon known as hypoxic pulmonary vasoconstriction. Blood vessels in the lungs constrict when they are exposed to low oxygen levels, this global vasoconstriction leads to increased right-sided heart pressures. The result of increased right-sided heart pressures causes fluid to leak from the vascular space into the alveoli, which results in impaired gas exchange. This has been demonstrated via Pulmonary Function Testing bronchoalveolar lavage, right sided-heart catheterization and arterial blood gas analysis showing decreased Vital Capacity, decreased Peak Expiratory Flows, pulmonary exudates rich in proteins and Increased Alveolar-Arterial oxygen Gradients. One study showed a 54% increase in estimated lung mass. It is not uncommon to hear crackles in all lung fields when auscultating asymptomatic individuals. The presence of intra-alveolar fluid inhibits the diffusion of oxygen

across the alveolar membrane, thereby reducing oxygenation across the pulmonary capillary membrane.

Signs and Symptoms of AMS. Symptoms tend to occur within the first 1- 3 days of exposure to altitude. Symptoms should resolve spontaneously within 15 - 24 hours, if ascent is arrested. The three most prominent symptoms are Headache, Nausea and Vomiting. (FMST.07.13c) Although nonspecific, headache must be present to make a diagnosis of AMS. 95% of AMS victims will complain of one or more of these three. Other common complaints include dizziness, anorexia, drowsiness, malaise, weakness, and insomnia. It is possible that findings of dyspnea on exertion and mental status changes along with ataxic gait may accompany the classic AMS signs and symptoms. These are clear indications for descent, as these are the hallmark signs of progression to HACE and the development of HAPE. Again, headache is the most common and prominent symptom of AMS. The Lake Louise Consensus criteria was created to give clinicians formal guidance in diagnosing AMS. The criteria require the presence of headache plus one or more symptoms from the above list. (i.e. nausea, vomiting, anorexia, insomnia, dizziness, fatigue, and dyspnea on exertion.)

c. Field Management (FMST.07.13d)

(1) Stop any further ascent. Light duty. No tobacco.

(2) Symptomatic Treatment. Mild analgesics such as Aspirin or Tylenol for the headache, Compazine or Phenergan for nausea/vomiting.

(3) Other Meds: Acetazolamide (Diamox), Dexamethasone (Decadron), and oxygen. The effects of Diamox and Decadron are additive when used together and are a very effective means for treating AMS.

(4) Hyperventilate: It is unrealistic to try to consciously control one's respiratory rate over any appreciable period of time. It is nearly impossible to maintain conscious hyperventilation while functioning in the field.

(5) Descend. If none of these methods help, then individuals should descend 1,000-3,000 feet. This will usually result in marked relief of symptoms. Most cases, however, improve in 1-2 days with symptomatic treatment.

(6) ALL PATIENTS WITH AMS MUST BE EVALUATED FOR HAPE AND HACE.

HOW?

- Field sobriety test.
- Mini-mental status exam.
- Auscultation of lung fields.

d. Preventive Measures: (FMST.07.13e)

(1) The best approach to high-altitude travel is staged, gradual ascent, when moving at altitudes greater than 3000 meters. Gradual ascent means no faster than 3000 ft/day up to 14,000 and no greater than 1,000 ft/day over 14,000 ft.

(2) Avoid any alcohol and/or sedatives during the first two nights.

(3) Maintain adequate fluid intake.

(4) Work high - Sleep low.

(5) High Carbohydrate Diet (approximately 70% of diet).

(6) Avoid overexertion, but remember that mild exercise assists acclimatization.

(7) No smoking.

(8) Pharmacological options: Prevention of AMS is possible with Acetazolamide (Diamox). It dramatically lowers the incidence of AMS when taken prophylactically. The usual dose is 125 mg PO BID starting 24-48 hours before ascent and should be continued for 3-4 days after the start of the mission. It is also effective when used as a treatment especially when combined with Decadron. In individuals with contraindications or allergies to Diamox use Decadron (dosed PO/IM/IV 4 mg Q6).

3. **HIGH ALTITUDE CEREBRAL EDEMA (HACE).**

a. Definition. HACE is a high-altitude illness characterized by swelling of the brain. (FMST.07.13f)

(1) HACE can occur as low as 8,000 feet, but typically occurs at more than 12,000 feet.

(2) HACE, considered by most, to be a progression of AMS, occurs as a failure to heed the warnings of AMS signs and symptoms and continue ascent.

b. Pathophysiology. Increased cerebral blood flow along with the fluid shifts mentioned above lead to vasogenic edema (leaky blood vessels) inside the confined space of the skull. It is also thought that inflammation plays a key role in the development of HACE and occurs as follows. Hypoxia triggers the release of inflammatory mediators (likely Transforming Growth Factor Beta, TGF-β and Vascular Endothelial Growth Factor, VEGF.) These mediators cause increased vascular permeability that result in fluid accumulation within the brain – edema and increased intracranial pressure. Once this pressure nears arterial pressure, cerebral blood flow will be impaired. Eventually, intra-cranial pressure will rise further with a concomitant increase in the risk of brain herniation leading to marked alterations in mental status, ataxia and even cranial nerve palsies. Increased CSF production further compounds the problem. (FMST.07.13g)

c. Differential Diagnosis: Other illnesses are possible in individuals ascending to altitude. Your differential should include the following when neurologic deficits are involved:

   (1) Cerebrovascular spasm and Transient Ischemic Attacks.

   (2) Cerebral vascular thromboembolic events.

   (3) Intra-cranial hemorrhage (aneurysm, AVM).

   (3) Hypothermia.

   (4) Meningitis

   (5) Encephalitis

   (6) "Hangover"

d. Signs and Symptoms:

   (1) Early signs and symptoms are those found in AMS. (i.e. H/A, NV)

   (2) When the above warning signs go unheeded the true signs and symptoms of HACE can quickly follow: (FMST.07.13h)

      (a) Ataxia (loss of muscle coordination leading to difficulty maintaining balance), especially prominent in the victim's gait.

      (b) Mental Status Changes, poor judgement, personality change which may progress to stupor, coma and death relatively rapidly without proper treatment.

      (c) Lassitude, confusion, hallucinations, convulsions, behavioral aberrations.

      (d) Paralysis of one or more extremities (see Differential Diagnosis).

   (3) The most important impediment to early recognition of HACE is its insidious onset. Early signs and symptoms frequently go unrecognized or are ignored by patients, as well as their companions, who also may be suffering to some degree from the effects of altitude.

e. Field Treatment (FMST.07.13i)

(1) Early recognition is the KEY. Look for Ataxia and Mental Status Changes.

(2) Treatment should be immediate since fatalities can occur within a few hours in severe cases.

(3) Once diagnosed, the patient should be placed in the most comfortable position possible, descent should be directed immediately and administration of high concentration O2 should be given, in route, if available.

(4) Decadron (early), 10mg PO/IM/IV followed by 4 mg Q6. Diamox can also be used 125-250 mg BID.

(5) Consider diuretics – use carefully.

(6) If loss of consciousness, think airway management and ABC's.

(7) Casevac to an advanced medical facility ASAP!

(8) A Gamow Bag can be a lifesaver if decent is not immediately possible. Perform 4-6 compressions per day each treatment should last 40-60 minutes.

NOTE: The Gamow Bag is a one-man portable hyperbaric chamber. It weighs 14.5 lbs., is operated by a foot-pump, and is made of nylon with an airtight zipper. With the HAPE or HACE victim inside, the bag is pressurized with the foot-pump to an internal pressure of 2 psi. This is the equivalent to a 5000 - 6000-foot descent in regards to the partial pressure of oxygen inside the bag. Air is circulated within the bag via the foot-pump. This can be a life-saving resource in the event a victim is caught at altitude and cannot descend due to weather or other uncontrollable variables.

f. Prevention

(1) Preventive measures are the same as those discussed for AMS.

(2) Decadron can be used as a prophylactic medication along with Diamox, in cases of unavoidable rapid ascent to extreme altitude. However one must continue this regimen, while at altitude, or face rapid onset of symptoms once the medication is stopped.

4. **HIGH ALTITUDE PULMONARY EDEMA (HAPE)**.

a. Definition. HAPE is a high-altitude illness that is characterized by filling of the lungs with edema fluid. (FMST.07.13j)

(1) HAPE rarely occurs below 12,000 feet.

(2) HAPE victims in the past were commonly described, after autopsy, as victims of "pneumonia" and "Inflammation of the lungs". Since 1960, medical science has had the technology to investigate the complexities of the physiological changes associated with hypobaric hypoxia.

(3) Incidence: Occurs in 1-2 % of people brought rapidly to 12,000ft.

(4) Commonly occurs within 2-4 days of ascent.

(5) Risk for developing HAPE remain the same as above:

(a) Rate of ascent
(b) Sleeping altitude
(c) Level of exertion
(d) Gender (M>F)
(f) Individual physiology

b. Pathophysiology of HAPE:

(1) Hypoxic Pulmonary Vasoconstriction

(a) Normal part of the acclimatization process (elevated PAP).
(b) Adaptive at sea level (i.e. lobar pneumonia).
(c) Maladaptive at altitude (global hypoxia).
(d) Tends to be uneven, leading to high pressure/high flow area.

(2) Overperfusion

(a) Localized areas exposed to high pressure and high flow.
(b) Shearing forces and stress failure at the microvascular level.
(c) Endothelial damage to capillaries leads to inflammatory response and increased vascular permeability.
(d) Elevated hydrostatic pressure combined with increased vascular permeability leads to leakage of protein rich fluids into the alveoli and decreased oxygenation

(3) Nitric Oxide deficiency

Individuals prone to HAPE may have a deficiency in the amount of Nitric Oxide produced in the vascular endothelium resulting in more profound hypoxic pulmonary vasoconstriction.

c. Pathologic Findings of HAPE: Postmortem studies of victims of HAPE have yielded some interesting results:

(1) Grossly, the lungs are congested and swollen, as seen with other causes of pulmonary edema. Average lung weight on autopsy is 2-4 times normal. Histologically, one sees pulmonary edema with a protein rich exudate filling the alveoli. It is also common to find the alveolar spaces filled with hyaline membranes.

(2) The left heart is completely normal in appearance. However, the right ventricle, right atrium, and pulmonary arteries are distended and dilated.

(3) Greater than 50% of HAPE victims are found to have evidence of HACE.

d. Signs and Symptoms. These tend to occur within 2-4 days of arrival at altitude. Usually the symptoms of AMS are present before or occur with the symptoms of HAPE. (FMST.07.13k)

(1) Early signs:

(a) Dry cough, frequently occurring at night.

(b) Dyspnea on Exertion (DOE), especially with ambulation uphill.

(c) Mild chest pain - usually perceived as an ache beneath the sternum.

(d) Decreased work performance and increased recovery time between events.

(e) Peripheral or Central Cyanosis.

(2) Later signs:

(a) Dyspnea at Rest.

(b) Productive cough which yields large amounts of pink, frothy sputum.

(c) Rapid pulse and respiratory rates.

(d) Audible crackles on auscultation.

(e) **Mental status changes, Ataxia**, Loss of Consciousness.

(f) CXR findings:

- Prominent pulmonary arteries.
- Normal heart size.
- Patchy infiltrates.

ECHO findings:

-Non-cardiogenic pulmonary edema -increased PAP
-Tricuspid regurgitation.
-Normal left ventricle
-Prominent pulmonary arteries -enlarged right atrium
-Enlarged right ventricle.

NOTE: Crackles heard on lung exam, as an isolated finding is NOT an indication for descent or grounds for the diagnosis of HAPE. It is common (up to 15% of climbers on Mt. Rainier) to find asymptomatic rales in climbers at altitudes of 10,000-12,000ft. Concern is heightened when crackles occur in constellation with other signs and symptoms.

e. Field Treatment (FMST.07.131)

(1) The most important emergency care measure is immediate descent to a lower altitude, since fatalities can occur within 6-12 hours in severe cases.

(a) A descent of at least 2,000-3,000 feet can be a definitive treatment if the condition is caught early enough.

(2) The patient should be placed in the most comfortable position (usually sitting) and given **high-flow O2 if available**. It is important to minimize activity, as exertion increases pulmonary artery pressures. Remember oxygen is Lasix for the lungs.

(3) Medication regimens include:

(a) Nifedipine (10mg PO initially then 30mg SR Q12) – calcium channel blocker - pulmonary artery/arteriole dilator. SR=sustained relief.

(b) Diamox and Decadron may be appropriate regimens, depending on your medication availability.

(c) Morphine (5mg IV initially decreased PAP) as a last resort.

(4) Oxygen is always appropriate, however, it is not always available.

(5) Gamow Bag can be a lifesaver if decent is not immediately possible.

f. Prevention Measures. Preventive measures are the same as those discussed for AMS.

(1) Acetazolamide (Diamox) has been proven to be successful in the prevention of HAPE. 125 - 250 mg PO bid helps prevent HAPE in individuals with recurrent episodes.

(2) Nifedipine (30 mg SRq12h) - also effective in individuals with recurrent HAPE.

(3) It should be noted that a history of HAPE is NOT a contraindication to the individual participating in high-altitude activities. Each case should be evaluated individually.

NOTE: Remember decent is the most important treatment for both HACE and HAPE. (FMST.07.13m)

5. **<u>SYSTEMIC EDEMA AT HIGH ALTITUDES</u>.**

Occasionally, certain individuals, especially women, will develop edema of the face, hands, and feet at altitude. This edema usually tends to occur in the absence of other symptoms. In susceptible persons, repeat episodes are common. The edema goes away upon return to lower altitude. There is no need for descent in these cases because the patient is usually able to tolerate the edema and continues to function well. However, they should be examined carefully to rule out the presence of HAPE or HACE. Diuretics are usually not warranted. It is also probably reasonable to restrict salt intake. The only indication for descent is if the edema is causing functional disability, i.e. inability to don one's boots, or facial edema severe enough to restrict vision.

6. **<u>SICKLE CRISIS</u>.**

Persons with sickle cell trait are at risk for developing sickling attacks when exposed to hypoxic stress, as might occur at high altitudes. However, sickling can occur at lower elevations. Individuals with Mediterranean family ties should be considered at risk for having sickle cell trait. The trait is more likely in black individuals as well. Any patient complaining of Left Upper Quadrant pain, SOB, and/or arthralgia at altitude should have sickling crisis included in the differential diagnosis regardless of race. It has been known to occur in white females as well. It has been recommended by some that individuals with sickle cell trait should be restricted to an altitude of 8,000 feet and below. However, more liberal recommendations exist, suggesting that the individual and their symptoms dictate ultimate altitude.

7. **<u>ALTITUDE THROAT</u>.**

At higher altitudes, mouth breathing is often increased during exertion. Inhalation of this cold, dry air can result in drying of the mucous membranes in the throat and upper airway. This condition responds to humidified air, salt-water gargles or lozenges. If exudate or edema is present, throat culture and/or antibiotics may be warranted.

8. **HIGH ALTITUDE FLATUS EXPULSION HAFE).**

Ascent to altitude can result in rapid expansion of intestinal gas, due to the decreased pressure of altitude. This produces abdominal discomfort and the passage of colonic gas. This condition is not associated with serious consequences; however, it may create tension amongst tent-team members. Treatment consists of oral administration of anti-flatulence such as simethicone, Mylanta II, or Gaviscon. Know and avoid your gas producing foods!

9. **HIGH ALTITUDE RETINAL HEMORRHAGE.**

Up to 60% of individuals ascending to altitudes greater than 18,000 feet will develop asymptomatic retinal hemorrhages. On fundoscopic exam, there will be hyperemia of the disc and dilation and tortuosity of the vessels. "Flame" hemorrhages and "cotton wool spots" can be seen on exam. The only indication for descent in an individual with retinal hemorrhages is loss of visual acuity. This usually results when a large hemorrhage involves the macular region of the retina. Remember that Carbon Monoxide poisoning can present with retinal hemorrhage as well.

10. **THROMBOEMBOLIC EVENTS.**

Thromboembolic disease occurs with increased frequency at higher altitudes. It is believed that the common circumstances of volume depletion due to dehydration and Polycythemia due to altitude exposure predispose individuals to coagulopathic events. In addition, extended periods of inactivity due to "going to ground" during storms contribute to this predisposition. There does not seem to be any direct influence of hypobaric hypoxia on the body's normal coagulation system.

11. **ACETAZOLAMIDE.**

Acetazolamide (Diamox) is a Carbonic Anhydrate Inhibitor, which is especially active in the brain, the lung, and the kidney.

a. Kidney:

(1) Uptake of bicarbonate ion is inhibited.

(2) Bicarbonate diuresis is induced.

(3) Metabolic acidosis ensues within one hour of medication.

(4) Sequence mimics and accelerates the natural acclimatization process.

(5) Enhanced HVR is the end result.

(6) Diuresis addresses problems associated with fluid retention.

b. Brain:

(1) Direct effects at the level of the Central Respiratory Center.

(2) Diminishes periodic breathing during sleep.

(3) Fewer episodes of apnea and extreme hypoxemia.

(4) CSF production is inhibited and CSF absorption is enhanced.

c. Side Effects:

(1) Polyuria.

(2) Flattens taste of carbonated beverages.

(3) Peripheral paraesthesias, myopia, impotence, nausea, and drowsiness.

(4) Contraindicated for Sulfa Allergic and G6PD-deficient individuals (reversible when medication is discontinued.)

UNITED STATES MARINE CORPS
Mountain Warfare Training Center
Bridgeport, California 93517-5001

FMST.07.35
03/13/02

STUDENT HANDOUT

HEAT ILLNESSES

TERMINAL LEARNING OBJECTIVE: Given a casualty in a wilderness environment, and necessary equipment and supplies, manage common heat related injuries to prevent death or further injury per the reference (FMST.07.35)

ENABLING LEARNING OBJECTIVES:

(1) Without the aid of references, given a list of heat illnesses match the definition to the correct condition, in accordance with the references. (FMST.07.35a)

(2) Without the aid of references, select from a given list the correct field treatment for heat cramps, in accordance with the references. (FMST.07.35b)

(3) Without the aid of references, select from a given list the correct field treatment for heat exhaustion, in accordance with the references. (FMST.07.35c)

(4) Without the aid of references, select from a given list the correct field treatment for heat stroke, in accordance with the references. (FMST.07.35d)

1. **CAUSES OF HEAT ILLNESSES**. Heat illness is caused by one of or a combination of the following factors:

a. Excessive environmental heat and humidity.

b. Faulty heat dissipation mechanisms within the body.

c. Stressed heat dissipation mechanisms.

d. Prolonged strenuous physical activity with inadequate fluid and electrolyte replacement.

2. **THERMOREGULATION**. The body is constantly adjusting itself in order to maintain its optimal temperature. This is called thermoregulation. A basic physics principle applies, in that heat is transferred from a higher to a lower temperature. What this means is that when the environmental temperature is higher than the skin temperature, the body gains heat, but when the environmental temperature is lower, the body loses heat. As a result, elevated environmental temperature adds to the heat burden of the body and can interfere with heat dissipation.

a. Heat Transfer Mechanisms. Keeping in mind the above principle we will now discuss the four methods of heat transfer.

1) Conduction: This is heat exchange between two surfaces in direct contact. An example would be lying on a hot or cold surface.

2) Convection: This is heat exchange from a surface to a gas or liquid, usually air or water. The amount of heat transferred depends on many variables, such as density, flow rate and surface area exposed.

3) Radiation: This is heat transfer between the body and its surroundings. All matter either absorbs heat or emits thermal radiation. The sun emits heat and the body absorbs it. A interesting fact is that highly pigmented skin while protective against UV radiation absorbs about 20% more heat than lightly pigmented skin.

4) Evaporation: This is the heat flow from the body surface to the environment by sweating.

b. If the body is unable to maintain thermal equilibrium by radiation, convection, and conduction and core temperature rises, sweating must occur to permit heat loss by vaporization of water. Since gastric emptying of water is about 1L/hr and sweat rates can exceed 1L/hr, a healthy diet along with constant replenishment of water lost is imperative to prevent illness.

3. **PREDISPOSING FACTORS**.

a. An underlying illness such as infection.

b. The temperature and humidity of previous days effect the likelihood of becoming a heat casualty.

c. Most heat illnesses occur during the morning hours.

d. Thirst is not an accurate indication of hydration; therefore you must drink even when you don't feel thirsty.

e. People usually do not voluntarily drink as much water as they lose.

f. All persons in the heat should be considered dehydrated, unless they have been forced to drink more than they desired.

4. **PREVENTIVE MEASURES**.

a. Drink 6-8 qts every day to replace fluids lost by sweating or from other bodily functions.

b. Consumption of a solute rich meal or a fluid replacement beverage about 2 hours before exercise can potentially counteract the sweat/salt losses.

5. **EARLY SIGNS OF HEAT RELATED ILLENSSES**. Changes in cognitive functions appear to occur frequently prior to the physical symptoms of heat illnesses. It is therefore imperative that medical personnel and leaders watch for confusion and impaired ability to function or work in the members of their group.

6. **SPECIFIC HEAT RELATED ILLNESSES**.

a. Heat Induced Syncope. (FMST.07.35a) This event is caused by the gravitational displacement of blood into the legs of an upright and stationary person combined with hemodynamic displacement of blood into peripheral circulation to support heat transfer at the body surface.

1) Prodromal symptoms.

(A) Restlessness

(B) Sighing

(C) Dysphoria

(D) Nausea

(E) Yawning

2) Treatment.

(A) The fainting is usually brief and responds to horizontal positioning.

(B) Rest in a cool, shaded area.

(C) Oral fluids.

(D) Persons should be cautioned against standing in hot environments and advised to flex leg muscles repeatedly while standing still to enhance venous return and warned to sit or lie down if warning signs reappear.

b. Heat Cramps. These are painful spasms of muscles, caused by sweating resulting in excessive loss of body salt. Generally, the spasms involve voluntary skeletal muscle subjected to prior exertion. This can occur even if a person is replacing water loss, but not salt losses.

1) Precipitating Factors: Cramps are preceded by several hours of sustained effort, combining heavy sweating in hot surroundings, with the ingestion of large volumes of water.

2) Symptoms: Painful muscle cramps in the arms, legs, and abdomen.

3) Treatment: (FMST.07.35b)

(A) Rest in a cool, shaded area.

(B) Gentle massage of affected muscles (may relieve spasms) or place the affected muscle in a stretch.

(C) Generally, heat cramps respond quickly to salt solutions.

1.) 1/4 to 1/2 teaspoon of salt in a quart of water (sip).

2.) In severe cases IV isotonic saline 0.9% NaCL.

NOTE: Salt tablets are not recommended due to the gastric irritation they cause.

c. Heat Exhaustion. This occurs when body salt losses and dehydration from sweating are so severe that a person can no longer maintain adequate blood pressure. It tends to develop over several days, that present ample opportunity exists for electrolyte and water imbalances to occur. Some 20% of cases are found to also have some form of gastroenteritis. Additionally, 60% of heat exhaustion casualties will also have symptoms of heat cramps.

1) Problems in recognition.

(A) Early signs may go unrecognized or be mistaken for malingering.

(B) Headache and visual disturbances may mimic prodromal of migraine headache.

(C) Dizziness and syncope may lead to an evaluation for arrythmia.

(D) Chills, ^ temperature, and sweating may lead to an evaluation for infection.

(E) Altered mental status or ataxia may lead to an evaluation for Meningitis or high altitude medical problem.

2. Symptoms:

(A) Headache

(B) Fatigue

(C) Piloerection

(D) Nausea/Vomiting

(E) Profuse sweating

(F) Dizziness

(G) Hyperirritability, Anxiety

(H) Chills

(I) Heat cramps

3. Clinical Findings:

(A) Tachycardia

(B) Hypotension

(C) Core temperature may be normal or elevated. Spontaneous cooling may occur, usually with profuse sweating. This will not happen with Heat stroke.

(D) Hyperventilation

(E) Syncope

(4) Treatment is based on symptoms and focuses on reduction of body temperature and rehydration. (FMST.07.35c)

(A) Lay down in a cool, shaded area.

(B) If body temperature is elevated, lower by any means available.

(C) Sprinkling with water and fanning.

(D) Ice bags over superficial great vessels and fanning.

(E) Periodic toweling of the skin renews the evaporative process.

(F) If patient is alert, orally rehydrate with cool water or commercially available rehydration mixture. Goal is 2L over 1st 2 hours.

(G) If patient is not alert, rehydrate via IV with LR.

(H) Victim must rest and continue to be rehydrated over next 24 to 36 hours.

d. Heatstroke (Hyperthermia). " A true medical emergency". This is a failure of the body's cooling mechanisms that rid the body of excessive heat build-up. Prior to 1950 the mortality rate was 40 to 75%. Long term survival of victims is directly related to the speed of effective treatment measures. Traditionally, the diagnosis of Heatstroke required the following 3 signs: a core temperature > 104°F(41°C), altered mental status, and cessation of sweating. Although, this symptom complex does represent full-blown Heatstroke adherence to this strict criteria may delay critical interventions. Unless an alternative etiology is obvious, the previously healthy person who collapses after physical exertion in hot weather should be considered to have exertional Heatstroke. Treat for Heatstroke whenever you suspect it and always suspect it in the heat.

1. Symptoms:

   (A) Headache

   (B) Shortness of breath

   (C) Nausea/ vomiting

   (D) Low BP

   (E) Skin may be wet or dry

   (F) Constricted pupils

   (G) Seizures

   (H) Dry mouth

   (I) Core temp > 104°F (41°C)

   (J) Weakness/ dizziness

   (K) Hot, red skin

   (L) Tachycardia

   (M) Confusion, Aggression, Combative

   (N) Coma

2. Clinical Findings.

   (A) CNS: altered mental status, agitation, ataxia, delirium, hallucinations, convulsions, seizures, coma, pupillary constriction.

   (B) Cardiovascular: hypotension (a late and ominous finding), shock, sinus tachycardia with ST segment & T wave abnormalities.

   (C) Pulmonary: hyperventilation, pulmonary edema.

   (D) Renal: acute renal failure, hematuria, pyuria, proteinuria, UA casts.

   (E) Gastrointestinal: nausea, vomiting, +/- disseminated intravascular coagulation, hematemesis, melena.

   (F) Hematologic: WBC 20-30 thousand, < platelets, < clotting factors.

   (G) Hepatic: increased liver function tests.

3. Emergency Treatment. (FMST.07.35d)

(A) Rapid reduction of body core temperature by any means available until core temperature is 102°F.

1) Ice water bath (controversial).
2) Cool water bath.
3) Pouring or hosing cool water over victim and fanning.
4) Placing ice bags at neck, armpits, groin, forehead, chest etc..

(B) Maintain ABC's.

(C) Monitor core temperature.

(D) IV access infuses RL or normal saline 500cc/hr x 4hrs. (watch for pulmonary edema).

(E) Control seizures with valium.

(F) Transport STAT. Cool enroute.

UNITED STATES MARINE CORPS
Mountain Warfare Training Center
Bridgeport, California 93517-5001

FMST.07.15
04/01/02

STUDENT HANDOUT

COMBAT CASUALTY CARE

TERMINAL LEARNING OBJECTIVE: Given a simulated casualty in a simulated combat environment, treat the casualty in accordance with the references. (FMST.07.15)

ENABLING LEARNING OBJECTIVES:

(1) Without the aid of references, choose from a given list the three correct phases of care of tactical casualty management, in accordance with the references. (FMST.07.15a)

(2) Without the aid of references, given a simulated casualty and the necessary equipment, conduct combat casualty care, in accordance with the references. (FMST.07.15b)

OUTLINE

1. Summary of Findings and Recommendations.

   a. Pre-hospital care of combat casualties is divided into three phases: (FMST.07.15a)
      1. Care under Fire.
      2. Tactical Field Care
      3. Combat Casualty Evacuation (CASEVAC) Care.

   b. Suppression of hostile fire may take temporary precedence over the rendering of care in the Care under Fire phase.

   c. Cervical spine immobilization is not required for victims of penetrating head or neck trauma.

   d. CPR should not be attempted on the battlefield for victims of blast or penetrating trauma who have no pulse, respirations, or other signs of life.

   e. The use of tourniquets is encouraged in the Care under Fire phase. The tourniquet is removed and bleeding controlled with direct pressure in subsequent phases when feasible.

f. The nasopharyngeal airway is the airway of first choice for unconscious patients until the CASEVAC phase.

g. For patients with airway obstruction in the Tactical Field Care phase unrelieved by a nasopharyneal airway, the next airway maneuver is cricothyroidotomy.

h. Endotracheal intubation is not the standard of care until the CASEVAC phase.

i. Evaluation of the laryngeal mask airway and the Combitube for use by SOF corpsmen and medics is recommended.

j. Progressive, severe respiratory distress in the setting of unilateral blunt or penetrating chest trauma on the battlefield should result in a presumptive diagnosis of tension pneumothorax and decompression of that hemithorax with needle thoracostomy.

k. Chest tube insertion is not indicated until the CASEVAC phase of care.

l. Oxygen is not expected to be carried onto the battlefield, but CASEVAC assets should have it.

m. No IV fluids are to be given until the CASEVAC phase unless the patient is in hypovolemic shock from a bleeding site which has now been controlled. Fluid resuscitation is not necessary for patients who are not in shock and not appropriate for patients in hypovolemic shock from uncontrolled hemorrhage (penetrating wounds of the chest and abdomen).

n. Saline locks may be used instead of IV's if fluid resuscitation is not required.

o. 18 gauge catheters may be used instead of 14 or 16 gauge.

p. Hespan (1000cc) is the fluid of choice for initial resuscitation of patients with controlled hemorrhagic shock.

q. Morphine is to be used IV (5mg) instead of IM.

r. IV cefoxitin is to be used as soon as possible for patients with penetrating abdominal trauma, grossly contaminated wounds, massive soft tissue trauma, open fractures, or any patient in whom a long delay until definitive treatment is expected.

s. Casualties should not be completely undressed for a secondary survey in the field. Removal of clothing should be limited to that necessary to expose known or suspected wounds.

t. More specific combat casualty care planning should be carried out and based on specific mission-oriented scenarios.

u. The establishment of Combat Casualty Transport Teams and their use on CASEVAC assets is recommended.

v. Electronic monitoring should be routinely used on CASEVAC assets unless conditions prohibit.

w. MAST trousers are not recommended as a standard of care in any phase.

2. Phases of Care:

a. Phase One: Care Under Fire

1. Return fire as directed or required.

2. Keep yourself from getting shot.

3. Try to keep the casualty from sustaining additional wounds.

4. Stop any massive external hemorrhage with a tourniquet.

5. Take the casualty with you when you leave.

b. Phase Two: Tactical Field Care

1. Airway management

- Unconscious casualty without airway obstruction: > Nasopharyngeal airway.

- Unconscious casualty with airway obstruction: > Cricothyroidotomy.

- Cervical spine immobilization is not necessary for casualties with penetrating head or neck trauma.

2. Breathing

- Consider tension pneumothorax and decompress with needle thoracostomy if a casualty has unilateral penetrating chest trauma and progressive respiratory disstress.

3. Bleeding

- Control any remaining bleeding with a tourniquet or direct pressure.

4. IV

- Start an 18 gauge IV or saline lock.

5. Fluid Resuscitation

- Controlled hemorrhage without shock: > No fluids necessary.

- Controlled hemorrhage with shock: > Hespan 1000cc

- Uncontrolled (infra-abdominal or thoracic) hemorrhage: > No IV fluid resuscitation.

6. Inspect and dress wound.

7. Check for additional wounds.

8. Analgesia as necessary: > Morphine (5mg) IV, wait 5minutes, repeat as necessary.

9. Splint fractures and recheck pulses.

10. Antibiotics

Cefoxitin 2gm slow IV push (over 3-5 minutes) for penetrating abdominal trauma, grossly contaminated wounds, massive soft tissue trauma, open fractures, or any patient in whom a long delay until CASEVAC or definitive treatment is expected.

11. Cardiopulmonary Resuscitation

CPR on the battlefield for victims of blast or penetrating trauma who have no pulse, no respirations, and no other signs of life will not be successful if attempted.

c. Phase Three: Combat Casualty Evacuation (CASEVAC) Care

1. Airway management

Unconscious casualty without airway obstruction: > Nasopharyngeal airway, endotracheal intubation, Combitube or laryngeal mask airway.

Unconscious casualty with airway obstruction: > Cricothyroidotomy if endotracheal intubation and/or other airway devices are unsuccessful.

2. Breathing

Consider tension pneumothorax or hemothorax and decompress with needle thoracostomy if a casualty has unilateral penetrating chest trauma and progressive respiratory distress.

Consider chest tube insertion if a suspected tension pneumothorax is not relieved by needle thoracostomy

Oxygen.

3. Bleeding

Consider removing tourniquets and using direct pressure to control bleeding if possible.

4. IV

Start an 18 gauge IV or saline lock if not already done.

5. Fluid resuscitation

No hemorrhage or controlled hemorrhage without shock: > Lactated Ringer's at 250cc/hr.

Controlled hemorrhage with shock: > Hespan 1000cc initially.

Uncontrolled (infra-abdominal or thoracic) hemorrhage: > No IV fluid resuscitation.

Head Wound patient: > Hespan at KVO unless there is concurrent controlled hemorrhagic shock.

UNITED STATES MARINE CORPS
Mountain Warfare Training Center
Bridgeport, California 93517-5001

FMST.07.37
04/01/02

STUDENT HANDOUT

BURN MANAGEMENT

TERMINAL LEARNING OBJECTIVE. Given a casualty in a wilderness environment, and necessary equipment and supplies, treat burn injuries to prevent death or further injury per the reference. (FMST.07.37)

ENABLING LEARNING OBJECTIVES

1. Without the aid of references, given a list of the depths and appearances of burns, match them to the type of burn, in accordance with the references. (FMST.07.37a)

2. Without the aid of references, given a list choose the major burns that require rapid casualty evacuation, in accordance with the references. (FMST.07.37b)

3. Without the aid of references, given a list choose the first step in the treatment of burn injuries, in accordance with the references. (FMST.07.37c)

OUTLINE

1. **GENERAL:** Each year there are about 2 million individuals burned severely enough to seek medical attention. Around 70,000 require hospitalization. Deaths are on the decrease largely due to the increase in the use of smoke detectors. It has been estimated that 90% of all burn injuries are preventable. Children and young adults are frequently affected by burn injuries.

2. **PHYSIOLOGY:** For non-chemical burns, the primary events of injury occur during the time of heat contact. Coagulation necrosis takes place within cells and denaturation of collagen in the dermis. Blood vessels are completely destroyed or endothelium damage is severe enough to cause clotting which leads to ischemic necrosis of remaining viable cells. Burn wounds are not static. Surrounding the "coagulation zone" is a zone of capillary and small vessel stasis. Circulation becomes stagnant from a strip formed by red blood cells, and aggregation of platelets and white blood cells. The fate of the burn wound depends on the progression of this zone of stasis.

7

3. In patients with burns less than 10% of the total body surface area (TBSA), the inflammatory process is generally limited to the burn site itself. As burns approach 20% TBSA, the local

response becomes systemic. Burns of 30% TBSA or greater, often lead to multiple organ system failure, which in a patient with a severe burn almost invariably leads to a fatal outcome, often due to acute renal failure.

Cardiovascular changes begin immediately after a burn. The degree of these changes depends initially on the size of the burn and to a lesser extent the depth of the burn. Burns less than 15% TBSA can be fluid resuscitated with crystalloid fluids. As burns pass 20% TBSA, massive shifts of fluid and electrolytes occurs. This results in massive fluid shift from the intravascular to the extravascular space. Unless intravascular volume is repleted, classic hypovolemic shock occurs.

3. **TYPES OF BURNS**

a. Scald Burns: Scalding is the most common cause of burns, usually resulting from hot water. Water at 140°F creates a deep partial-thickness or full-thickness burn in three seconds. At 156°F the same burn occurs in one second. Freshly brewed coffee is generally about 180°F. Boiling water always causes a deep burn. Soups and sauces are thicker, remaining in contact with the skin longer. Exposed areas tend to burn less deeply than areas covered with a thin layer of clothes. Clothing retains heat and keeps liquid in contact with the skin longer. Immersion scalds are deep and cause severe burns resulting from longer contact time. These burns are more common in children and elderly patients because of their thin skin. Burns from hot oils are generally deep partial-thickness or full-thickness. Temperatures for cooking oils reach around 400°F, and tar can be up to 500°F.

b. Flame Burns: The next most common burn injuries are flame burns. Although household injuries have gone down with the advent of smoke detectors, a high number of burn injuries still occur from careless smoking, improper use of flammable liquids, automobile accidents, and clothing ignited from stoves or space heaters.

c. Flash Burns: Flash burns result from intense heat for a brief time caused by explosions of natural gas, propane, gasoline, and other flammable liquids.

d. Contact Burns: These burns usually result from direct contact of hot metals, plastics, glass, or coals. Contact burns are usually limited in size, but are deep wounds. With the increase of wood stove usage, this type of burn injury is on the rise in children. Contact burns, dealing with an unconscious patient or involving molten materials, are usually fourth degree. The most common contact burn in the wilderness is from hot coals.

e. Electrical Burns: Electrical burns are actually thermal burns from very high-intensity heat. When electricity encounters resistance from the body it turns into heat. Severity of the burn depends on the portion of the body affected relative to the amperage. The smaller the body part, the more intense the heat is, and the part is less able to dissipate the heat. Consequently the fingers and toes are almost always destroyed with severe damage to the foreams. Larger areas can dissipate the current enough to prevent extensive damage. Arc burns take the most direct route, not the path of least resistance. These are deep and destuctive burns which mostly occur at the joints.

Electrical burns may have other associated injuries that should not be overlooked on the initial assessment. Fall related injuries are common. The intense muscle contraction resulting from electrocution can cause fractures of the lumbar vertebrae, humerus, or femur, and also may cause dislocation of the hips and shoulders. The casualty may have cardiac symptoms of a myocardial contusion or infarction. There can be a conduction system failure or an actual rupture of the heart wall. If there are no signs present of cardiac problems intially after a shock of 110 or 220 volts there is a minute chance they will appear later. Nervous system damage is possible whenever a current passes from one side of the body to the other more than likely affecting the spinal cord.

f. Chemical Burns: Usually caused by strong acids or alkalis and are most often the result of industrial accidents, home use of drain cleaners, assault, and improper use of harsh solvents. Chemical burns continue to burn until they are inactivated by reaction with tissue or dilution by flushing with water. Acids tend to tan the skin, creating an impermeable barrier that limits futher penetration. Alkalis combine with cutaneous lipids and saponify[1] the skin until they are neutralized. The appearance of the wound can be misleading. Unless the observer can be 100% sure, all chemical burns should be considered full thickness.

g. NBC Burns: Nuclear weapons cause burns in two ways: by direct absorption of thermal radiation through exposed surfaces (flash burns); or by the indirect action of fires caused by the weapon (flash burns). The first step in treatment is to prevent contamination to the rescuers and other patients.Vesicants such as mustard gas cause skin burns with edema and blister formation. Burns resulting from vesicants should be flushed with copious amounts of water. Absorbent powder (flour, talcum powder, Fuller's earth) can be used if water is scarce. The powders should be wiped away with a moist towel. The military uses M258A1 kits for skin decontamination. White Phosphorus ignites spontaneously when exposed to air. Remove the air with an air-tight seal; mud is optional. A brief rinse with 1% copper sulfate[2] will impead further oxidation.

4. **CLINICAL PRESENTATION:** The severity of the burn injury is related to the size of the burn, the depth of the burn, and the part of the body that is burned.

a. Burn Size: Burns are the only quantifiable form of trauma. The burn size is the single most important factor in predicting mortality, need for specialized care, and the complications expected from the burn. We can measure burn size with the "rule of nines". The "rule of nines" is where areas of the body represent approximately 9% of a person's total body surface area (TBSA), each leg 18% (the front of the leg 9%, and the back of the leg 9%), the front of the torso represents 18%, the back of the trunk is 18%, the head represents 9%, and the perineum 1%. For infants and small children under four years of age, the head represents a larger percentage 18%, and the legs a smaller percentage 13%, each arm 9%, front troso 13%, back troso 13%, and the perineum 1%. The percentage of the burn area and that of the unburned area need to add up to equal 100%. For smaller burn areas, use the Rule of Palmar Surface: the patient's palmar surface equals about 1% TBSA.

[1] Saponify: to convert fat into soap by treating it with an alkali.
[2] Navy uses sodium bicarbonate instead of copper sulfate

b. Burn Depth: The depth of the burn is either described in degree: first, second, third or fourth, or by the depth of the injury. The following describes burns in terms of partial-thickness or full-thickness. While these descriptions appear to separate burns into defined categories, many burns have a mixture of characteristics making a precise diagnosis difficult.

(1) First-Degree: These burns involve the epidermis only and do not blister. They are most commonly caused by ultraviolet light. An example would be a sunburn. The burned skin is painful and red. It should heal in 7 days without scarring.

(2) Superficial Second-Degree, or Superficial Partial-Thickness: These burns include the epidermis and upper layers of the dermis. They characteristically form blisters, under which the skin is red and moist, these burns are painful to the touch. Blisters can take up to two days to appear. These burns are usually caused by hot liquids. Wounds should heal in 14 to 21 days; they may scar depending on the extent of the burn.

(3) Deep Second-Degree, or Deep Partial-Thickness: Burns that involve deeper layers of the dermis. There is damage to hair follicles and sweat glands. They are characterized with a molted pink and white color with blisters forming immediately. These wounds may be less sensitive to touch than the surrounding normal skin, or maybe tender to touch. The patient often complains of discomfort rather than pain. Capillary refill may be slow or absent, when pressure is applied to the wound. These burns are caused by hot liquids, oil, steam, or flame. They may be difficult to distinguish from Third-Degree/Full-Thickness burns. Healing takes 3-9 weeks. Scarring is probable, the degree is related to the amount and the depth of dermal injury. Surgical grafting may be required.

(4) Third-Degree, or Full-Thickness: Involves the entire thickness of the skin, epidermis through the dermis, down to subcutaneous fat. All structures of the epidermis and dermis are destroyed. The wound is classically described as leathery, firm, and depressed when compared to normal skin. The tissue is charred, pale, and insensitive to light and touch. It is often misdiagnosed as a Deep Second-Degree/Deep Partial-Thickness due to the similar clinical findings. This burn will not spontaneously heal. Surgical repair and/or skin grafts are necessary. There is significant scarring.

(5) Fourth-Degree: These burns involve all layers of the skin, subcutaneous fat, muscle, and bone. These are devastating, life threatening injuries. They almost always have a charred appearance, and often only the cause of the burn gives a clue to the amount of underlying tissue destruction. (FMST.07.37a)

c. Burn Classification: The American Burn Association has devised a classification of burns, dividing them into major, moderate, and minor burns. Patients are placed in groups related to their risk. Low-risk patients are between the ages of 10 and 50 years old. High-risk patients are less than 10 and greater than 50 years old. Poor-risk are patients with underlying medical illnesses such as heart disease, diabetes, or chronic pulmonary problems.

(1) Minor Burns: Imply outpatient treatment. Minor burns involve TBSA of less than 15% in the low-risk group, or less than 10% TBSA in the high-risk group; full-thickness burns of less than 2% TBSA in anyone; and no other injuries.

(2) Moderate Burns: Are partial-thickness burns of 15-25% TBSA in the low-risk group, 10-20% TBSA in the high-risk group; full-thickness burns of equal to or less than 10% TBSA in anyone; burns not involving the hands, face, feet, perineum, or circumferential limbs.

(3) Major Burns: Defined as partial-thickness burns greater than 25% TBSA in the low-risk group, or greater than 20% TBSA in the high risk group; full-thickness burns of greater than 10% TBSA in anyone; burns involving the hands, face, feet, or perineum; burns crossing major joints; circumferential limb burns; burns complicated by inhalation injury; electrical burns; burns complicated by fractures or other forms of trauma; and burns in poor-risk patients. These burns require rapid casualty evacuation. (FMST.07.37b)

5. **TREATMENT:**

a. Initial Care:

(1) The first step in the treatment of burn injuries is to stop the burning process. (FMST.07.37c) Remove patients from heat, remove all their clothes that are affected by hot liquids, shut off electricity (very important), remove chemicals by flushing the affected area for 10 to 15 minutes, and decontamination of NBC. Heat can continue to injure tissue for a surprisingly long time. No first aid will be effective until the burning process has stopped.

(2) Manage the ABC's. Once an airway has been established give oxygen. If qualified, consider intubation earlier rather than later.

(3) Assess for associated injuries such as fractures or lacerations and inhalation injury.

(4) Remove clothing and jewelry from the burn area. This will help prevent constriction from swelling of the burn tissue. Do not try to remove anything that has adhered to the wound.

(5) Evaluate the Burn (depth, extent, pain).

b. General Treatment for the Patient:

(1) Gently wash the burn with tepid water and mild soap to remove any debris and to clean the skin surface around the burn site. Pat dry. Remove the skin from blisters that have popped open (do not open blisters unless necessary for function of hands or feet).

(2) Dress burn with a thin layer of antibiotic ointment.

(3) Cover the burn with Second Skin (if the burn is small enough) or cover with a thin layer of gauze, or clean, dry clothing. Burns on the face, neck and hands may be left open to the air after applying silvadene ointment. Covering the wound reduces pain and evaporative losses, but do not use an occlusive dressing.

(4) When evacuation is imminent, do not redress or reexamine the injury. If evacuation is prolonged, redress once a day. Remove old dressings, reclean (removing the old ointment), and apply fresh ointment and a clean dry covering. (Note: if stuck, soak off old dressings with clean, tepid water.)

(5) Do not pack wounds, which are larger than 20% of the body surface area, in ice. Do not leave wet coverings on burns for more than two hours at a time to reduce the risk of hypothermia.

(6) Stabilize the body temperature. When skin is lost, so is the patient's ability to thermoregulate.

(7) Have the patient drink as much fluid as he or she can tolerate without vomiting. Include some salt in the oral fluids, but do not make these solutions stronger than . 9%. This is the equivalent of a pinch of salt per 8 ounce glass.

(8) IV therapy as follows, Baxters (Parkland) Formula: In the first 24 hours give Ringers Lactate, 4ml times the weight in kilograms times the percent TBSA burned. Half of the solution is given in the first eight hours. The second half is given over the next 16 hours. The second 24 hours should be given fluids to maintain blood pressure, colloids (usually in the form of albumin) are the preferred choice.

4 x wt (kg) x TBSA= amount LR given in the first 24 hours

Example a 70kg patient with 50% TBSA this is the formula for IV therapy for the first 24 hours.

4 x 70kg x 50% = 14,000ml in 1ST 24 hours
7,000ml hours 1-8
3,500ml hours 9-16
3,500ml hours 17-24

c. General Management of the Burn: Caring for the wound itself is often the least important aspect of burn care. All burn wounds are sterile for the first 24 to 48 hours. Burn management is aimed at keeping the wound clean and reducing the pain.

(1) Elevate burned extremities to minimize swelling. Swelling retards healing and encourages infection. Have the patient gently and regularly move burned areas as much as possible.

(2) Burns are tetanus-prone wounds. Check on the last known tetanus shot. The patient may need a renewal.

(3) Ibuprofen is probably the best over the counter analgesic for burn pain (including sunburn).

(4) If you have no ointment or dressings, leave the burn alone. The burn's surface will dry into a scab like covering that provides a significant amount of protection.

6. **INHALATION INJURIES:** Half of all fire-related deaths are from smoke inhalation. Smoke inhalation doubles the mortality rate for any burn. Smoke inhalation is a general term for carbon monoxide poisoning, thermal airway injury, and smoke poisoning.

a. Carbon Monoxide Poisoning: Carbon monoxide (CO) is a colorless, odorless, tasteless gas that has an affinity for hemoglobin 200 times greater than that of oxygen. Carboxyhemoglobin (COHb) levels are measured as the percentage of hemoglobin bound to carbon monoxide. The following COHb levels correlate with clinical symptoms:

<10%	No symptoms
20%	Headache, nausea, vomiting, loss of dexterity
30%	Confusion, lethargy, ECG ST depression
40%	Coma
50%	Death

The half-life of COHb is 4 to 5 hours at room air, 90 minutes with 100% oxygen, and 20 to 25 minutes at 3 ata in a hyperbaric chamber. Treatment for less than 40% COHb is a 100% oxygen. At 40% or greater the hyperbaric chamber should be used. If patient is still in a coma after COHb levels are normal, prognosis is poor, and it has been experience that they rarely awaken.

b. Thermal Airway Injury: Injury to the respiratory tract from inhalation of hot gas or steam. The lower respiratory tract is usually not damaged, unless there was an inhalation of flame or steam. Any patients who were in an explosion, with burns of the hands, face, and upper torso are at risk. Signs of carbon such as soot in the pharnyx should alert the caregiver to airway injury. Maintaining the airway is the important issue here.

c. Smoke Poisoning: Is the inhalation of noxious gases that are the products of combustion. Cyanide poisoning is known to occur in victims of smoke inhalation. Other toxic gases released are sulfur dioxide, hydrogen chloride, phosgene, and ammonia. These gases cause damage to the respiratory tract similar to chemical burns. Physical signs include facial burns, intraoral or pharyngeal burns, singed nasal hairs, soot in mouth or nose, hoarseness, carbonaceous sputum, and expiratory wheezing.

UNITED STATES MARINE CORPS
Mountain Warfare Training Center
Bridgeport, California 93517-5001

WMC
04/09/02

LESSON PLAN

NOMENCLATURE AND CARE OF MOUNTAINEERING EQUIPMENT

LESSON PURPOSE. The purpose of this period of instruction is to introduce you to the types of equipment used here and how to care for it so as to prevent its untimely failure. This lesson relates to all climbing and installation work that you perform here.

1. **ROPES.** All ropes used in the military must meet UIAA standards or U. S. Federal Test Standard 191A. Most ropes have a 5-year shelf life and maximum 2-year service life.

a. Static. Black in color.

(1) Construction. Kernmantle

(2) Minimum tensile strength. 7500 lbs.

(3) Maximum elongation. 1.5%

(4) Diameter. 11mm

(5) Sizes.

(a) 165 ft ± 5 ft

(b) 300 ft ± 10 ft

(6) Usage. Rescue operations and bridging where a low amount of elongation is desirable under a working load.

b. Dynamic. Olive Drab in color.

(1) Construction. Water-resistant treated Kernmantle to reduce friction.

(2) Minimum tensile strength. 6500 lbs.

(3) Maximum elongation. 6%

(4) Diameter. 10.5mm and 11mm

(5) Size. 165 ft ± 5 ft.

(6) Usage. For lead climbing/party climbing.

c. Maxim Dry Rope. Olive Drab or Multi-Colored.

(1) Construction. Water-repellent treated Kernmantle.

(2) Minimum tensile strength. 3472 lbs.

(3) Maximum elongation. 6%

(4) Diameter. 9mm

(5) Size. 150 ft ± 5 ft

(6) Usage. For glacier travel/ice climbing

d. Gold Line II.

(1) Construction. Eight strand braided nylon plymor.

(2) Minimum tensile strength. 4500 lbs.

(3) Maximum elongation. 20%

(4) Diameter. 11 mm

(5) Size. 300 ft or 600 ft spools

(6) Usage. Sling Ropes and litters only.

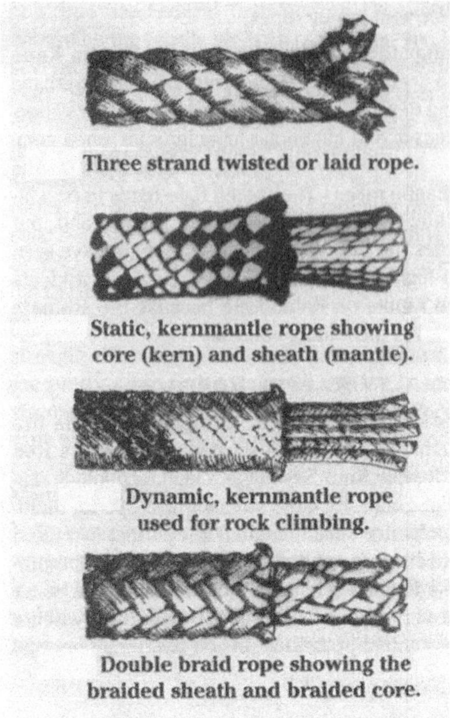

<u>INSIDES AND OUTSIDE SHEATHS OF VARIOUS TYPES OF ROPES</u>

NOTE: Sling ropes are made from 15 foot lengths of plymor or dynamic rope ONLY. Twenty-five foot practice coils should be constructed with static rope, but dynamic rope can be used.

2. (5 Min) **<u>ADVANTAGES/DISADVANTAGES</u>**

a. <u>Advantages of Nylon Rope</u>

(1) High strength to weight ratio.

(2) Good energy absorption in dynamic ropes.

(3) Flexible.

(4) Rot resistant, not affected by frost.

b. <u>Disadvantages of Nylon Rope</u>

(1) Low melting point. Nylon fuses at 400°F and melts at 480°F.

(2) Susceptible to abrasions and cuts.

(3) Affected by chemicals and light.

c. Advantages of Manila Rope

(1) Easily gripped.

(2) Hard wearing.

(3) Does not deteriorate in heat.

d. Disadvantages of Manila Rode

(1) Heavy, kinks, especially when wet. Absorbs water and swells.

(2) Burns at +300°F.

(3) Edible by rodents.

3. (3 Min) **GENERAL INFORMATION**

a. Nylon rope stretches under tension and will rupture at between 30% and 70% elongation, depending on construction.

b. Nylon rope loses as much as 30% strength when wet.

c. Temperatures as low as 250°F will damage a nylon rope.

4. (5 Min) **NYLON WEBBING**

a. The type of nylon webbing available is tubular. Tubular nylon webbing is very strong and flexible. All rules that apply to nylon rope apply to tubular nylon webbing. The size of nylon webbing used is:

(1) 1 inch tubular nylon. Tensile strength approximately 4,000 - 4,500 lbs., depending on the manufacturer.

b. Pre-sewn Spectra Runners. Tensile strength approximately 5,500 lbs.

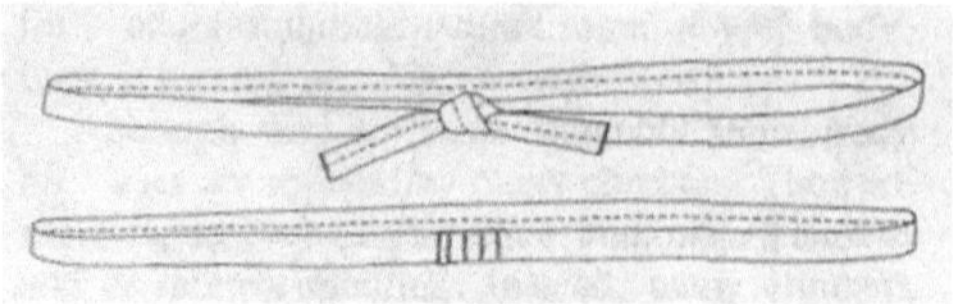

NYLON WEBBING

NOTE: These are minimum strengths. Some manufactures make even stronger webbing.

TRANSITION: We have just discussed general information in nylon webbing, are there any questions? Not only do we use ropes, but we also use carabiners in our installations, we will discuss the types of carabiners used:

5. (5 Min) **CARABINERS**. Also commonly known as snaplinks. Both locking and non-locking are used.

   a. Purpose. Carabiners are used for the following purposes:

      (1) To attach ropes or runners to pieces of protection.

      (2) To attach the rappel rope to the rappel seat for seat-hip rappels or for crossing rope bridges.

      (3) To attach the individuals safety rope to a safety line on a rope installation.

      (4) To form field expedient pulley systems.

   b. Nomenclature of a non-locking carabiner

(1) Gate

(2) Gate pivot pin.

(3) Locking pin.

(4) Body.

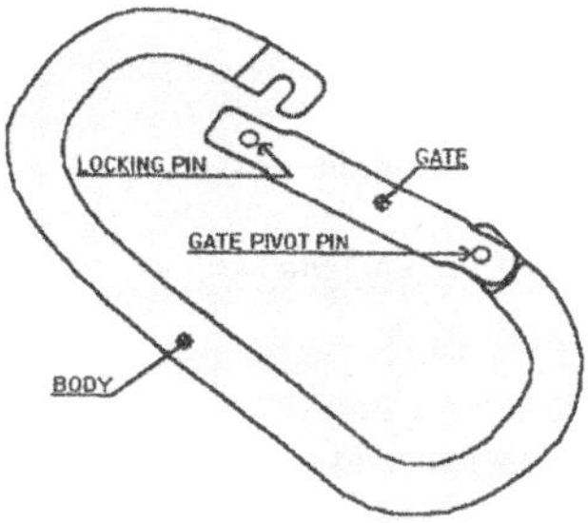

c. Nomenclature of a locking carabiner

(1) Gate.

(2) Gate pivot pin.

(3) Locking notch.

(4) Locking nut.

(5) Body.

NON LOCKI[illegible]UM CARABINER

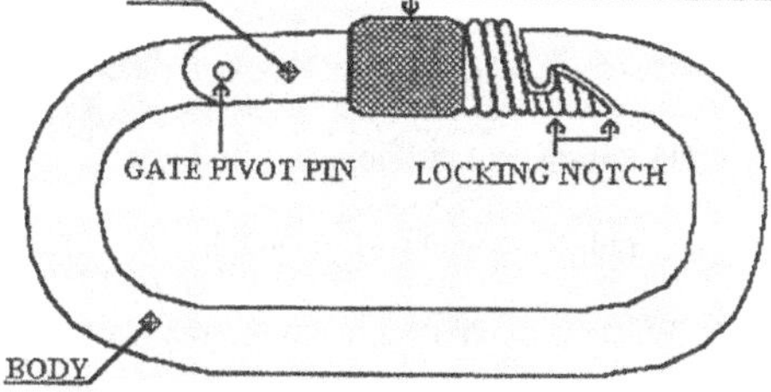

STEEL LOCKING CARABINER

d. There are two types of carabiners used. The two types and their characteristics are:

(1) Steel locking carabiners

(a) Large steel locking "D" (various manufacturers): Minimum tensile strength of 5,500 lbs.

(b) Steel-locking oval Stubai 82 is not in the MAC Kit and obsolete. However, it is being used at MWTC to save money, even though they are beyond the service life. Tensile strength of only 3,300 lbs.

(2) Aluminum non-locking carabiners

(a) Aluminum non-locking oval (various manufacturers): minimum tensile strength of 4,200 lbs.

e. Serviceability Check for a Carabiner. The following steps are used for you to check a carabiner for serviceability:

(1) The gate snaps shut with no friction and with no gap between the locking pin and locking notch.

(2) There is no excessive side to side movement of the gate.

(3) The pivot pin is tight.

(4) The locking pin is tight.

(5) The locking nut travels freely and locks securely.

(6) There are no cracks or flaws in the metal.

NOTE: The weakest part of a carabiner is the gate. If an engraver is used to mark a carabiner, it should be applied to the gate and not the load bearing side.

f. Preventive Maintenance for a Carabiner.

(1) Remove all dirt, moisture and grime.

(2) Lubricate with tri-flow graphite and clean off thoroughly.

NOTE: Whenever you use a locking carabiner ensure that the locking nut is always locked down (tightened).

6. (2 Min) **CARE OF THE CARABINER**. Do not drop the carabiner as this may result in either actual damage to the carabiner or in dirt getting into the workings of the carabiner and damaging it.

UNITED STATES MARINE CORPS
Mountain Warfare Training Center
Bridgeport, California 93517-5001

FMST.07.11
05/01/01

STUDENT HANDOUT

HYPOTHERMIA/REWARMING

TERMINAL LEARNING OBJECTIVE: Given a hypothermia casualty in a cold weather environment and necessary equipment and supplies, manage a hypothermia casualty to prevent death or further injury per the reference. (FMST.07.11)

ENABLING LEARNING OBJECTIVES:

(1) Without the aid of references, select from a given list the definition of hypothermia, in accordance with the references. (FMST.07.11a)

(2) Without the aid of references, select from a given list three of the five mechanisms by which the body loses heat, in accordance with the references. (FMST.07.11b)

(3) Without the aid of references, select from a given list the reason why gentle handling of a hypothermia victim is important, in accordance with the references. (FMST.07.11c)

(4) Without the aid of references, select from a given list the four rewarming techniques used to rewarm a hypothermia victim, in accordance with the references. (FMST.07.11d)

(5) Without the aid of references, select from a given list the proper steps of the sleeping bag rewarming method of rewarming a hypothermia victim, in accordance with the references. (FMST.07.11e)

(6) Without the aid of references, select from a given list the temperature of the water when rewarming a hypothermia casualty in a warm water bath, in accordance with the references. (FMST.07.11f)

8

1. **HYPOTHERMIA**:

a. Definition: A fall in core body temperature to 95°F (35°C) or less. (FMST.07.11a)

b. Hypothermia is potentially lethal. The mortality rate is greater than 50% in severe cases and cases complicated by trauma where the mortality can approach 100%. The most common cause of death in hypothermia is ventricular fibrillation.

c. When speaking of hypothermia, core temperature values must be used. In a controlled environment of an MTF or ER the gold standard is a core temperature taken by rectal or esophageal thermometry. The term "exposure" has often been used to speak of hypothermia but this term is inexact and has numerous other connotations. It should not be used as a synonym for hypothermia.

d. Incidence.

(1) Classically, hypothermia has been thought to occur in extremely cold environments. However, it can occur in temperate climates and at Southern latitudes. In fact, greater than 50% of cases of hypothermia occur at ambient temperatures of 50°F or greater. Contrary to popular belief, the highest incidence of hypothermia is seen in urban areas.

(2) Hypothermia tends tends to occur more frequently in males with a male to female ratio of 2.8:1.0. It's also more common in the elderly, those with drug and alcohol dependency and those with psychiatric diagnoses.

Many standard clinical thermometers don't register a low enough temperature to accurately diagnose hypothermia. Currently, a low-reading rectal thermometer is available through the Federal Stock System (NSN 6515-00-139-4593). All medical personnel attached to cold weather units or deploying to cold weather environments should carry this thermometer. In a high altitude cold weather environment medical providers should be adept at correlating signs and symptoms with specific temperature ranges so as not to unduly expose the patients to the elements. The more accurate, and invasive temperature determination should only be performed when the patient can be adequately protected from the elements.

2. **PATHOPHYSIOLOGY**:

a. Heat Conservation/Loss.

(1) The body is a furnace that generates heat. The ability of the body to thermo regulate depends on the ratio of heat production to heat loss. An imbalance of these two processes can lead to hypothermia. About 75% of food energy consumed goes toward generating body heat while the other 25% goes toward maintaining basic metabolic function.

(2) Thermogenesis can be divided into two main categories:

(A) Shivering thermogenesis creates heat by muscle activity with subsequent breakdown of glycogen. Shivering usually begins at a core temperature of around 97°F (36°C) and finally ceases at core temperatures of 85°- 90°F (30°- 34°C).

(B) Non-shivering thermogenesis creates heat by hormonal interaction with target organs such as the heart. Epinephrine, norepinephrine, growth hormone and thyroid hormone have all been implicated in playing a role in non-shivering thermogenesis.

(3) Changes in hormone secretion patterns are extremely important in acclimatization to cold weather.

(4) Once heat is generated it is retained by three main mechanisms: Peripheral vasoconstriction, abolition of sweating and piloerection. Peripheral vasoconstriction is by far the most important and efficient mechanism of heat conservation. The main internal temperature sensors in the body are located in the hypothalamus in the brain. When these areas sense a drop in core temperature they initiate peripheral vasoconstriction at the surface of the skin to prevent a loss of core body heat.

(5) The body loses heat by five main mechanisms: (FMST.07.11b)

(A) **Radiation** is heat transferred to the surrounding environment via infrared radiation.

(B) **Convection** is the loss of heat via air/water current.

(C) **Conduction** is the direct contact-transfer of heat from a hot to a cold object.

(D) **Evaporation** is the conversion of water from a liquid phase to a gaseous phase (perspiration).

(E) **Respiratory heat loss** takes place when warm humidified air from the lungs is expelled into the atmosphere.

(F) The mechanism accounting for the greatest percentage of heat loss is highly variable and depends on ambient conditions, i.e., a nude man standing in a cold room will lose approximately 65% of his heat by radiation. The same man standing in calm cold water will lose most of his heat by conduction and at a rate 25 times faster.

**NOTE: The liver and brain are the most thermogenic organs on a per weight basis. However, due to their greater mass, skeletal muscle and skin produce the largest percentage of body heat.

(6) Numerous factors predispose to the development of hypothermia. These include but are not limited to:

(A) Malnutrition.
(B) Lack of adaptation.
(C) Inactivity.
(D) Sedatives.
(E) Physical exhaustion.
(F) Burns.
(G) Extremes of age.
(G) Dehydration.
(H) Comorbid illness.

3. **SIGNS AND SYMPTOMS**:

a. Clinical manifestations of hypothermia reflect CNS, and cardiorespiratory involvement.

(1) 98°-95°F (37°-35.6°C): Mild shivering, cold sensation and impaired fine motor coordination.

(2) 95°- 92°F (35°-32.8°C): Violent shivering, difficulty speaking, sluggish thinking, and amnesia. Large muscle in coordination.

(3) 92°-86°F (32.2°-30°C): Shivering is replaced by muscular rigidity. Exposed skin is blue or puffy. Movements are jerky. Dulled sensorium, but victim may be able to maintain posture and the appearance of being in contact with surroundings. Possible atrial arrhythmias.

(4) 85°-81°F (29.4°-27.2°C): Coma, lack of reflexes, possible ventricular arrhythmias. Cessation of shivering below 86oF.

(5) Below 78°F (25.6°C): Failure of cardiac and respiratory centers, pulmonary edema, ventricular fibrillation, and asystole.

b. Some patients with hypothermia will exhibit a phenomenon that is called, "paradoxical undressing."

(1) In these instances the hypothermia victim is found nude with their cloths beside them. If the victims are women they are often misdiagnosed as sexual assault cases. This phenomena results from loss of hypothalamic control over peripheral vasoconstriction which leads to flushing of the periphery with relatively warm blood causing the individual to feel hot. Subsequently the individual undresses to cool down.

c. Again it must be stated that the only way to correctly diagnose hypothermia is with a core temperature reading.

4. **FIELD TREATMENT**:

a. ABC's, C-Spine precautions, remove wet clothing, replace with dry clothes and insulate with a vapor barrier system. (Wrapping the victim in two plastic garbage bags or bubble wrap is ideal. Further insulation may be necessary). Handle the victim carefully to prevent ventricular fibrillation caused by unnecessary jarring and rough handling. (FMST.07.11c) CASEVAC.

b. Victims with altered consciousness should be NPO. Under no circumstances should ETOH be given. This is a peripheral vasodilator and will only enhance heat loss.

c. If CASEVAC is not feasible for a long period of time, or if the victim is only mildly hypothermic, it may be necessary to rewarm in the field by using a proper sleeping bag rewarming technique. Additionally, insulated hot packs placed at the high heat loss areas will help.

d. The hypothermic victim often has no discernible vital signs.

(1) Before beginning CPR thoroughly assess pulses for 30-45 seconds to detect viable cardiovascular activity.
(2) Severely hypothermic victims in ventricular fibrillation and asystole have been successfully resuscitated even after periods as long as 4 hours.

5. **REWARMING**: The re-warming of a hypothermia casualty in a BAS setting is less than optimal since the AMAL for a BAS does not include monitoring devices, laboratory facilities, etc., but rewarming of a casualty can be successfully conducted even in these fairly primitive conditions. The following are three basic techniques of re-warming:

- Passive external re-warming: a sleeping bag or blanket that is not pre-warmed.
- Active external re-warming: a pre-warmed sleeping bag, bear hugger or warm water bath.
- Active internal or core re-warming: IV fluids, warmed/humidified O2, lavage, extracorporal re-warming.

a. Four ideal re-warming strategies that may be available in the field include: (FMST.07.11d)

(1) Sleeping bag rewarming.

(2) Water bath/counter Current rewarming.

(3) Warm/humidified airway rewarming.

(3) Warmed I.V. solutions. NS is preferable as a cold liver does not like to metabolize Lactate.

6. **REWARMING TECHNIQUES (BAS)**:

a. Sleeping Bag Rewarming. (FMST.07.11e) This is the simplest, but least effective method.

(1) Make the diagnosis. Take a rectal temperature to determine if the casualty is clinically hypothermic (95°F or less). Use this temperature as a baseline to determine stabilization.

(2) Warm sleeping bag (pre-warm sleeping bag with two volunteers). Placing a hypothermic casualty into a cold sleeping bag will cause further heat loss by conduction.

(3) Strip casualty. Remove all wet clothing avoiding unnecessary handling. Muscular movement will pump cold blood to the core.

(a) Place the casualty between the two volunteers, if space permits. Their body heat is transferred from the two volunteers to the cold casualty.

(4) Monitor core temperature frequently throughout transport to ensure that the casualty's core temperature is not continuing to drop.

(5) Adequate insulation is required to prevent further heat loss to the environment.

(6) Augment heat by placing insulated heating pads in the high heat loss areas: head, axilla, groin, popliteal region and antecubital fossa. (Be careful not to cause burns)

**NOTE: It should be noted that an evacuation bag can be used; this will provide more room for the casualty and volunteers.

b. Water Bath Re-warming. (FMST.07.11f) This method has a long history of success in rewarming even severely hypothermic casualties. The casualty is placed in a portable/field expedient bathtub or a life raft. The extremities and head are kept out of the water to avoid significant afterdrop. The water bath temperature should be 104°-108°F. An anesthesia temperature probe should be used to constantly monitor the core temperature. Since the casualty will cool the bath water around him, it will be necessary to continue to add warm water to maintain the proper water bath temperature. Keep in mind however that this technique has definite drawbacks. In the event of ventricular fibrillation caused by re-warming shock, unsynchronized cardioversion can be very dangerous in a wet environment. Furthermore, if the casualty in the water is not on a firm surface CPR, if needed, can be very difficult to perform ineffective.

**NOTE: Individual in heat distress can be rapidly cooled using this method with "cool" water instead.

c. Warm Airway Re-warming. A warmed and humidified air/oxygen mixture is used to provide a warming media within the lungs. These devices consist of a method of generating warm humidified air and are usually portable. The warm humidified air modestly increases the amount of heat that can be delivered to core. There should be a thermometer in the airway tubing to monitor the temperature. To prevent injury to the bronchi, the temperature should not exceed 115°F (108° to 115°F is ideal). Warm airway rewarming should be used in conjunction with other rewarming methods. This method, by itself, will probably not deliver sufficient heat to rewarm a severely hypothermic casualty by itself and may in fact blunt the shivering response robbing the body of an important endogenous mechanism for producing heat.

Caveat: When possible, active external re-warming should always be combined with active core re-warming techniques so as to minimize the effects of afterdrop. This occurs as a result of peripheral vasodilatation causing warm core blood to move to the surface while relatively cold surface blood moves to the core and may lower the fibrillation threshold.

**NOTE: Hypothermic submersion incident casualties cannot tolerate humidified air for any length of time.

d. Warmed I.V. Solutions. Hypothermia is a common response to I.V. therapy. Solutions that have been prewarmed have been shown to prevent this complication. There are various methods for warming I.V. solutions:

(1) Crystalloid solutions can be warmed by warm water bath and microwaved with no adverse changes to its integrity.

(2) Fresh Frozen Plasma (FFP; -20°C): Studies have shown that microwaving for 30 seconds followed by gentle manipulation of the bag for 10 seconds, repeated five times will result in only 2.6% cell destruction.

(3) Packed Red Blood Cells (PRBC 4°C): Using conventional rewarming methods, studies have shown that the outer five millimeters of the bag can have isolated areas of hemolysis. Until a device is made that can uniformly shake up PRBC during warming, we do not recommend microwaving of blood. Rewarming of blood by dilution with warm, calcium-free crystalloid solution is useful.

WIND CHILL CHART

WIND	AMBIENT TEMPERATURE (FAHRENHEIT/CENTIGRADE)													
CALM F	35	30	25	20	15	10	5	0	-5	-10	-15	-20	-25	-30
CALM C	27	-1	-4	-7	-9	-12	-15	-18	-21	-23	-26	-29	-32	-34
	EQUIVALENT WIND CHILL FACTOR (FAHRENHEIT/CENTIGRADE)													
5 MPH F	33	27	21	16	12	7	1	-6	-11	-15	-20	-26	-31	-35
8 KPH C	1	-3	-6	-9	-11	-14	-17	-21	-24	-26	-29	-32	-35	-37
10 MPH F	21	16	9	2	-2	-9	-15	-22	-27	-31	-38	-45	-52	-58
16 KPH C	-6	-9	-13	-17	-19	-23	-26	-30	-33	-35	-39	-43	-47	-50
15 MPH F	16	11	1	-6	-11	-18	-25	-33	-40	-45	-51	-60	-65	-70
23 KPH C	-9	-12	-17	-21	-24	-28	-32	-36	-40	-43	-46	-51	-54	-57
20 MPH F	12	3	-4	-9	-17	-24	-32	-40	-46	-52	-60	-68	-76	-81

32 KPH C	-11	-16	-20	-23	-27	-31	-36	-40	-43	-47	-51	-56	-60	-63
25 MPH F	7	0	-7	-15	-22	-29	-37	-45	-52	-58	-67	-75	-83	-89
40 KPH C	-14	-18	-22	-26	-30	-34	-38	-43	-47	-50	-55	-59	-64	-67
30 MPH F	5	-2	-11	-18	-26	-33	-41	-49	-56	-63	-70	-78	-87	-94
48 KPH C	-15	-19	-24	-28	-32	-36	-41	-45	-51	-53	-57	-61	-66	-70
35 MPH F	3	-4	-13	-20	-27	-35	-43	-52	-60	-67	-72	-83	-90	-98
56 KPH C	-16	-20	-25	-31	-33	-37	-42	-47	-51	-55	-58	-64	-68	-72
40 MPH F	1	-4	-15	-22	-29	-36	-45	-54	-62	-69	-76	-87	-94	-101
64 KPH C	-17	-20	-26	-30	-34	-38	-43	-48	-52	-56	-60	-66	-70	-74

7 REWARMING TECHNIQUES (HOSPITAL):

a. Optimal re-warming techniques for higher echelon care facilities still tend to employ rather simple therapeutic modalities. The three safest and most effective are listed below:

- Warmed IVF.
- Warmed/humidified oxygen.
- Bear Hugger.

b. Many texts advocate the use of extracorporal blood re-warming. However, this technique has some disadvantages. It requires a high level of technical and medical support, which are often not available. Since patients undergoing bypass must be anti-coagulated, patients with multiple traumatic injuries may not be good candidates. Lastly, research indicates that anticoagulation with heparin may actually worsen the severity of frostbite that may occur in patients with hypothermia.

c. Lavage: peritoneal, thoracic and mediastinal lavage have all been suggested for use in severe hypothermia. However, the use of these techniques in inexperienced hands tends to carry a much higher complication rate. If performed correctly, peritoneal lavage can be very beneficial with a lower complication rate than thoracic lavage. Gastric, colonic and bladder lavage are indeed safer but are of limited efficacy, as they do not offer access to areas with significant surface area.

d. ACLS: At core temperatures below 86°F attempt three progressive unsynchronized shocks at 200J, 300J, and 360Js. If no response continue CPR and REWARM THE VICTIM. Re-attempt counter-shocks when core temperature rises above 86°F. Above 86°F, perform counter shocks in 2° increments above 86°F, i.e.88°F, 90°F and 92°F. Employ ACLS drug protocols above 93°F along with conventional normothermic ACLS protocol.

UNITED STATES MARINE CORPS
Mountain Warfare Training Center
Bridgeport, California 93517-5001

FMST.07.12
03/21/02

STUDENT HANDOUT

SUBMERSION INCIDENTS

TERMINAL LEARNING OBJECTIVE: Given a casualty in a cold weather environment and the necessary equipment and supplies, manage a submersion incident casualty to prevent death or further injury per the reference. (FMST.07.12)

ENABLING LEARNING OBJECTIVES:

(1) Without the aid of references, given a list of the events, which take place to a submersion victim, select the correct sequence of events, in accordance with the references. (FMST.07.12a)

(2) Without the aid of references, select from a given list, three of the five methods that can be used to increase survival time during cold water immersion, in accordance with the references. (FMST.07.12b)

(3) Without the aid of references, select from a given list, three of the six favorable prognostic signs post submersion incident, in accordance with the references. (FMST.07.12c)

(4) Without the aid of references, select from a given list, three of the four unfavorable prognostic signs post submersion incident, in accordance with the references. (FMST.07.12d)

(5) Without the aid of references, select from a given list the correct field management priorities of a submersion victim, in accordance with the references. (FMST.07.12e)

10

OUTLINE.

1. **TERMINOLOGY**.

a. Definition: A submersion incident is that medical problem that occurs after a casualty has been submersed under water.

(1) Drowning refers to death by suffocation following submersion in water.

(2) Near drowning refers to submersion in water with at least temporary survival (greater than 24 hours).

NOTE: An effort is being made to standardize the terminology used regarding submersion incidents, in order to clarify communication between care-givers. Terms like secondary drowning and other definitions are being discouraged.

2. **PATHOPHYSIOLOGY**.

a. Sequence of events: (FMST.07.12a)

(1) Panic initially, followed by a violent struggle.

(2) Gulping and swallowing air and water to avoid aspiration.

(3) Breath holding until hypoxia leads to unconsciousness.

(4) Once consciousness is lost, the gag reflex relaxes and passive influx of water into the lungs occurs. A small percentage of victims (10-15%) have significant laryngospasm, which prevents any appreciable volume of water to enter the lungs. This is referred to as "dry drowning".

(5) Aspiration is more likely in submersion victims who enter very cold water (10°C). An involuntary gasp reflex occurs, with a marked increase in respiratory rate and depth. This individual is at increased risk of aspiration, and subsequent drowning.

b. Submersion in very cold water will quickly results in hypothermia.

(1) Rapid induction of total body hypothermia (core temperatures of 89.6°F or less) may be protective in submersion incidents. Significant decreases in oxygen requirement result, especially in the case of the brain itself. Marked decreases in cerebral blood flow and oxygen demand have been demonstrated (see Hypothermia lesson). This protective effect of hypothermia has been used to explain the many cases of complete recovery after extended periods of submersion, which have only occurred in cold water.

(2) The "Mammalian Diving Reflex" slows the heart rate, shunts blood to the brain and closes the airway when more primitive marine mammals are exposed to extremely cold water. General consensus holds that this reflex is not a significant response in human beings. However, it has been found to be more prominent in new-born human infants. It has also been speculated that this "Reflex" may be more significant in those individuals who have survived extended submersion times.

c. Fatalities in submersion victims associated with hypothermia are primarily due to three known mechanisms:

(1) Immersion syndrome. Sudden death due to cardiac arrest secondary to massive vagal stimulation. Usually associated with sudden total body exposure to frigid water.

(2) Excessive fatigue and confusion leading to fatal errors of judgment.

(3) Direct hypothermic effects. Thermal conductivity in water is approximately 25 times that of air. Despite the evidence showing hypothermia's protective effects, the complications associated with hypothermia (cardiac arrhythmias, CNS depression, loss of motor control) can lead to submersion and drowning.

d. Some authorities emphasize distinguishing between "wet" and "dry" drowning. In wet drowning, aspiration of water into the lung occurs. (85-90% of casualties). In dry drowning, laryngospasm prevents aspiration of water (10-15% of cases). Ultimately, the end result is the same: obstruction to respiration, oxygen depletion, and asphyxiation. The morbidity and mortality associated with the two types of drowning are not significantly different.

e. Fresh-water and salt-water drowning are somewhat different.

(1) In fresh-water aspiration, the osmotic difference between hypotonic fresh-water and the relatively hypertonic blood leads to net water movement into the vasculature.

This flow across the alveolar membrane also leads to "washout" of surfactant, which leads to collapse of the alveolar spaces. Hypoxemia follows.

(2) In sea-water aspiration, the osmotic shift is reversed, leading to fluid outpouring from the blood stream flooding the alveoli and the interstitium. This leads to decreased lung compliance as well as direct damage to pulmonary capillaries. Again, hypoxemia results.

(3) Hypoxemia is the end result in both types of drowning. No significant difference in morbidity in survivors has been shown between fresh-water and salt-water drowning. The only bodies of water shown to have a worsening effect on prognosis are the hyper-saline bodies, like the Dead Sea and the Great Salt Lake. Otherwise, prognosis is similar.

f. Hypoxemia is the primary insult after submersion. Efforts to correct this problem should be vigorous and sustained. All clinically dead submersion casualties should receive CPR, or ACLS if available, immediately upon removal from the water, regardless of water temperature or duration of submersion. Since the overwhelming majority of naturally occurring water bodies in the world have temperatures below 92°F, all submersion victims should be treated aggressively as hypothermic casualties. Survivals after submersion times up to 66 minutes have been documented. Remember: No one is dead until they are warm and dead.

g. In the case of prolonged submersion or significant aspiration, pulmonary derangements occur rapidly. In lesser insults, the onset of symptoms may be delayed for as long as 24 hours. Physiological changes include:

(1) Cardiovascular. Casualties develop respiratory acidosis and hypoxemia, which may trigger dysrhythmias. Electrolyte changes are not usually significant in survivors of submersion incidents and should correct once spontaneous respirations are re-established.

(2) CNS. C-Spine injuries are common in submersion incidents, most commonly a result of shallow-water diving. Injury patterns include wedge, tear-drop and burst fractures of the upper C-Spine, frequently resulting in tetraplegia. Concussive head injuries also occur. Drug and alcohol use should always be considered.

(3) Metabolic. Respiratory acidosis is the initial problem. Acid/base derangement should correct after resuscitation of the casualty without specific intervention.

3. **SURVIVAL TECHNIQUES IN COLD WATER**.

a. Swimming motion increases heat loss more than it increases heat production. The direct hypothermic effects of cold-water immersion make it impossible to swim for long periods of time. The average swimmer is unlikely to swim more than a kilometer in 50°F water. Swimming is advised only when a rescue vehicle is nearby, or the distance to shore is less than ½ mile.

b. Treading Water and Drown Proofing.

(1) Treading water results in reduced survival time due to increased heat loss secondary to increased activity.

(2) "Drown Proofing" is a technique whereby the swimmer relaxes in the water, floating facedown, arms spread, allowing his face to submerge between breaths. "Drown proofing" is the least desirable survival technique in cold water since the head and neck (high heat loss areas) are maximally exposed.

c. Personal Floatation Device (PFD). Without the aid of the PFD, the swimmer caught in cold water is forced to use less desirable methods to stay afloat.

(1) If wearing a PFD, the individual can utilize the "Heat Escape Lessening Posture" (HELP). This position minimizes the exposure of the swimmer's high heat loss areas because the arms are folded across the chest and pressed to the sides, the knees are drawn up to the chest and the legs are crossed at the ankles. Unfortunately, this position is fatiguing and becomes difficult to hold.

(2) If there is more than one person in the water, the "Huddle" position should be used. Similar to a football huddle, the chest, abdomen and groin area should be pressed together. Children or injured individuals should be placed in the middle of the huddle, since they become hypothermic more quickly.

d. Increasing Survival Time: (FMST.07.12b)

(1) Try to enter the water in a lifeboat or raft.

(2) Locate and don a PFD as quickly as possible.

(3) Wear several layers of clothes, especially to protect high heat loss areas.

(4) If you must enter the water, enter gradually rather than jumping or diving.

(5) If afloat in the water, avoid unnecessary movement. Assume the HELP or Huddle position.

(6) Don't try to swim.

(7) Remember the situation is not hopeless. Have the will to SURVIVE!

4. **RESCUE TECHNIQUES**.

a. Reaching the Casualty. The basic order for water rescue is: Reach and Pull, Throw, Tow and, as a last resort, Go.

(1) Reach and pull. When the casualty is responsive and close to the shore, begin by holding out an object for him to grab then pull him from the water. Floating line is considered the best choice, but if it is not available, a tree branch, fishing pole, oar or other object will do as well.

(2) Throw. Should the casualty be alert, but too far away for you to reach and pull, throw an object that will float to them. A PFD or ring buoy is best, if available. The primary course of action is to throw anything that will float, and to do it as quickly and as accurately as possible.

(3) Tow. Once the conscious casualty has an object to hold onto, try to find a way to tow him to shore. From a secure position, tow the casualty ashore.

(4) Go. Swim to the casualty as a last resort. You should be a strong swimmer, trained in water rescue, in order to avoid the dangers associated with this form of rescue. If the casualty is unconscious or unresponsive, you must go to the casualty.

5. **SIGNS AND SYMPTOMS**.

a. History of Incident. Several questions need to be answered, as they will help determine the approach to the casualty.

(1) When a person is found floating face down on the water surface or is brought to the surface by a rescuer, there usually is little question about the cause of unconsciousness.

(2) When near-drowning is unwitnessed or occurs after a dive or other water activity (surfing), a head injury and a C-spine injury should be presumed, and C-Spine precautions should be instituted.

(3) Determine how long the casualty was submerged. This helps determine prognosis.

(4) How cold was the water? Very cold water (less than 70°F/21°C) may be protective.

(5) What was the casualty's health prior to submersion? Drowning could have been secondary to alcohol ingestion, cardiac arrest or other illness/injury.

b. Casualties may be symptomatic or asymptomatic at the scene:

(1) Respiratory signs and symptoms may include dyspnea, tachypnea (SOB), sore throat, substemal burning, pleuritic chest pain, persistent cough, frothy sputum, cyanosis, rales, rhonchi and wheezes.

(2) Cardiovascular signs may include tachycardia and hypotension.

(3) Neurological findings may include anxiety, restlessness, lethargy, confusion, convulsions, incontinence, hyporeflexia and coma.

(4) Abdominal distention is common as is fever and hypothermia.

c. Favorable prognostic signs: (FMST.07.12c)

(1) Casualty is alert.

(2) Basic or advanced life support is received at the scene.

(3) Good response to initial resuscitation.

(4) Hypothermic.

(5) Older child or adult.

(6) Brief submersion time.

d. Unfavorable prognostic signs: (FMST.07.12d)

(1) Age under three years.

(2) Fixed dilated pupils.

(3) Submersed five minutes or longer.

(4) No resuscitation attempts for the first ten minutes.

(5) PH less than 7.10.

(6) Cardio-pulmonary arrest as initial presentation.

(7) Resuscitation time greater than 30 minutes.

6. **FIELD MANAGEMENT**.

a. The scene of a submersion incident is frequently chaotic and first consideration should be the safety of the rescuers, then retrieval of the casualty from the water.

b. Priorities: (FMST.07.12e)

(1) Establish an airway and check breathing and circulation (ABC's). Rescue breathing, can be performed by a trained rescuer, while the casualty is still in the water.

(2) Observe C-Spine precautions, since cervical spine and head injuries are common, especially in diving accidents.

(3) Begin CPR as soon as possible, if necessary, when the casualty reaches the shore, or suitable rescue platform.

(4) The symptomatic patient should be given high-flow 02 and IV access should be established, if available.

(5) Medevac ASAP.

(6) It is recommended that all victims of submersion be observed in a medical facility for a minimum of 6-8 hours. It is not uncommon for asymptomatic near-drowning victims to refuse further care. All that can be done is attempt to get a contact number in order to follow-up with the victim via telephone 8 hours later.

c. Vomiting is common and should be anticipated by turning the patient's head and having suction ready, if available. Frequently, the cause of vomiting is gastric distention.

(1) The American Heart Association recommends that gastric distention be relieved only when it interferes with CPR.

(2) Gastric decompression can be achieved by "breaking" the patient. This maneuver is performed by placing the patient in the prone position and lifting upwards at the abdomen.

(3) Another method is to turn the patient to a lateral recumbent position and compress the abdomen with the palm of one hand.

(4) The preferred method of gastric drainage is to pass an NG tube and relieve distention with gentle suction.

(5) Gastric drainage should not be attempted without suction equipment available, due to the risk of aspirating vomitus.

7. **PROGNOSTIC PREDICTORS**.

a. 0 - 2 "Unfavorable Prognostic Indicators" yield a greater than 90% survival rate (see above).

b. 3 or more "Unfavorable Prognostic Indicators" yields a less than 5% survival rate.

c. Corm's Criteria: Based on the status of the patient 1 hour after initial resuscitation.

(1) A - Awake	All survive (99+%)
(2) B - Blunted	Most survive (95+%)
(3) C – Comatose	
C1 - Flexor Response	Good survival with ICU care (80+%)
C2 - Extensor Response	Good survival with ICU care (80+%)
03 - Flaccid	Poor survival
C4 - Arrest	Dismal survival

UNITED STATES MARINE CORPS
Mountain Warfare Training Center
Bridgeport, California 93517-5001

FMST.07.14

04/02/02

STUDENT HANDOUT

WILDERNESS ORTHOPAEDIC/ TRAUMA INJURIES

TERMINAL LEARNING OBJECTIVE: Given a casualty in a mountainous enviornment and, necessary equipment and supplies, manage common orthopedic injuries in a mountainous enviornment to prevent death or further injury per the reference. (FMST.07.14)

EABLING LEARNING OBJECTIVES:

(1) Without the aid of references, select from a given list the phases of clearing a cervical spine in the field in accordance with the references. (FMST.07.14a)

(2) Without the aid of references, select from a given list, the correct treatment for an open book pelvic fracture in accordance with the references. (FMST.07.14b)

(3) Without the aid of references, select from a given list, six considerations for orthopedic patient evacuation in accordance with the references. (FMST.07.14c)

OUTLINE

1. INTRODUCTION.

Orthopedic injuries occuring in the wilderness are not significantly different from those suffered in garrison. Bones break, ligaments are torn and strained, and joints are dislocated in both settings. The key to preparing for and managing wilderness orthopedic injuries is understanding that the provider will have significant limitations with regard to diagnostic tools and initial management resources at his/her disposal. The personal risks each expedition member is exposed to can rise dramatically if the team is needed to transport a victim out of the back-country. Determination of a casualty's mobility becomes a major consideration for the wilderness medical provider. Assessment of the terrain, creative use of available natural resources, and effective leadership are all necessary components of caring for orthopedic injuries in the wilderness.

10

←

2. GENERAL: Making the diagnosis.

The wilderness "doc", whatever his/her training background, needs to have good physical exam skills. NO X-RAY and NO LABS will be available in the back-country. These basic tools are not available in the field. The provider will need to rely on his/her observation, palpation, and auscultation skills to make a diagnosis. This person will also need to use the patient, if possible, to uncover the likely injury by taking a thorough history which includes careful review of the mechanism of injury.

A. HISTORY: Critical information can be gained from talking to your patient.

- Mechanism of Injury - Determine:
  1. High vs. low-velocity accident
  2. Direction of force
  3. Nature of activity
  4. Loss of consciousness

- Ask the casualty:
  1. Did they hear a pop, snap, crack, or breaking sound?
  2. How long ago did the injury occur?
  3. Can they move the injured body part?
  4. Was there swelling within 15 minutes of injury or did it take longer?
  5. Was there immediate disability or was there a gradual loss of function?
  6. Was there immediate discoloration of a joint?
  7. Can they bear weight on the injured extremity?

- Other important information:
  1. Evaluate the environment. (i.e. Hot/Cold, Wet/Dry)
  2. Allergies.
  3. Medications currently being taken.
  4. Previous injuries/surgeries to the site in question.

B. PHYSICAL EXAMINATION

Your powers of observation become crucial.

- Look for:
  1. Swelling.
  2. Discoloration and bruising.
  3. Obvious deformity/angulation.
  4. Open wounds. (with or without protruding bone fragments)
  5. Differences right vs. left.

6. Muscle spasm surrounding injured site.
7. Point tenderness.
8. Crepitus.

- Beware:

Consider the environment while assessing your patient
- Don't undress the casualty in the cold!
- Protect your patient from the elements. (Sun, Rain, Cold)
- Talor your physical exam to meet the constraints of the environment.
    * palpate under clothing.
    * visualize one region at a time, then re-dress.
    * set up temporary shelter from the elements.

3. Head Injuries:

a. A blow to the head may lead to increased intracranial pressure (ICP) or intracranial bleeding neither of which are managable in the wilderness. The job of the clinician is to diferentiate a serious life threatening injury from a minor one.

1. Minor injury: No loss of consiousness (LOC), or LOC of less than 15 seconds with immediate return to full alertness. The casualty can not be on medications which increase risk of bleeding or have a history of bleeding disorders. Patient may be monitored every two hours for mental status changes, lethargy, irritibility, persistent nausea and vomiting, changes in speech or visual changes.

2. Serious Injury: LOC greater than 15 seconds, and/or persisitent confusion or memory loss; signs or symptoms of increased ICP: Debilitating headache, mental status changes, persistent nausea and vomiting, appearance of clear fluid in external auditory canal, Battle sign, raccoon eyes, or seizures.

3. Field treatment of serious head injury: Suspect injury of C-spine, manage airway-be able to clear vomit, elevate head 30 degrees. Evacuate to treatment facility ASAP.

4. Spinal Injuries.

A. Cervical Spine

1. High Risk Activities:

C-spine injuries in the wilderness usually occur after either a fall from a significant height or high-velocity accident from skiing. Common winter activities predisposing participants to C-spine injuries are skiing, snow-boarding, and snow-mobiling.

2. Anatomy:

The cervical spine consists of seven cervical vertebrae interposed between the base of the skull and the thoracic spine. The cervical spine has a great deal of mobility to allow maximal range of motion for the skull. This increased mobility comes at a high price; this portion of the spine is less stable than the rest of the vertebral column, and hence more subject to injury. The C-Spine is stabilized by three longitudinal ligaments: The anterior longitudinal ligament (ALL) runs longitudinally along the anterior surface of the vertebral bodies. It is broad and very strong and helps to prevent hyper-extension of the head and neck. The posterior longitudinal ligament (PLL) also runs longitudinally, this time along the posterior surface of the vertebral bodies, within the vertebral canal itself. This ligament is relatively narrow and somewhat weaker than the ALL, and helps to prevent hyper-flexion at the neck. Third is the supraspinous ligament, connecting the spinous processes. This, too, is a strong ligament, acting to inhibit hyper-flexion.

The peripheral nerves innervating the muscles and the sensory nerves of the upper extremities originate from the cervical portion of the spinal cord and these are some of the nerves likely to be affected in the event of a C-Spine injury. Hence, it is important to assess the neurologic status of the upper extremities if a cervical injury is suspected.

3. C-Spine Injury Statistics:

-Most common injury: Flexion injury at C5/6
(so look for a deficit in the C6 distribution)

-28% of C-Spine injuries have another spinal fracture associated with it.
(so examine entire spine)

-10% of Head injuries/Facial fractures also have a C-Spine injury, especially if there was LOC.

4. Remember:

a. Head injury -> Assume C-Spine injury.
b. Perform complete Neurovascular exam
-Motor, Sensory, Reflexes, Pulses, Babinski reflex.

B. Clear a C-Spine. (FMST.07.14a)

When faced with a possible C-Spine injury in the Wilderness, clearing it without X-ray should be considered only in extreme situations. However, there are situations which demand that every effort be made to make a casualty ambulatory, because the movement of the individual would demand herculean effort from the rest of the team, potentially placing more individuals at risk of injury or prolonging exposure to extreme environmental conditions.

Clearing a C-Spine in the field should follow a three-phased assessment. If the patient fails the assessment at any one of the phases, then the team is obligated to maintain spinal precautions and transport the patient. The phases are outlined below, and expanded upon in the lecture.

PHASE ONE:

-ensure no alteration of mental status.
-maintain in-line traction.
-perform complete neurovascular exam (focusing on the upper extremities)
-palpate spinous processes one-by-one while asking casualty whether palpation causes pain (supraspinous ligament)
-distinguish between pain over spinous process and muscle soreness associated with paraspinal musculature.
-only if palpation was completely pain-free can you move on to ...

PHASE TWO:

-continue to maintain in-line traction
-palpate anteriorly to the left and right of trachea. along the anterior vertebral bodies (anterior longitudinal ligament).
-Assess for bony tenderness under the angle of the jaw to the clavicle.
-Distinguish between tenderness of the sternocleidomastoid muscle and actual vertebral pain.
-only if palpation was pain-free can you progress to...

PHASE THREE:

-while loosely maintaining in-line traction, have patient move head through active range of motion

*first flexion/extension
*followed by rotational movement

-stop test at the first indication of pain with movement

Note: If you are able to move the patient's head through a full range of motion without pain, you have effectively cleared his C-Spine. Have the patient resume activity slowly and cautiously; evaluate any other complaints noted.

B. Thoracic and Lumbar Spine:

-Look for a T-L Spine injury with calcaneal fractures. Approximtely 10% of patients with a calcaneal fracture will also have an associated lumbar fracture.

C. Pelvis:

-Think HEMORRHAGE/SHOCK if faced with a Pelvic Fracture
Up to 6 liters of blood loss possible – (internal iliac areteries)

Place gentle constricting wrap around pelvis if open book fracture is suspected. (FMST.07.14b)

4. FRACTURES.

Some common wilderness fractures:

a. Metacarpal/Phalangeal -the hands are your means of interfacing with the environment, sometimes it's a forceful meeting.

b. Distal Radius -Very common wilderness fracture.

c. Scaphoid -check for snuffbox tenderness, pain with axial load to thumb.

d. Lunate -also common, pain dorsally at base of 3rd MC.

f. Clavicle -need good NV exam, check for pneumothorax.

g. Long bone -Radius/Ulna: if proximal, check elbow
- -Humerus: radial nerve runs in spiral groove
- -Tib/Fib: think compartment syndrome- Fib fx may ambulate with cane
- -Femur: think Hemorrhage, think traction.

f. Ankle -commonly a fracture/dislocation, check NV

g. Hip - Leg will typically be externally rotated

REMEMBER:
- Always think HEMORRHAGE with Long Bone fractures
- Always perform NEUROVASCULAR Exam before and after treatment
- Always SPLINT AND PAD for stability, comfort, function and transport.
- Indications for REDUCTION are: NV deficit , severe angulation, severe pain, if angulation prediposes to open fx or makes transport difficult.
- OPEN FRACTURES: Have 6-8 hours to get to surgery. Field Tx- Gently wash off with betadine/ iodine solution to get off obvious dirt- don't scrub. Wrap in sterile gauze. Don't place exposed bone back under skin. Give antibiotics: Augmentin, 2nd or 3rd generation cephalosporin, a quinolone or tetracycline.

THE REASONS WE PLACE FEMUR FRACTURES UNDER TRACTION.

1. Re-establishing length tightens fascia and tamponades bleeding.
2. Dramatic pain relief.
3. Helps prevent open fractures.

4. Helps reduce secondary soft tissue damage.
5. Increases ease and safety of transport.

SOME BASIC RULES OF EXTREMITY SPLINTING:

- Splint all fractures before moving the casualty, unless in imminent danger.
- Splint all fractures as they are found, unless severe angulation complicates transport or is causing neurovascular deficits.
- Splint to include the joint above and below.
- Construct splints using uninjured extremity, then splint in the position of function, comfort or stability.

5. DISLOCATIONS.

a. Rapid Diagnosis and Reduction is imperative, if evacuation time is > 1-2 hours.

- Easier to reduce immediately after injury. (muscle spasm)
- Makes transport much easier. (increased patient comfort)
- Dramatic pain relief.
- Early reduction reduces risk of Neurovascular injury.
- Reduction could make difference between an ambulatory vs. litter patient
  the safety of the entire expedition could be placed at risk

b. Signs and Symptoms: Remember, NO X-RAYS!

- Decreased Range of Motion.
- Obvious Deformity. (compare right vs. left)
- Usually NO Crepitus.
- Typical Posture.

c. Reduction Techniques:

- Phalanges	Linear traction, buddy tape. Can not reduce volar displacement of 1st phalanyx.
- Shoulder	Bedsheet method- counter traction. Prone with weights method. External rotation. Snowbird technique.
- Patella	Extension with medially directed pressure.
- Ankle	Linear traction with knee flexion.

- Hip — Anchor pelvis, hip flexed at 30 degrees, upward traction.

- Sternoclavicular — Roll between scapulae, and apply forceful downward pressure.

-Knee — Should only be reduced as an absolute last resort. Apply anterior traction gently.

REMEMBER: Analgesia and Sedation should be utilized when available before attempting joint or fracture reduction, unless multiple traumatic injuries are present.

6. EVACUATION CRITERIA:

Conditions demanding expeditious evacuation: (FMST.07.14c)

1. Suspected cervical, thoracic, and/or lumbar spine injuries.
2. Pelvic injury with instability and/or significant blood loss.
3. Any open fractures. (6-8 hour window, antibiotics early)
4. Compartment syndromes. (pain with passive motion, pallor, pulseless, paresthesia)
5. Hip or Knee dislocations.
6. Any other injury the medical provider feels he/she is not prepared to manage.

7. OTHER TRAUMA.

Heavy bleeding: Usually can be treated by direct pressure, if a tourniquet is required,release pressure every five minutes to reassess.

Contusion: Apply ice 1st 48 hours which will provide pain relief and limit expansion of hematoma.

Lacerations and avulsions: High pressure irrigation with clean water, may remove debris with flame treated forceps. Use steri-strips for minor clean lacs. Keep dirty or puncture wounds open and give antibiotics.

Animal Bites: Copious irrigation, and antibiotic, do not close. Consider rabies if bite was unprovoked and animal was acting strange. (bats, raccoons, skunks, canines, and felines) Has never been seen in livestock, rabbits, squirrels, rats or mice. Most common zoonotic source in North America is the skunk.

Severed part:	Can be reattached up to 24 hours after injury if the cut is clean. Gently clean part, wrap in gauze and keep cool without freezing.

UNITED STATES MARINE CORPS
Mountain Warfare Training Center
Bridgeport, California 93517-5001

FMST.07.38
04/02/02

STUDENT HANDOUT

REPTILE AND ARTHROPOD ENVENOMATION BITES AND STINGS

TERMINAL LEARNING OBJECTIVE: Given a simulated casualty, in a wilderness environment, treat reptile and arthropod envenomation, in accordance with the references. (FMST.07.38)

ENABLING LEARNING OBJECTIVES:

1) Without the aid of references, select from a given list the first aid measures used for pit viper envenomation, in accordance with the references. (FMST.07.38a)

2) Without the aid of references, select from a given list the first aid measures used for coral snake envenomation, in accordance with the references. (FMST.07.38b)

3) Without the aid of references, select from a given list the first aid measures used for widow spiders (lactordectus species) bites, in accordance with the references. (FMST.07.38c)

4) Without the aid of references, select from a given list the first aid measures used for brown recluse spider (loxosceles species) bites, in accordance with the references. (FMST.07.38d)

5) Without the aid of references, select from a given list choose the first aid measures used for bee, wasp, and hornet stings, in accordance with the references. (FMST.07.38e)

12

OUTLINE.

1. **PIT VIPERS**. Crotalidae have a triangular head, cat-like vertical pupils, hinged fangs and a heat-sensitive "pit" on each side of the head between the tip of the nose and the eye. Rattlesnakes have a variable number of rattles depending upon age and number of molts. They sometimes strike without rattling. About 60% of this country's venomous bites are attributed to rattlesnakes. Cottonmouths (water moccasins) and copperheads are the other two commonly encountered North American pit vipers. Copperhead and cottonmouth venom are quite similar but weaker than most rattlesnake venom.

a. Anatomy of Pit Vipers.

(1) Venom apparatus: Venom gland; Compressor glandular muscle; Primary duct; Accessory gland; Secondary duct; Fang sheath; and fangs.

(2) Facial pit organs: Highly sensitive paired receptors of infrared radiation that can detect temperature changes of less than 0.2°C; used to detect warm-blooded prey/predators and aim strike; also pit organs may have a role in determining volume of venom injected.

(3) Rattles or buttons: Interlocking keratin rings at tip of tail in rattlesnakes; a new rattle is added with each molting cycle (every 50-400 days); some are lost due to trauma.

(4) The strike: Aimed primarily by facial pits; rarely strike farther than 1/2 their body length; speed - approximately 8 feet/second.

(a) Rattlesnakes generally inject 25-75% of their venom when they bite humans.

(b) Can bite without injecting any venom ("dry bite").

1. 20% - no envenomation.
2. 20-30% - mild envenomation.
3. 30-50% - moderate to severe envenomation.

b. Signs and Symptoms of Snake Venom Poisoning:

(1) Pit Vipers:

(a) Local:

1. Puncture wounds/scratches:

A. Pattern can be misleading (venomous vs. nonvenomous).

B. Must differentiate from other animal/insect bites or plant puncture wounds- usually venomous snakebite wounds are larger, with more bleeding.

C. May be one or many fang marks present; range from a few mm to 4cm apart.

2. Pain, edema, erythema:

A. Usually presents within 30 minutes if envenomation.

B. More severe envenomation generally yields more rapid progression and more severe pain.

C. Pain is usually characterized as burning and immediate in onset.

D. Pain usually confined to bite site.

E. Edema limited to subcutaneous tissues (with no increase in intra-compartmental pressures).

F. Usually starts within 5 minutes (if none in 10 minutes, probably no significant envenomation unless dealing with Mojave rattlesnake).

G. Gradually spreads up extremity for 36 hours.

H. Danger of airway obstruction in bites to head and/or neck.

3. Ecchymosis:

A. Ecchymosis starts within several hours.

B. May involve entire extremity.

4. Lymphangitis:

A. Venom absorbed rapidly through lymphatic system results in lymphatic inflammation.

5. Petechiae, vesicles, hemorrhagic bullae:

A. Will occur in untreated rattlesnake envenomation.

B. Rarely seen when treated early with adequate antivenin.

C. Usually occurs 6-36 hr after bite.

6. Necrosis, tissue destruction:

A. Due to direct venom effects.

B, Can be prevented if adequate antivenin is given within 2 hours of bite.

(b) Systemic:

1. Nausea, vomiting:

A. Common (early onset may indicate severe envenomation).

2. Weakness, diaphoresis, chills, dizziness/vertigo.

A. Syncope is common with all pit viper bites.

B. Frequency is proportional to severity of bite.

3. Change in taste:

A. May complain within minutes of rubbery or metallic taste.

4. Increased salivation.

5. Fever.

6. Tingling, numbness in scalp/face/fingers/toes:

A. Can occur within 10 minutes and indicates moderate to severe envenomation.

7. Fasciculations in face/neck/back/ or other involved extremity:

A. Can occur early and indicates severe envenomation.

8. Visual disturbances:

A. Blurred vision, yellowing of vision and blindness.

9. Tachycardia, bradycardia.

10. Hemorrhage, hemolysis, thrombosis and DIC:

A. Bleeding from wounds, gingival bleeding, epistaxis, Hematuria, hematemesis, Melena, lower GI bleeding, hemoptysis, peritoneal hemorrhage and cerebral hemorrhage.

B. Can occur as early as 6 hours; more common 12-72 hours after bite.

C. Systemic bleeding only occurs in moderate to severe envenomation.

D. Most coagulopathic effects secondary to proteases acting at various sites in the coagulation cascade.

11. Hypotension, shock:

A. Can occur rapidly in severe envenomation.

B. Early, due to pooling of blood.

C. Late may be due to loss of volume.

D. Generally little effect from decreased cardiac contractility.

12. Pulmonary edema:

A. Common in severe envenomation.

B. Secondary to toxin and inflammatory injury of pulmonary capillary membranes and pooling of blood in major vessels and capillary beds of the lung.

C. Compounded by direct cardiodepressant factor in some venoms.

13. Oliguria, anuria:

A. Hypotension with resultant decreased GFR. (#1).

B. Hemoglobinuria and myoglobinuria with renal tubular obstruction.

C. Direct toxic effect of venom on kidneys.

D. Acute tubular necrosis from rhabdomyolysis.

14. Paresis, Paralysis:

A. Seen with Eastern Diamondback and Mojave rattlesnakes.

B. Uncommon with other rattlesnakes.

15. Respiratory failure:

A. Uncommon with pit viper bites (except Mojave).

B. Complicated by cardiovascular failure.

16. Seizures:

A. Probably due to hypertension and hypoxia.

17. Coma:

A. Usually sensorium is clear.

B. When coma occurs, it is secondary to cerebral anoxia.

18. Death:

A. Generally occurs in 6 to 48 hours in untreated severe envenomation.

B. Can occur sooner with IV envenomation.

c. First Aid Measures for Pit Viper Envenomation. (FMST.07.38a)

(1) Do not try to kill the snake and bring it in for identification.

(a) Only vital to identify the snake when Mojave rattlesnake or coral snake bites are a possibility, as management is altered.

(b) Risk of a second bite to the victim or rescuer.

(c) Should never delay transportation of the victim.

(2) Remove any jewelry, which could become a tourniquet with progression of edema.

(3) Calm and reassure patient to decrease heart rate and circulation of venom.

(4) Constriction band.

(a) Most widely accepted first aid measure in the literature.

(b) Should be more than 1/2 inch wide.

(c) Apply to proximal end of extremity affected and wrap to approximately 5 cm above swelling (site of envenomation). Then wrap back towards proximal end of extremity from where the wrap started.

(d) Apply only tight enough to occlude lymphatic and superficial venous return.

(e) Probably no benefit when applied 30 min or more after bite.

(f) Maintain until antivenin is started.

(5) Incision and suction.

(a) Very controversial; should only be used by experienced medical personnel.

(b) Do not incise wound.

(c) Suction best applied by mechanical means (Sawyer extractor).

(d) Do not use mouth (unknown oral lesions may expose rescuer to risk).

(e) Should be started within 30 minutes of envenomation.

(f) Continue suction for 30 to 60 min.

(6) Rest and immobilization.

(a) Put victim at rest and splint extremity in position of function, at or just below heart level.

(b) Allow room for swelling in splint.

(7) Ice.

(a) Avoid any method of cooling.

(8) Watch for adverse reactions.

(a) Rare documented cases of anaphylaxis caused by snake venom in patients previously bitten or otherwise exposed to snake venom.

(b) Some evidence that venom can activate the alternate-complement system directly and cause a similar reaction.

(9) Rapid transport to a medical facility.

(a) The most important measure in first aid.

(b) No measures should delay this; antivenin is the only proven definitive therapy.

2. **CORAL SNAKES**. The elapidae species of coral snakes are brightly colored, with black noses and alternating red-yellow-red-black bands around their bodies (remember "red on yellow, kill a fellow"). They have relatively small mouths with fixed fangs. From southern Mexico through tropical South America, the rules for distinguishing coral snakes are highly unreliable. Unless you are a knowledgeable herpetologist, it is best not to pick up colorful snakes in the tropics.

a. Signs and Symptoms of Coral Snake Envenomation:

(1) Fang marks may be hard to see.

(2) Frequent delay in onset of symptoms, followed by extremely rapid progression.

(3) Little or no pain at bite site.

(4) No local necrosis.

(5) Earliest evidence may be drowsiness or euphoria.

(6) N/V, increased salivation, paraesthesias at bite site.

(7) Bulbar paralysis progressing to peripheral paralysis.

(8) Paraesthesias, and fasciculation.

(9) Occasional seizures.

(10) Mild Hypotension.

(11) Death is due to respiratory and cardiovascular failure.

b. First Aid Measures For Coral Snake Envenomation. (FMST.07.38b)

(1) None of proven benefit except rest, reassurance and rapid transportation to medical facility for antivenom administration.

3. **GILA MONSTERS**. No direct injection mechanism for their venom, but have powerful jaws that allow them to chew and tear at their victims. Drooling venom produces a substantial amount of pain and possible complications resembling pit viper envenomation. Follow the recommendations for pit viper envenomation. Fatal Gila monster encounters are extremely rare.

4. **VENOMOUS ARTHROPODS**. Arthropods represent the largest phylum in the animal kingdom. It contains spiders, scorpions, insects, ticks, kissing bugs, water

bugs, caterpillars, moths, butterflies, grasshoppers, centipedes, and millipedes, among others. Some arthropods sting (bees, ants, scorpions, etc.), others bite (spiders, centipedes, kissing bugs, etc.), while still others discharge a secretion that is toxic (millipedes, caterpillars, etc.). Although the number of arthropods that are sufficiently venomous to endanger humans is not known, these animals are implicated in far more bites and stings than all other phyla combined. Almost all of the 20,000 species of spiders are venomous, but luckily for, man, only relatively small numbers have fangs long and strong enough to penetrate the human skin. There are some 500 species of scorpions and all are venomous, although only a small number pose a significant danger to humans.

a. The number of deaths from arthropod stings and bites is not known, nor do most countries keep records of the incidence of such injuries. In Mexico, parts of Central and South America, and in North Africa, deaths from scorpion stings may exceed several thousand per year. Spider bites probably account for no more than 200 deaths per year, worldwide. The number of deaths from arthropod bites or stings in the temperate countries is far greater than the number of deaths from snakebite. Almost all of these deaths, however, are from anaphylactic reactions.

5. **SPIDERS**. -There are at least 200 species of spiders that have been implicated in significant bites to humans (Russell, 1988a). Spiders got the misplaced blame for the bites seen in one large series of case studies. Of some 600 suspected spider bites, 80% were found to be caused by other arthropods. The arthropods most frequently involved in these mis-diagnoses were ticks (including their embedded mouth parts), kissing bugs, mites, bedbugs, fleas, flies, beetles, water bugs and various other stinging arthropods.

a. Lactordectus species (Widow spiders): It is most commonly referred to as the Black Widow, Brown Widow or Frog-legged spider depending on the species. There are many other commons names such as: hourglass, poison lady, deadly spider, red-bottom spider, T-spider, Gray Lady spider, or shoe-button spider. Although both male and female widow spiders are venomous, only the latter have fangs large and strong enough to penetrate the human skin. Mature females range in body length from 10 to 18mm, whereas males are from 3 to 5mm.

(1) Signs and Symptoms of spider envenomation.

(a) Sharp, pinprick-like bite, but in some cases the incident is so minor that it goes unnoticed.

(b) Initial bite-pain is often followed by dull-numbing pain in the affected extremity, and cramps in one of the large muscle masses.

(c) Muscle fasciculations can be seen within 30 minutes of the bite.

(d) Sweating is common.

(e) Complaints of weakness and pain in the regional lymph nodes.

(f) Pain in the lower back, thighs, or abdomen is common and rigidity of the abdominal muscles is seen in severe envenomations.

(g) Severe paroxysmal muscle cramps may occur accompanied by arthralgias.

(h) Tissue reaction around the bite may vary, depending on the species involved. In most cases in the United States, there is no reaction at the bite site and the puncture wounds may be impossible to find. There may be an immediate 2 to 4 mm blanched area around the puncture wounds, surrounded by a slightly erythematous area, which in time becomes pallid. This may persist for 20 to 30 minutes.

(i) Hypertension is a characteristic finding in moderate to severe poisonings. It usually appears 2 to 3 hours post envenomation.

(j) Elevation of body temperature.

(k) Increased salivation.

(1) Anorexia.

(2) Treatment for Widow Spider envenomation.
(FMST.07.38c)

(a) First aid measures are supportive in nature. If the pain is intense, ice can be placed over the wound until the patient arrives at a medical facility.

(b) Antivenin is given in severe cases at medical facilities.

b. Loxosceles species (Violin spiders): Are known in the United States as fiddle-back, or brown recluse spiders. There are over 100 species of Loxosceles. The abdomen of these spiders varies in color from gray to orange or reddish-brown to dark brown. The violin on the cephalothorax is brown to blackish and distinct from the pale yellow to reddish-brown background of the cephalothorax. This spider has three pairs of eyes, forming a curved row. Females average 8 to 15 mm in body length whereas males average 6 to 10 mm. Both are venomous.

(1) Signs and Symptoms of Violin spider envenomation.

(a) Pain; about the same degree of a bee sting but patients may be completely unaware of the bite.

(b) Localized burning sensation around the site of injury.

(c) Pruritus is often present and the area begins to appear red, with a small-blanched area around the immediate bite site.

(d) The reddened area enlarges during the subsequent 1- 8 hours. It often becomes irregular in shape and, as time passes, hemorrhages may develop throughout the area.

(e) A small bleb or vesicle forms at the bite site and increases in size.

(f) Vesicle subsequently ruptures and a pustule forms.

(g) The whole area becomes swollen and painful.

(h) Lymphadenopathy may develop.

(i) Necrosis can develop from superficial to deep layers.

(j) Rare systemic signs and symptoms include:

1. Malaise.
2. Stomach cramps.
3. Nausea and vomiting.
4. Jaundice.
5. Spleen enlargement.
6. Hemolysis.
7. Hematuria.
8. Thrombocytopenia.

(2) Treatment for Loxosceles spider envenomation: (FMST.07.38d)

(a) There are no first-aid measures of value. In fact, all first-aid measures should be avoided, because the natural appearance of the lesion is most important in determining the diagnosis.

(b) An ice pack may be placed over the wound if the pain is severe.

(c) Excision of the bite site is no longer recommended as this has been demonstrated to cause worsening of necrosis and tissue damage. Definitive therapy includes use of dapsone, antibiotics, tetanus prophylaxis and possible steroid use.

c. There are many other species of spiders that include Steatoda species (cobweb spiders), Phidippus species (jumping spiders), Chiracanthium species (running spiders), Lycosa species (wolf spiders), and the newest member to the United States is the Hobo spider. There have been several reported deaths in the Northwestern United States. Originally a European species of brown spider, the Hobo spider was introduced into the U.S. in 1936, and little research has been done on it to date.

6. **SCORPIONS**. Approximately 75 of the 650 species of scorpions can be considered of sufficient importance to warrant medical attention. In the United States, members of the genera Hadrurus and Vejovis are capable of inflicting a painful and erythematous, but non-lethal lesion. However, species of the genus Centruroides are sufficiently dangerous to warrant definitive medical care. Most stings are inflicted by the Vejovis species.

a. Centruroides species. There are approximately 30 species of this genus confined to the New World. Of these, eight are of considerable medical importance, and most of these species are found in Mexico. In the United States, they are commonly referred to as "bark scorpions", because of the preference for hiding under the loose bark of trees, in dead trees, or in logs. Their general color is straw to yellowish-brown or reddish-brown, and they are often easily distinguishable from other scorpions in the same habitat by their long thin tail with its "telson" on the end, the pedipalps, or pincer-like claws. They can reach lengths up to 55 mm.

(1) Signs and symptoms of envenomation.

(a) Initial pain, rarely severe.

(b) Area becomes sensitive to touch (tap test).

(c) Tachycardia may ensue after one hour.

(d) Respiratory rate increases.

(e) Fasciculations may be seen over the face or large muscle masses.

(f) Complaints of generalized muscle weakness.

(g) Respiratory distress may progress to paralysis.

(h) Slurring of speech may be present.

(i) Convulsions have been reported.

(2) Treatment for scorpion envenomation.

(a) There is no first-aid measure of value for scorpion stings.

(b) Ice may be placed over the wound to reduce pain.

(c) In severe cases, antivenin may be recommended.

b. Vejovis species is closely related to the Urectonus. Both genera are ground scorpions. Whereas Urectonus usually inhabits mountain habitats from southern California to Oregon, Bejovis has a wide distribution from the southern portions of Canada, south through Wyoming and Colorado to Texas and west to California. Some species measure up to 85mm, but most are from 30 to 35mm. Vejovis spinigerus, a common western U.S. species, is often called the striped-tail scorpion.

c. Hadrurus species is native to North America. These are the longest and most stout-bodied of our scorpions, generally referred to as the giant hairy scorpions because of their size and the conspicuous bristles on their legs, pedipalps and caudal segments. Adults may can measure up to 135 mm. These are burrowing species and may be found as deep as 2 feet in sandy soil. They are native to Arizona, California and parts of Utah, Nevada and Idaho, as well as Mexico.

7. **BEES, WASPS, HORNETS AND ANTS**. While it may take over 100 bees to inflict a lethal dose of venom in most adults, one sting can cause a fatal anaphylactic reaction in a hypersensitive person. There are 3 to 4 times more deaths in the United States from bees than from snakebites. In the few fatalities that have resulted from multiple bee stings, death has been attributed to acute cardiovascular collapse.

a. Treatment for bee stings: (FMST.07.38e)

(1) The stingers of many bees may remain in the skin and should be removed by teasing or scraping rather than pulling, to avoid even more venom being injected into the patient.

(2) Ice may be applied to the sting site.

(3) Persons with known hypersensitivity to stings should carry a Bee-Sting Kit, containing an antihistamine and epinephrine when in endemic areas.

(4) Desensitization can be carried out using whole-body antigens or, preferably, whole-venom antigens.

8. **TICKS AND MITES**. Carry many diseases. In North America, some species of Dermacentor and Amblyomma cause tick paralysis.

a. Signs and symptoms of Tick Paralysis.

(1) Restlessness and irritability.

(2) Lethargy and anorexia.

(3) Generalized muscle weakness.

(4) Incoordination and ataxia.

(5) Nystagmus.

(6) Ascending flaccid paralysis with loss of deep tendon reflexes.

(7) Bulbar or respiratory paralysis may develop.

b. Treatment for Tick Paralysis:

(1) Remove the tick.

(2) Treat symptomatically (ventilator support may be necessary).

(3) Lesions should be cleaned.

(4) Antivenin and/or corticosteroids are used in severe cases.

9. **LEPIDOPTERA (CATERPILLARS)**. Caterpillars commonly cause envenomation in humans. There are at least ten families of caterpillars that are venomous.

a. MEGALOPYGE OPERCULARIS (Puss Caterpillar or Woolly Slug) The most commonly seen venomous caterpillar in the United States.

(1) Description: Hairy, flat, ovoid shaped: 30-35 millimeters in length.

(2) Location: Texas north to Maryland and Missouri.

(3) Stinging apparatus: Consists of spines intermingled with the hairs of the body:

b. SIGNS AND SYMPTOMS.

(1) Needling pain (Can be intense).

(2) Redness.

(3) Swelling.

(4) Nausea.

(5) Headache.

(6) Fever.

(7) Vomiting.

(8) Lymphadenopathy.

(9) Hypotension.

(10) Shock.

(11) Bleeding of the mucus membranes.

a. Treatment

(1) Apply contact tape to affected area, and remove tape. This will effectively remove spines from affected area.

(2) Group I corticosteroid cream and ointment.

(3) Codeine or Demerol may be required for pain.

(4) Antiemitics may be needed to control vomiting.

Note: Anaphylactic reactions have been reported, but are rare.

UNITED STATES MARINE CORPS
Mountain Warfare Training Center
Bridgeport, California 93517-5001

FMST.07.10

04/02/02

STUDENT HANDOUT

FREEZING AND NEAR FREEZING TISSUE INJURIES

TERMINAL LEARNING OBJECTIVE: Given a casualty in a cold weather environment, and necessary equipment and supplies manage common cold weather injuries in a cold weather environment to prevent death or further injury per the reference. (FMST.07.10)

ENABLING LEARNING OBJECTIVES.

1) Without the aid of references, from a given list select the correct definition of frostbite, in accordance with the references. (FMST.07.10a)

2) Without the aid of references, from a given list select the correct two mechanisms of injury for frostbite, in accordance with the references. (FMST.07.10b)

3) Without the aid of references, from a given list select the correct signs and symptoms of frostbite, in accordance with the references. (FMST.07.10c)

4) Without the aid of references, from a given list select the correct favorable post thaw signs of a frostbite injury, in accordance with the references. (FMST.07.10d)

5) Without the aid of references, from a given list select the correct unfavorable post thaw signs of a frostbite injury, in accordance with the references. (FMST.07.10e)

6) Without the aid of references, from a given list select the correct field treatment for frostbite, in accordance with the references. (FMST.07.10f)

12

7) Without the aid of references, from a given list select the correct primary field treatment consideration when treating frostbite in the field, in accordance with the references. (FMST.07.10g)

8) Without the aid of references, from a given list select the correct four methods to reduce the chance of becoming a frostbite casualty, in accordance with the references. (FMST.07.10h)

9) Without the aid of references, from a given list select the correct definition of immersion foot, in accordance with the references. (FMST.07.10i)

10) Without the aid of references, from a given list select the correct mechanism of injury for immersion foot, in accordance with the references. (FMST.07.10j)

11) Without the aid of references, given a list of categories and characteristics match the category of immersion foot to its characteristics, in accordance with the references. (FMST.07.10k)

12) Without the aid of references, from a given list select the correct field treatment of immersion foot, in accordance with the references. (FMST.07.10l)

OUTLINE.

1. **FROSTBITE**

a. Definition: Frostbite is the actual freezing of tissue. (FMST.07.10a)

(1) Frostbite usually occurs as a result of exposure to subfreezing temperatures for a long period of time. However, brief exposure to high wind-chill or very cold temperatures can also cause frostbite. Frostbite usually occurs after more than twelve hours of exposure to subfreezing temperatures. Skin temperature of the extremities is dependent on heat production and is more a function of underlying muscle mass than of circulation. Thus, the areas of the body most prone to frostbite (hands, feet, ears, nose, chin and molar regions) have small amounts of underlying muscle tissue.

(2) Incidence. Frostbite, unlike hypothermia tends to spare the extremes of age. Thus, it is most prevalent in the middle years of life and like hypothermia, the greatest incidence occurs in urban areas. Other factors predisposing a person to frostbite include: poor nutritional status, dehydration, prior cold injury, individuals from the southeastern part of the United States, diabetics, individuals with atherosclerotic disease, and use of tobacco products.

(3) Although the actual numbers are not known, persons born in the south are 3.7 times more susceptible to frostbite. Persons with a history of prior cold injury are 1.7 times more prone to have a second injury. This includes parts of the body other than the originally frostbitten region.

b. Mechanisms of Tissue Injury. Tissue injury resulting from frostbite is produced in two ways: (FMST.07.10b)

(1) *Actual freezing of the tissues.* Ice crystal formation in the extra-cellular space leads to an osmotic gradient across the cell membrane drawing water out of the cell. This process, known as crenation, causes a marked change in intra-cellular electrolyte and enzyme concentrations, which can lead to cell death or recovery depending on the duration of exposure to freezing conditions and post thaw care of the tissue. Intra-cellular ice crystal formation is much more destructive, however, this requires much faster freezing rates.

(2) O*bstruction of blood supply to the tissues.* Cell injury and death caused by ice crystal formation leads to the release of inflammatory mediators, which damage vascular endothelial cells. The net result is increased vascular permeability. Fluid leaks out of the vascular space and into the intercellular space causing tissue edema and increased blood viscosity. Elevated viscosity combined with platelet aggregation, to the walls of damaged blood vessels, lead to vascular obstruction with resultant tissue damage.

c. Signs/Symptoms. (FMST.07.10c) Initially there is discomfort and pain in the affected tissues, which can progress to loss of sensation. The skin appears white and bloodless. There is decreased or no capillary refill. The skin may feel waxy and will not glide freely over a bony prominence. With minor injury, blisters form, usually in the first 24 hours, and edema may also be present. Deep injury is not associated with substantial formation of blisters. There is also little edema seen in deep injury.

(1) Favorable prognostic signs of frostbitten tissue. These signs indicate probably little or no tissue loss, post thaw. (FMST.07.10d)

(A) Warmth of tissue.

(B) Normal tissue color.

(C) Preservation of sensation.

(D) Blister formation. *A* blister extending the full distal length of the injured digit is a good prognostic sign. However, a blister limited to the line of demarcation between normal and frostbitten tissue is a less favorable sign.

(E) Edema persisting more than 24 hours.

(F) Rapid capillary refill.

(2) Unfavorable prognostic signs. These indicate a poor prognosis with probable tissue loss, post thaw: (FMST.07.10e)

(A) Complete absence of edema.

(B) Cyanotic tissue.

(C) Continued loss of sensation.

(D) Affected areas continue to stay cold.

(3) Since it is extremely difficult to predict ultimate tissue loss from initial examination, any classification must be done retrospectively. Classically, frostbite is divided into four degrees of severity:

(A) First degree - Erythema, edema, transient tingling and/or burning.

(B) Second degree - Blisters, edema, anesthesia and/or paresthesias.

(C) Third degree - Involves the entire thickness of skin and extends into subcutaneous tissue.

(D) Fourth degree - Involves the entire thickness of the skin along with muscle and bone.

(F) Frostnip is a completely reversible injury that is technically not considered frostbite. Frostnip classically presents as a white patch and numbness and is readily warmed in the field by placing the affected extremity in one's axilla or groin. No resultant skin changes occur.

d. Field Management (FMST.07.10f)

(1) The correct treatment for frostbite is as follows:

(A) Once an extremity has suffered frostbite, the primary consideration of treatment is not to rewarm the region if there is a chance of refreezing. (FMST.07.10g) This occurs because the amount of tissue damage is greatly increased if the part is thawed then allowed to re-freeze. Generally, the sooner a frostbitten region is thawed the lower the probability and extent of damage. Rapid rewarming is the preferred method of thawing a frostbitten region. This consists of immersion of an affected part in circulating warm water of 104°-108°F until the tissues are soft and pliable, which usually takes about 40-60 minutes.

(B) If slow rewarming methods (such as allowing a part to sit at room temperature) is used, some refreezing of the melted fluid occurs and the new ice crystals are actually larger than those formed during the original freezing causing further tissue destruction.

(C) Out dated methods such as rubbing with snow or rewarming next to an open fire, etc., only enhance tissue damage.

(D) Once thawing has taken place no benefit of continued rapid rewarming can be demonstrated.

(E) Cold injuries that are old and have thawed spontaneously should not be rapidly re-warmed again.

(F) Nicotine use should be prohibited since it is a vasoconstrictor and may enhance tissue loss.

(G) There is no need for ointments, salves, etc.

(H) Blisters should be allowed to remain intact.

(I) Medevac all cases of frostbite, as soon as the tactical situation permits.

(J) One final note; in military tactical situations such as ours, the victim often must walk down below the snow line to reach a BAS or base camp where rapid rewarming can be carried out. In this situation let the victim walk if he has frostbitten feet. Once his feet have thawed, he will be unable to walk and will have to be carried out.

(K) Hospital and long-term care: After re-warming the frostbitten tissue, attention must turn to prevention of infection and functional restoration. The patient's tetanus status should be determined and meticulous care should be taken to keep the injury clean. Daily whirlpool therapy for 30-45 minutes has been shown to be beneficial. The use of post-thaw medications is controversial but there is evidence that topical aloe vera and NSAIDS to control inflammation are probably beneficial. Some studies have demonstrated benefit from the use of allopurinol as a free radical scavenger. Also, animal studies have shown a definite benefit from the use of 90% DMSO gel, however, this use is not FDA approved on humans and it may be difficult or impossible to obtain this medication. The wide scale use of heparin and dextran 40 has been suggested but a definitive recommendation that supports their use still cannot be made. Studies suggesting the use of thrombolytics are very small and not well controlled therefore there is no current recommendation for use of thrombolytics. Lastly, it has been shown that there is a clear benefit for the use of early and aggressive physical therapy to preserve function and prevent contracture.

e. Prevention (FMST.07.10h)

(1) Some general principles for the prevention of frostbite are listed below

(a) Dress to keep comfortably cool. The head radiates 30-50% or more of all body heat and so goes the old adage, "If your feet are cold put on your hat".

(b) Nutrition. Eat the right type and quantity of foods.

(c) Avoid tight fitting and constrictive clothing. This decreases blood flow and increases the chance of frostbite. Rings, jewelry, etc., conduct heat away from the skin and increase the chance of frostbite.

(d) Wear mittens instead of gloves. They are much warmer. For work involving a high level of dexterity use contact gloves.

(e) Do not touch metals with bare hands at extremely low temperatures. Your flesh will stick to the surface and this will result in subsequent frostbite and tissue loss. If you find yourself in such a predicament pour a warm fluid over the metal to thaw your flesh off the metal.

(f) Avoid nicotine.

2. **IMMERSION FOOT**.

a. Definition. (FMST.07.10i) Immersion foot (AKA trenchfoot) is a non-freezing injury of the extremities in which the tissues are damaged.

b. Usually a disease of the feet (but it can occur in the hands), it occurs when the extremity is exposed to a cold-wet environment (external temperatures above freezing and less than

$50^{o}F$) for long periods of time (usually greater than 12 hours). Prolonged standing enhances the pathologic potential for injury due to wet and cold conditions.

c. Mechanism of Injury. (FMST.07.10j) Even though this injury is called immersion foot, actual immersion of the feet in water is not necessary. This injury can also occur in the hands.

(1) The circulation of blood to the affected extremity is reduced because the victim's extremity is cold, wet and the peripheral blood vessels have constricted to reduce overall heat loss.

(2) Because prolonged cooling damages sensitive tissues, nerve injuries are also involved. This nerve damage is responsible for the severe pain and paresthesias. Damaged skin and subcutaneous tissues result in erythema, edema and possible blister formation.

d. Signs/Symptoms Early signs include numbness and paresthesias followed by complete numbness and sometimes leg/arm cramps. The skin at first appears erythematous but becomes progressively pale and mottled then grayish-blue as the condition progresses. Often there is a large amount of edema and blister formation. In severe cases the arterial pulses are absent. Immersion foot is classified by severity: (FMST.07.10k)

CATEGORY	CHARACTERISTICS
Minimal	Reddening of skin; slight sensory changes
Mild	Edema; sensory changes (reversible)
Moderate	Edema, redness, blisters, intra-cutaneous hemorrhage; irreversible nerve damage
Severe	Severe edema, massive intra-cutaneous hemorrhage; necrosis, gangrene

e. Field Management (FMST.07.10l) Treatment consists of:

(1) Pat drying of extremity.

(2) Gentle rewarming.

(3) Elevation of affected extremity.

(4) Bed rest.

(5) Treatment after this is supportive. For more severe injuries, debridement or resection of a portion of the foot may be necessary. However, this is best accomplished at a higher echelon of care. Do not be fooled by this injury. These patients can get

extremely ill from wet-gangrene which may necessitate amputation of tissue. However, the initial treatment should be as conservative as possible. DO NOT TREAT THESE PATIENT'S WITH WATER IMMERSION THERAPY. If infection is present obtain surgical consult early, and plan on small sequential amputations.

3. **CHILBLAINS (PERNIO)**.

a. Definition. Chilblains are a non-freezing injury that leads to red inflamed, pruritic skin.

b. Mechanism of Injury. Pernio is caused by intermittent exposure to above freezing temperatures and high humidity. (More extreme cold is usually associated with lower humidity). Classically, chilblains occur in young individuals and most commonly on the anterior surface of the legs and the dorsum of the hands. The cause of pernio is not clearly known, but it is thought to reflect an abnormal vascular response to the environment.

c. Signs/Symptoms. Skin is red, tender, and pruritic, edematous and blanches with pressure. Itching may be severe and hyperhydrosis can occur.

d. Management. The field treatment is massage and gentle heat. This treatment is usually self-limiting and will not recur without re-exposure. Topical steroids are effective for itching and burning.

4. **CORNEAL FROSTBITE**

a. Definition. Actual freezing of the cornea.

b. Mechanism of injury. This occurs when large amounts of cold air pass over and cool the cornea, resulting in corneal clouding and loss of vision. This can occur as the result of even short periods of exposure to activities associated with high wind chill i.e. snowmobiling.

c. Signs/Symptoms. On examination the cornea will be clouded over.

d. Treatment and Prevention. Antibiotic ointment and DO NOT PATCH THE EYES. Prevention is the key.

UNITED STATES MARINE CORPS
Mountain Warfare Training
Center Bridgeport, California
93517-5001

FMST.07.22
04/03/02

STUDENT HANDOUT

LAND NAVIGATION REVIEW

TERMINAL LEARNING OBJECTIVE In a mountainous environment, navigate in mountainous terrain, in accordance with the references. (FMST.07.22)

ENABLING LEARNING OBJECTIVES

1) Given a map and compass, orientate the map, in accordance with the references. (FMST.07.22a)

2) Given a map and a protractor locate a six-digit grid coordinate, in accordance with the references. (FMST.07.22b)

3) Given a map, compass, and protractor perform a resection to locate an unknown point from three known points, in accordance with the references. (FMST.07.22c)

4) Without the aid of references, select from a given list the correct definition for true north, grid north, and magnetic north, in accordance with the references. (FMST.07.22d)

OUTLINE.

1. **THE MAP**. Before you can properly use a map, there is some basic information you will need to understand.

a. Definition. A map is a reduced or scale drawing of the ground and important things on the ground as seen from the air. Essentially it is a picture of the surface of the earth, as it would appear looking at it from an aircraft.

b. Marginal Information. As you take a quick look at the borders, you will notice that there is a large amount of information. When using a map, the instructions are placed around the outer edges of the map and are known as marginal information. All maps are not the same, some having more marginal information than others, and the information being located in different places around the border. It is important that you know how to use this information if you plan to get the most out of your map. The different information on the margins is as follows:

(1) Sheet name. The sheet name is found in two places: the center of the upper margin and lower right margin of the map. The map is usually named after it's outstanding cultural or geographical features. Whenever possible, the name of the largest city is used.

(2) The sheet number. The sheet number is found in the upper right margin of your map and is used as a reference number for the map sheet.

(3) Series name and scale.

(a) The series name is found in the upper left margin of your map. A map series usually comprises a group of similar maps on the same scale and on the same sheet lines or format designed to cover a particular geographical area. It may also be a group of maps designed to serve a common purpose. The name given a series is of the most prominent area.

(b) The scale notation represents the ratio of the maps distance to the corresponding (actual) distance on the earth's surface. For example, the scale notation of 1:50,000 indicates that one unit of measure on the map equals 50,000 of the same units of measure on the actual surface of the ground (1 map inch = 50,000 earth inches).

(4) Series number. The series number appears in the upper right margin and lower left margin. It is a comprehensive reference expressed either as a four digit number or as a letter followed by three or four numbers.

(5) Index to adjoining sheets. The index to adjoining sheets is found in the lower right margin of your map.

(6) Bar scales. The bar scale is located in the center lower margin of your map.

(7) Contour interval. The contour interval is located in the center lower margin of your map, just below the bar scales. It states the vertical distance between adjacent contour lines on the map. For example, every contour line represents an increase in elevation as indicated by the distance of the contour interval.

(8) Northern References. There are 3 different northern references located on the declination diagram. It is located in the right portion of the lower margin of your map.

(a) True North. A line from any position on the earth's surface to the North Pole, which is the exact top of the earth.

(b) Grid North. The north direction established by the vertical grid lines on the map.

(c) Magnetic North. The direction to the North Magnetic Pole.

(9) Legend. Is located in the lower left margin. It illustrates and identifies some of the symbols on the map. There is not enough room on a map sheet to show the true outlines of objects and land features so the mapmaker uses a set of standard symbols to represent them. Sometimes the symbols are not the same, or standard types of symbols typically used. This is because of the type of map, the scale, or the origin. Because of this possible difference, to prevent errors in symbol identification, the legend should be referred to every time a map is used. (The example on the legend may state, "unimproved road, hard surface road, jeep trail," etc. The right side of the symbol will have three symbols; the first being "unimproved road", the second "hard surface road", and the third being "jeep trail".)

(10) Map colors. The colors you see on your map are not there to make the map pretty or enjoyable to the user. They have important meaning. The colors vary with different types of maps, but on a standard, large scale map, there are five basic colors:

(a) Black. Represents most of the man-made features.

(b) Red. Represents major roads.

(c) Blue. Represents water.

(d) Green. Represents vegetation.

(e) Brown. Represents all terrain features, such as hills, draws, fingers, depressions, and saddles by the use of contour lines. These contour lines are shown in brown.

(f) Other Colors. Occasionally other colors are used to show special information. These, as a rule, will be indicated in the marginal information. Currently, there are changes being made to the color system, which may result in changes of the marginal information in regards to map colors.

2. **HORIZONTAL SCALES**.

a. Bar Scale. This special ruler, located at the bottom of the map sheet, enables you to measure a distance between 2 points on your map and change that map distance into the actual ground distance.

(1) The primary scale. The primary scale is the part of the scale that starts at the zero (0) and extends to the right.

(2) The extension scale. The second part of the bar scale is the extension scale. It starts at the zero (0) and extends left.

b. Measuring Straight-Line Distance. If you are given a mission that requires you to move from your present position to a different position some distance away and you plan to travel in a straight line between these two points, the following method is used to determine the distance you will have to travel:

(1) First locate your present position and your destination on your map. Then you will lay a straight edged piece of paper on the map so that the edge of the paper touches the location of your present position and the location of your destination. Make a tick mark on the edge of the paper at each point. Now move the paper down to the bar scale. Place the left tick mark on the zero (0) of the scale so that the other tick mark will fall somewhere on the primary scale.

c. Measuring a Curved Line Distance. If you are given a mission that requires you to move from your present position to a different position and this time you choose to move along a road to reach your destination, the following method is used to work out the ground distance you will have to travel:

(1) Find your present position, your destination, and the road you will follow, on your map. Position a straight edged piece of paper at your location and make a tick mark,. Now, rotate the paper so that it runs along the road until the road turns, and make a tick mark on both the paper and the map. Keeping these 2 tick marks together, rotate the paper again until the road turns, and again mark the paper and the map. Continue this procedure until you have reached your destination. Now using your bar scale as previously stated, find the actual ground distance your squad will be traveling.

3. **ELEVATION AND RELIEF**. In order to begin learning the subject of elevation and relief, let's first look at some definitions.

a. Contour lines. Of the several ways to indicate elevation and relief on maps, the most common is through the use of contour lines. A contour line represents an imaginary line on the ground along which all points on this line are the same elevation above sea level.

b. Characteristics of Contour Lines. There are a few characteristics of a contour line that you should know that will help you understand what they represent on your map. These characteristics are as follows:

(1) Contour lines never cross.

(2) Contour lines connect on themselves.

c. Types of Contour Lines. In addition to knowing the definition, purpose, and characteristics of a contour line, you need to realize that there are different types of contour lines.

(1) Index contour lines. Starting at zero elevation, every 5th contour line is drawn with a heavier line.

(2) Intermediate contour lines. These are the contour lines falling between the Index contour lines.

(3) Supplementary contour lines. These lines are shown as dashed brown lines.

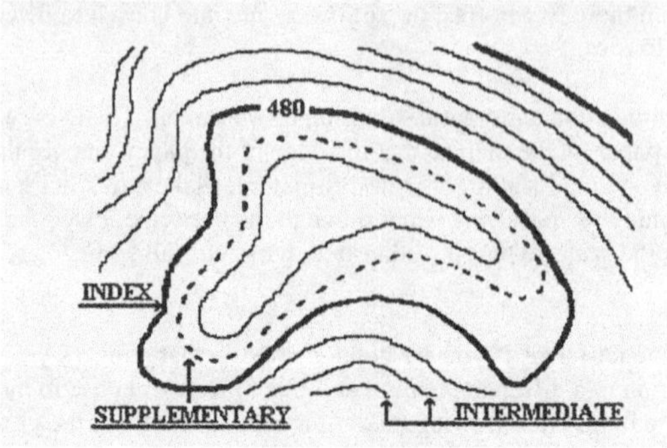

d. Determining Elevation of a Point. Now, you are ready to learn how to determine the elevation of any point of the ground represented on your map.

(1) Given elevation nearest the point, examine the area around your point until you find the nearest point that has a marked elevation. There are a few different ways that contour lines are marked, and they are:

(a) Labeled index contour line. Index contour lines are broken with the elevation of the line printed in the break.

(b) Spot elevations. At various places on your map, usually on prominent landforms such as road junctions or hilltops, a spot elevation will be given. It will be printed in either brown or black.

(c) Bench marks. Similar to Spot elevations, but more accurate. They are marked on your map by a black "x", with the letters "BM", and the elevation.

(d) A point on a contour line. If a point is located on a contour line, its elevation will be the same as that contour line.

(e) A point between contour lines. If a desired point is less than 1/4 the distance between 2 lines, its elevation is considered to be the same as the closest contour line. If the point is from 1/4 to 3/4 the distance, its elevation is considered to be in the middle of the 2 lines.

(f) A point on top of a hill. To estimate the elevation on the top of an unmarked hill, add 1/2 of the contour interval to the elevation of the highest contour line around the hill.)

4. **TERRAIN FEATURES**. There are many types of terrain features that you must be able to identify. By interpreting the arrangement of contour lines, you should be able to get a mental picture of how these terrain features would appear if you were actually out there on the ground looking at them. Some of the features you need to be able to recognize are as follows:

a. Hills. Defined as a point or small area of high ground. A hill is represented on your map by a number of contour lines each of which circles around and connects onto itself. These rough circles get smaller as they progress toward the top of the hill. You need to carefully compare the arrangement of these contour lines to the actual appearance of the hill.

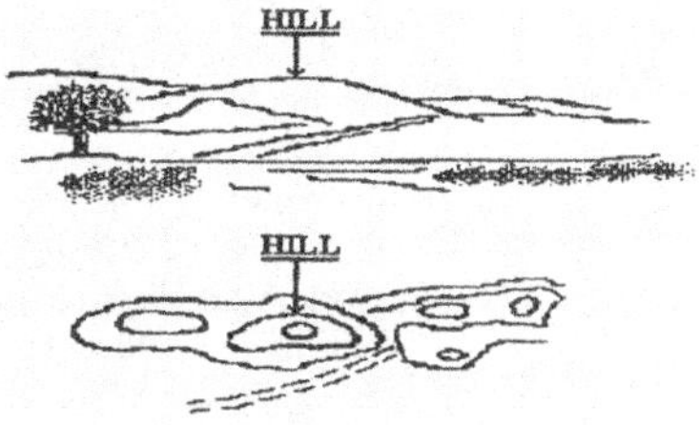

b. Spur. Defined as a line of high ground. The points along the top or crest of the spur are higher than the ground on both sides. A spur is represented on the map by a number of contour lines, each of which form a U-Shaped or V-Shaped design. The end or bottom of the U or V-Shaped lines point down to a lower elevation.

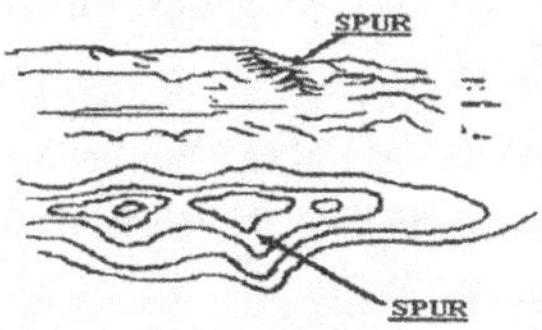

c. Draw. Defined as a line of low ground. The points located in the bottom of a draw are at a lower elevation than the ground to either side. A draw usually runs between 2 spurs. It is sometimes difficult to tell the difference between the contour lines that represent a draw from those that represent a spur because they are both U or V-Shaped lines. Remember that the curved portion representing a spur points down to ground of lower elevation, and the curved portion representing a draw points to ground of higher elevation. Now by looking closely at the area on your map, you will be able to tell which of the 2 you are dealing with. In many cases, draws will contain a stream, while spurs usually never have a stream running down their centers.

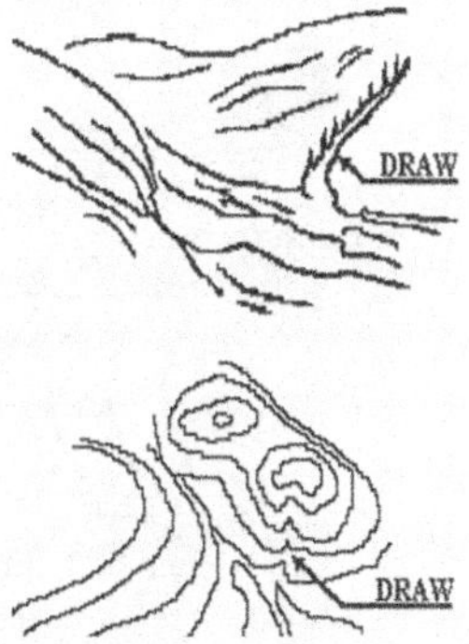

d. Saddle. Defined as the low ground between 2 hilltops, or a dip along the crest of a spur. On your map a saddle is represented by contour lines that form 2 hilltops that are located closely to each other. The area between these 2 hilltops being the location of the saddle.

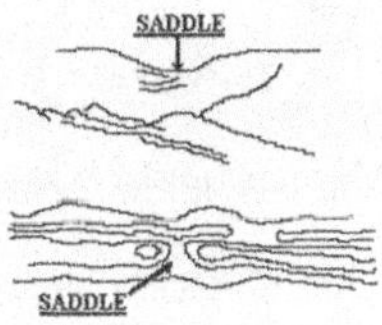

e. Depression. Defined as a point or area of low ground surrounded on all sides by higher ground. A depression is represented by contour lines forming circles within a small area on the map. These rough circles look identical to hilltops except for the presence of hachures. These are short stubby lines that are connected to the contour line at one end and point away from the contour lines at right angles in a downhill direction.

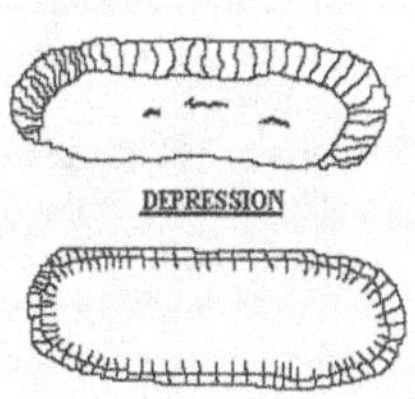

f. Cut. A cut is a man-made feature by which the bed of a road or railroad is leveled by cutting through high ground.

A cut is represented by a sudden ending of contour lines, the ends being connected by a straight line. Hachures are connected to this straight line and point downhill toward the roadbed. This sudden ending of contour lines occurs on both sides of the road bed.

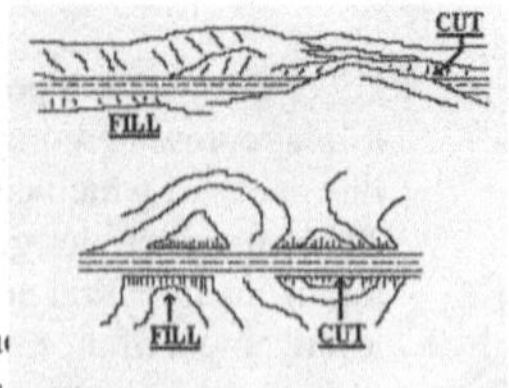

g. Fill. A fill is also a man-made feature by which the bed of a road is raised by filling in the low area. A sudden ending of contour lines can also represent a fill. A straight line connects these ends with Hachures pointing downhill away from the road bed. Again, this occurs on both sides of the road bed.

5. **SLOPES.** The rate of rise or fall of a ground form is known as its slope. In combat, this factor of slope must be considered when making plans for an operation. You must have the ability to recognize the ground forms in your area of operation by studying your map and the spacing between the contour lines on your map.

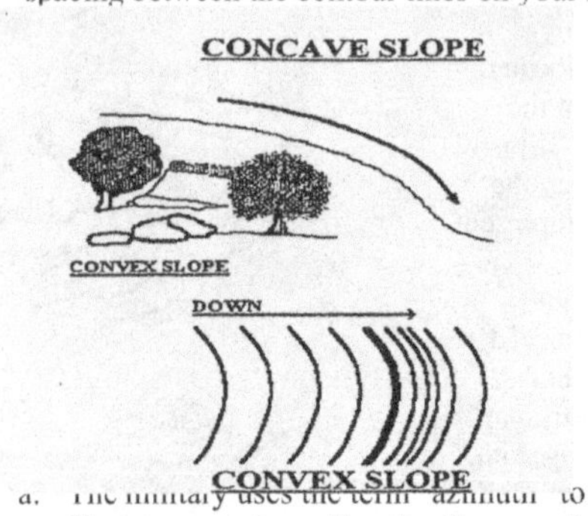

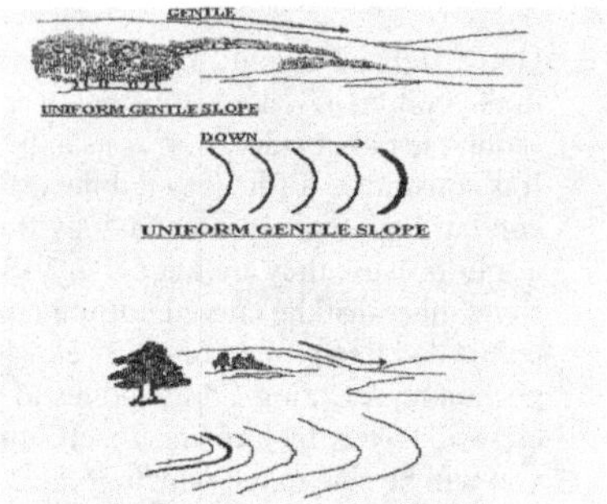

a. The military uses the term "azimuth" to describe the direction from one point to another. There are two base direction lines used to measure the value of an azimuth. They are as follows:

(1) Grid north. It is used when measuring direction from your map.

(2) Magnetic north. Used when measuring direction from your compass.

b. Units of Measurement. Now that you have learned about the two base direction lines that are used to measure direction, we need to learn the units of measurement that are used to measure the angle from one of these direction lines to the direction line.

(1) Degrees. The degree is the unit of measurement that is used most often. Anytime you need an azimuth to move from one point to another, you will use degrees as your unit of measurement. The base direction line, grid north or magnetic north, will be zero (0) degrees and measuring clockwise from this base direction line, the angle increases until you reach 360 degrees which brings you back around in a full circle to the base direction line.

(2) Mils. The mil is also used as a unit of measurement for direction. It is used when working with indirect fire weapons such as mortars and artillery. The azimuth you send will have to

be in mils, not in degrees. Remember, you are still measuring the angle between the base direction line and the line of sight from you to the distant object only you are using a different unit to measure the angle. Moving clockwise from the base direction line, which would be zero (0) mils, the angle back to the direction line. Now you know there are 360 degrees and also 6400 mils in the circle. This means that each degree equals approximately 17.5 mils.

c. Converting Map and Compass Directions. (WMC.5.2d) Earlier you learned that an azimuth was always measured from a base direction line. You also learned that we can use two different base direction lines. Grid north is the base direction line used when you are working with the map and getting a map direction or a grid azimuth. Magnetic north is the base direction line used when you are using the compass to get a compass direction or a magnetic azimuth. Because these two different base direction lines are used and because inmost cases there is a difference of one or more degrees between them, it is necessary for you to learn how to convert from a grid azimuth to a magnetic azimuth or from a magnetic azimuth to a grid azimuth. In order to learn this procedure we must take another look at the declination diagram.

(1) Declination diagram. A diagram that shows the relation between three different norths true north, grid north and magnetic north. It is the difference between grid north and magnetic north that we need to look at. The difference between these two north's is known as the grid-magnetic angle or G-M angle.

(a) Grid-Magnetic angle. The grid-magnetic angle is the angular difference between grid north and magnetic north. As you can see on this diagram a line that connects grid north and magnetic north represents the G-M angle. The size of the angle is stated off to the left side of the diagram in both degrees and mils. The G-M angle on your maps is 18 degrees or 320 mils. It is important to know that the G-M angle will change depending on where you are located on the earth's surface.

(2) Converting. When you receive a mission as a squad leader to move your squad from your present position to some distant point, you take out our map, a protractor and figure out the map direction or the grid azimuth to that distant point. However, before you can use your compass to direct you to your destination, you will have to convert the grid azimuth you have from your map, to a magnetic azimuth that you can set on your compass.

(a) The first step in making this conversion is to examine the declination diagram on your map. The G-M angle on your map is 18 degrees. You know that you will have to add the G-M angle to, or subtract it from the azimuth you measured on your map, or shot on your compass.

(b) The instructions as to whether you will add or subtract this angle are written on the map depending on whether you are going from grid to magnetic or vise-versa.

d. Orientation of a Map. (WMC.6.14a) In order to gain the best use of your map and to avoid confusion while navigating across terrain, you will need to know how to orient your map to the ground. *A* map is oriented when it is lying flat and level with it's north and south pointing in the same directions as north and south on the ground. There are two methods you can use to orient a map.

(1) Inspection. The first method used to orient the map is by inspection. By carefully examining your map and the ground around the area in which you are located to find linear features such as roads, railroads and powerless, you can align the feature on the map with the same feature on the ground. If there is only one linear terrain feature in the area in which you are located, you will have to be careful not to reverse your directions. Even though you have a road running north-south on your map aligned with the same road on the ground, the north end of your map may be pointed in a southerly direction down the road. To avoid this from occurring, find another terrain feature in the area that you can also identify on the map. Once you have aligned your map along the linear feature, check to insure that the second feature is positioned on the same side of the linear feature on your map as it is on the ground.

(2) Compass. This is the second method of orienting a map. You begin by placing your map on the ground so it lies flat and level. Open your compass so that the cover and eyepiece are completely open. Align the graduated straight edge of the compass along one of the north-south grid lines on your map with the cover end of the compass pointing toward the top of the map. This will place the stationary index line parallel to the grid north. Since the north seeking arrow on the dial of your compass will always point to magnetic north, you set your compass up just like a declination diagram. Grid north being the stationary index line and magnetic north being your north seeking arrow. Now rotate your map, insuring that the graduated straight edge of your compass remains aligned with the north-south grid line, until the face of the compass appears exactly as the declination diagram in the lower margin of the map. Your map is now oriented. For the Sonora Pass map sheet you would rotate the map until the north seeking arrow on the compass lies 16.5 degrees to the right of the stationary index line.

7. **GRID SYSTEM.** (WMC.6.14b) In most combat situations it is very important to be able to pinpoint your location, the location of another unit, or the location of an objective. In order to do this, it is necessary to have a location system that will allow you to accurately determine a specific location and state that location in such a way that will allow others to find that location quickly. This system is known as a grid system.

a. Grid Box. This box gives basic instructions on reading grids in determination of specific points on the map.

b. Grid Lines. The straight black lines that run across, and up and down the surface of your maps form the 1000 meter grid squares. These lines are called grid lines. If you look around the border of your map you will see that both the horizontal and vertical grid lines are identified by a two digit number.

c. Grid Square Identification. Each 1000 meter grid square on your map has it's own unique number. This number is called a grid coordinate. A four digit number is the result of combining the two digit number identification of the vertical grid line and the two digit number identification of the horizontal grid line that cross to form the lower left corner of the grid square. To find the grid coordinate of any grid square on your map you will always read your map right and up. You find the two digit number of that vertical grid line and you will have the first 2 digits of your coordinate. Next, find the two digit number of the horizontal grid line and place it after your vertical number and you now have the 4 digit grid coordinate that identifies your grid square.

d. Locating Terrain Features Within a Grid Square. If you are given a 4 digit grid coordinate to find a certain point on the earth's surface, that grid coordinate would put you within 1000 meters of that point. If you are told to find a certain feature within that grid square and you are good at reading your map, a 4-digit coordinate may get you close enough to your point to enable you to find it. However, there are many situations in which a 4 digit number will not get you close enough to a given position to allow you to accomplish your task. It is necessary to be able to locate points within a grid square and using a six-digit grid coordinate does this.

(1) Six Digit Grid Coordinate. The 1000-meter grid square will have to be broken down further into 100-meter grid squares. Because the surface of your map would become too cluttered by all these additional lines, a special instrument has been made to help you divide up a 1000-meter grid square. This instrument is called a coordinate scale or protractor.

8. **RESECTION**. (WMC.6.14c)

a. You will face a situation where you need to pinpoint the location of your position. This may not be possible by associating the terrain in your area with the map because you may have a poor map or the terrain in your immediate area is flat and has no distinguishing features. When you find yourself in this situation you will need to know how to perform a resection.

(1) Examine the terrain around your position and find two terrain features or objects that you can also identify on your map.

(2) Use your compass to find the magnetic azimuth to one of the distant terrain features you have identified.

(3) Convert the magnetic azimuth you found into a grid azimuth.

(4) Change the grid azimuth you found into a back azimuth.

(5) Drawing a line on your map from a distant point you identified in step one back toward your present position. Place a small mark on your map at this point. Using a straight edge, draw a line from where you plotted the distant object to the mark you made by the edge of your protractor. This line should be extended to insure it continues through the area where you are located.

(6) Repeat, finding the magnetic azimuth to the second distant object. That is at least 60 degrees from the first.

(7) When you have drawn your second line on the map, the point where the second line crosses the first line is the location of your position. The process that has just been covered is called a two-point resection.

(8) If your position is located on a linear line feature such as a road, power line, railroad or river, then it is only necessary to perform a one-point resection in order to pinpoint your location. Instead of identifying two distant points, you will only need to select one. Where the single line that you drew on your map crosses the linear feature on which you are located will be the location of your position.

9. **DETOURING AROUND AN OBSTACLE**.

a. There will be occasions when it will become necessary to move off your plotted route in order to avoid an obstacle such as a small lake or swamp. You must be able to move around these obstacles and return to the route you were following before you took your detour. This can be accomplished by using one of three detour methods: Far side landmark method, nearside landmark method, and the 90-degree offset method.

(1) Far side landmark. Used during hours of daylight when you are able to identify some feature on the far side of the obstacle that lies on your line of march and is obvious enough so that you can find it once you have moved around to the far side.

(2) Nearside landmark. Used during hours of daylight when you cannot find a landmark on the far side. However, you can identify a feature on the nearside that is on your line of march and you will still be able to see once you have moved around to the far side. Again, let us use the example of being on an azimuth of 198 degrees, and it brings you to a small lake. From the nearside you cannot locate any features on the far side. There is, however, a small pier on the nearside that is located on your line of march. You can now move around the lake using the easiest and safest route. Move to an area on the far side that is as close to being opposite of the nearside landmark as possible. You then figure out the back azimuth of your route azimuth, which in this case would be 18 degrees. Shoot an azimuth back across the lake to the pier. Move to the right or left until you are in a position where the pier lies at an azimuth of 18 degrees. This puts you back on your line of march and you can continue on following your original azimuth of 198 degrees.

(3) 90 degree offset. Used during hours of reduced visibility or during daylight hours when there are no visible landmarks along the line of march on either side of the obstacle. This time you are moving along a 360 degree azimuth and you encounter a large, impassable area of thick brush. The brush is too high to see any landmarks from either side of the thicket. The following procedure should be taken to detour around the thicket.

(a) Stop the pace count you have been making for movement along your assigned route.

(b) Decide which way you will detour around the thicket. Let's say in this case the right flank appears to be the easiest and safest. Add 90 degrees to your original azimuth of 360 degrees in order to move your patrol to the right.

(c) Move on this 90 degrees azimuth until you have reached a point where you can clear the right flank of the thicket. You will keep a new pace count starting from the point where you made your 90 degrees turn to the right to the point where you can clear the right flank of the thicket.

(d) Make another change of direction. This time you want to return to a 360 degrees azimuth so you will parallel your original course and move by the right flank of the thicket. In order to return to the 360 degrees azimuth you will now subtract 90 degrees.

(e) Move on the 360-degree azimuth until you have by passed the thicket. At the point of your second 90 degree turn, you will pick up the pace count you were making before you began making your detour around the obstacle. Even though you are not on your original line of march, you are moving parallel to your original line of march and therefore must include the distance you cover during this leg of the offset in your pace count for your route.

(f) Once you have bypassed the thicket, you will make another change of direction. This time, you want to move back toward your original line of march. Since that means you will be turning toward the left, you will subtract 90 degrees from the 360 degrees azimuth you have just completed. Your new direction will be 270 degrees.

(g) Pace back toward the original line of march following the 270-degree azimuth, the same number of paces as you made off the original line of march. In doing this you will have returned to your original line of march.

(h) Once you have moved back to your original line of march you will make your last change of direction. By adding 90 degrees to the original azimuth of the last leg of the offset, which in this case was 270 degrees, you will be back on the original azimuth of 360 degrees.

(i) Return to your original line of march pace count.

UNITED STATES MARINE CORPS
Mountain Warfare Training Center
Bridgeport, California 93517-5001

FMST.07.39
04/03/02

STUDENT HANDOUT

SEARCH AND RESCUE MANAGEMENT

TERMINAL LEARNING OBJECTIVE: In a mountainous environment, execute a search and rescue, in accordance with the references. (FMST.07.39)

ENABLING LEARNING OBJECTIVES:

(1) Without the aid of references and from a given list, choose the definition of a search, in accordance with the references. (FMST.07.39a)

(2) Without the aid of references and from a given list, choose the definition of a rescue, in accordance with the references. (FMST.07.39b)

(3) Without the aid of references and from a given list, choose one the four methods used to determine a search area, in accordance with the references. (FMST.07.39c)

(4) Without the aid of references and from a given list, choose one of the five methods for confining a search area, in accordance with the references. (FMST.07.39d)

(5) Given a lost individual scenario in a mountainous environment, plan a search and rescue in accordance with the references. (FMST.07.39e)

OUTLINE.

1. **GENERAL**. When discussing a search and rescue situation we must first be able to differentiate between the two facets of the operation. A "search" means that you are looking for an overdue missing person (FMST.07.39a). A "rescue" 'is the extraction of a known subject from a known location (FMST.07.39b). In either of these situations time is critical for several reasons:

a. Subject may need emergency medical care.

b. Size of the search area is limited.

c. Time and weather destroys clues.

d. Subject may need protection from the environment or from himself.

2. **SEARCH METHODS**. When conducting a search you will be searching for clues, instead of searching for the victim. There are two reasons for this:

a. There are more clues than victims. Anyone moving through an area will leave signs.

b. Clue detection greatly reduces search difficulty by reducing the search area.

c. This determination will assist the search manager in the use of assets, the justification for requesting outside agency assistance, and other management decisions. There are eight points that must be considered in determining the search urgency:

1. Subject profile.
2. Weather profile.
3. Equipment available to subject.
4. Subject's training, experience, and mobility.
5. Terrain hazards.
6. History of incidents in the area.
7. Time.
8. Political sensitivity.

3. **SUBJECT PROFILE**. The subject profile is based on the subject's capabilities and mobility in the terrain and conditions of the search area. The three points are age, medical condition, and number of subjects. Discussing this will assist us is rating the relative urgency:

a. Age: Very young/old, other.

b. Medical condition: Injured, ill, mental problems, healthy, or dead.

c. Number of subjects: Single or group.

d. Subject experience profile: No experience/no knowledge of area, no experience/knows area, experienced/no knowledge of area, or experienced/knows area.

e. Weather profile: Past or current bad weather, bad weather in next 8 hours, bad weather in next 24 hours, or no bad weather predicted. Temperature and visibility.

f. Equipment profile: Inadequate, questionable, or adequate for environment or weather.

g. Terrain/Hazard profile: Known hazardous terrain or, hazards or few or no hazards.

h. History of incidents in the area: Outcome of other searches for individuals with similar profiles in similar situations.

i. Time: Time since the subject went missing and the effects of time on clues.

j. Political sensitivity: External sources; VIPs, politicians, relatives, media, or higher authority in your chain of command.

4. **DETERMINING THE SEARCH AREA**. There are four methods used to determine the search area by rescue managers. The best application is to use a combination of these methods by personnel with search experience. (FMST.07.39c)

a. Theoretical: This is the distance that the victim could have traveled in the elapsed time since he/she was reported missing. (PI x R Squared)

b. Statistical: Historical data on searches for similar individuals in similar conditions.

c. Subjective: Evaluation of limiting factors for this specific incident and location.

d. Deductive reasoning: Sherlock Holmes version, based on analysis of circumstances involving the specific incident.

5. **BEHAVIOR ASPECTS OF A LOST PERSON**. General aspects of lost person behavior will give you an awareness of the actions of the individuals you are searching for by separating them into different categories. Some factors that could effect the search strategy are the general health of the subject, his/her past experiences, physiological effects of the environment, and the victims' mental state or biological cycle. Also, looking at the subject's ability to assist in his/her own survival with simple skills such as fire, shelter, signaling, and land navigation will determine strategy. The psychological impact on the victim should also be taken into account. Will the subject be overcome by Stress? Fear? Panic? Embarrassment? Or will the subject be calm and develop a positive "I'll be OK attitude? A further point to consider is the circumstances under which the subject became lost. Was the subject at a known location such as a campground? Enroute to a specific location or on a trail? Traveling in a wilderness area with no specific location or destination in mind? Or does the subject fall under one of the following categories:

a. Children age 1 to 3: Unaware of concept of being lost, no navigation skills or sense of direction, wander aimlessly, might seek out a convenient place to take a nap. i.e. hollow log, under bush/rock/picnic table.

b. Children age 3 to 6: More mobile, more inquisitive and easily distracted, will attempt to "go home" once they decide that they are lost, taught to avoid strangers and may hide from searchers.

c. Children 6 to 12: Good navigation skills and sense of direction in familiar terrain but become disoriented in unknown surroundings, may have intentionally run away and do not want to be found, may not answer when called, same fears as adults but no experiences to fall back on, once night sets in more likely to respond to searchers and except assistance.

d. Mentally retarded (all ages): Act similar to children 6 to 12, generally will not respond to name, will hide from searchers or seek shelter, once shelter found might remain in place, will do little or nothing to assist in their own survival or rescue.

e. Elderly (above 65): Possibly suffering from senility or Alzheimer's, easily confused or attracted to something that strikes their fancy, may not be oriented to the present, likely to over-extend themselves physically, and often have hearing impairments.

f. Despondent: Often are seeking solitude and will not respond to searchers, usually found near prominent terrain/scenic views, within sight and sound of civilization, rarely found in the underbrush.

g. Hikers: Rely on trails and have a set destination in mind, have poor land navigation once they loose the trail, often get confused at trail intersections or in springtime when trails are covered with snow, hiking parties of different ability levels get separated, will miss trails and head down wrong drainages. Hunters: Concentrate on game more than navigation, often pursue game into drainages/deadfalls/underbrush/deep snow, tend to over-extend themselves, typically not prepared for foul weather.

h. Cone pickers/Berry pickers/Photographers/ rockhounds/etc: Intention to stay or go to a specific location, carry little provisions or survival gear, go in good weather so they don't carry foul weather gear, often are not aware of surrounding as they are intent on pictures or the ground if pickers, these people are an extremely high risk group.

i. Fishermen: Well oriented once they arrive at lake or river, most often late due to accident, these are often recovery operations.

j. Climbers: Generally well equipped and self sufficient, tend to remain on marked routes and leave good plans with friends, primary factors are weather and hazardous conditions such as falling debris or avalanches, once located these searches often become technical rescues. 1. Marines: Subject profile will depend on individual's MOS, mission, training, and abilities learned as a civilian.

6. **BEGINNING THE SEARCH**. Once we have determined that a subject is missing we must determine whether or not he/she is in our search area. The sheriffs department will initiate a "Bastard Search", this is to ensure that the subject is not simply at the local tavern, shacked up with an old friend, or was not in an automobile accident enroute to the area. Next we will begin our search at one of the four following area:

a. Last Known Point (LKP): This is the last place that you have physical evidence to support that the subject was there. i.e. Subject's vehicle at the trailhead, name in a logbook, campsite with gear, etc.

b. Place Last Seen (PLS): The last place he was physically seen. i.e. On the trail by other hikers, at the hunting cabin, at a restaurant or gas station, or back home in San Diego.

c. Departure Point: The place where the subject was supposed to enter the search area. Usually a trailhead or specific point relating to the individual's activity.

d. Clues: As clues are discovered, and you confirm that they belong to the subject, you will update or adjust the search area.

7. **MATTSON**. Once we have determined that we do have an overdue subject and can assume that he/she is in our search area we need to break the search area into manageable sized segments. Depending upon the area to be searched, special training required to search high hazard areas, and assets available, we will divide the area by hazard or by terrain features. The lines of division should follow terrain features so that the searchers on the ground can easily navigate and stay within their zones. We will then determine the POA using the Mattson approach. This is done with each of the five or six members of the management team with the most search experience assigning a percentage to each of the segments using his individual judgment on the likelihood of the subject being in that area. Always include an extra column to represent the "rest of the world". Once all personnel have made their determinations, divide the totals and this will give you your areas to search in order of precedence:

NAME	A	B	C	D	E*	TOTAL
DUNSTON	40%	20%	10%	20%	10%	100%
KEOGH	50%	30%	5%	10%	5%	100%
KOFFEL	40%	25%	15%	10%	10%	100%
AYRES	60%	20%	10%	5%	5%	100%
TOTAL	190%	95%	40%	45%	30%	400%
AVERAGE	47.50%	23.75%	10.00%	11.25%	7.5%	
PRIORITY	1 ST	2ND	4TH	3RD	5TH	

8. **COMMAND SYSTEM**. The Office of Emergency Service (OES), has developed a command system that they use for all emergency situations. This system is called the

Incident Command System (ICS). The breakdown is very similar to the military chain of command and the duties correspond to those of the various "S" shops in the battalion. The personnel in each duty area will complete their assigned duties as you would in a military unit with the Incident Commander (IC) being overall responsible for the entire action. Normally, in California, the IC is the County Sheriff.

a. Incident Commander. CO

b. Operations-Tactics, coordination, insert resources. S-3

c. Plans-Strategy, intelligence, briefing and debriefing teams. S-2

d. Logistics-Beans, Band-Aids, beds, heads, motor-T, Special equipment. S-4

e. Communication-COC and field, landlines, frequency coordinator. S-6

f. Finance-Documentation of all man-hours, damages, legal issues.
S-1

g. Liaisons-inter agency, media, safety, family. PAO/Chaplain

9. **<u>CONFINING A SEARCH AREA</u>**. There are several reasons to confine the search area. The most important is that the longer the subject has free movement, the farther he/she can travel. (Remember the Theoretical Method) Additionally, this will ensure that the subject does not walk out without you being aware of it. It's real dumb to be conducting a search for a subject that is on his/her way home. The use of confinement will not interfere with the other searchers, decreases the chances of a more massive search and if promptly initiated has been shown to be highly effective. There are several methods that we can use. (FMST.07.39d)

a. Sign-cutters - Searchers that hike or ski the perimeter of the entire search area, then the perimeter of the segments. This will let you know if he/she moves from one area to another and allows you to concentrate limited assets in a specific area.

b. Trail-watchers/Road Blocks/Camp-ins - Searchers that camp at trailheads. They can alert you if the subject walks out and interview other individuals concerning possible sightings of the subject.

c. Look-outs - These personnel are already in place in fire towers. Simply notify them of the situation.

d. Track Traps - Clearing areas of the trail that are checked periodically for fresh tracks from the subject.

e. String-Lines - Spools of line with paper arrows attached that point to a safe area of a trail-watcher camp.

10. **USING CLUES**. The use of clues cannot be underestimated. They can tell you if a subject was or was not in the area, by freshness you can tell how long ago the subject was there, by the type of clue you can determine physical and mental condition, and by following clues you can determine a direction and speed of travel. Once you have located a CONCLUSIVE clue, you will re-mattson the area, using the clue as the LKP. The clues we will be looking for fall under four categories:

a. Physical - Footprints, equipment, candy wrappers, cigarette butts, etc.

b. Recorded - Summit logs or trail registers.

c. People - Witnesses who have seen the subject.

d. Event - Mirror flashes, camp fires, whistle signals, etc.

11. **ADDITIONAL ASSETS**. There are a variety of assets available to you in the search and rescue business. Some are volunteers, some charge money and some (Like the military) require higher authority to grant permission for their use. Remember that in addition to active searchers you can also use passive measures such as lights, noise, balloons, fires, smoke generators, sirens, or P.A. systems. Some examples of active searchers are:

a. Scent dogs.

b. Tracking dogs.

c. Human trackers.

d. Hasty teams.

e. Grid teams.

f. Helicopters.

g. Civil Air Patrol.

h. Snowmobile clubs.

i. Horse teams.

UNITED STATES MARINE CORPS
Mountain Warfare Training Center
Bridgeport, California 93517-5001

FMST.07.16
04/03/02

STUDENT HANDOUT

TRIAGE

TERMINAL LEARNING OBJECTIVE: Given simulated casualties, in a wilderness environment, triage casualties, in accordance with the references. (FMST.07.16)

ENABLING LEARNING OBJECTIVES:

1) Without the aid of references and from a given list, choose when actual triage most often begins, in accordance with the references. (FMST.07.16a)

2) Without the aid of references and from a given list, identify one of the four-triage categories used by NATO, in accordance with the references. (FMST.07.16b)

3) Without the aid of references and from a given list, choose the primary cause of death in a mass casualty situation, in accordance with the references. (FMST.07.16c)

OUTLINE.

1. **DEFINITION**: Triage or sorting (often used interchangeably) is the evaluation and classification of casualties for the purpose of establishing the priority for treatment and evacuation.

NOTE: Casualty sorting is one of the most important tasks for the entire medical service. No other job requires more informed judgment, hard work and courage. The Officer and the Corpsman responsible for triaging wounded Sailors and Marines must exercise sound judgment so that casualties receive proper, expedient care. Triage is a continuing process and any individual assigned should be the most capable and experienced available.

a. Decision Factors. The factors concerning casualty triage are based upon the following:

(1) The need for immediate resuscitation.

(2) The need for emergency surgery.

(3) The futility of surgery due to obviously lethal wounds.

(4) Time needed to perform surgery compared with:

(a) Probability of success.

(b) Number of other casualties needing treatment.

b. The actual triage begins most often with the casualty himself. (FMST.07.16a). This usually occurs when the injured individual determines that "Buddy-Aid" or "Self-Aid" will not be adequate.

(1) The corpsman, once presented with the casualty, must make several determinations:

(a) Is the casualty a walking-wounded or a litter case?

(b) If severely wounded, are resuscitative efforts required immediately?

(c) Does the casualty need any other immediate treatment (bleeding, asphyxia or pain control)?

(2) Once the casualty has been transported to a Battalion Aid Station (BAS), triage continues. The dental officer will commonly perform the duties of the Triage Officer at the echelon I or echelon II level, if present.

2. **NATO TRIAGE CATEGORY CODES**. (FMST.07.16b)

a. At each medical treatment facility (MTF) in the area of operations (AO), incoming casualties are classified by level of treatment required. Four triage category groups have been universally adopted for use by both United States and NATO Forces. These categories are listed and defined as they appear in standardization agreement (STANAG) No 2879:

(1) Immediate Treatment (Group T1) (RED Tag). To include those requiring emergency life-saving surgery. These procedures should not be time consuming and should concern only those patients with high chances for survival. Some examples are:

(a) Respiratory obstruction.

(b) Accessible hemorrhage.

(c) Emergency amputation.

(2) Delayed Treatment (Group T2) (YELLOW Tag). To include those badly in need of time-consuming major surgery, but whose general condition permits delay in surgical treatment without unduly endangering life. To mitigate the often, critical effects of delay in surgery, sustaining treatment, such as stabilizing I.V. fluids, splinting, administration of antibiotics, catheterization, gastric decompression, and relief of pain will be required. Examples are:

(a) Large muscle wounds.

(b) Fractures of major bones.

(c) Intra-abdominal and/or thoracic, head or spinal injuries.

(d) Uncomplicated major burns.

(3) Minimal Treatment (Group T3) (GREEN Tag). To include those with relatively minor injuries who can effectively care for themselves or who can be helped by untrained personnel. Some examples are:

(a) Minor lacerations.

(b) Abrasions.

(c) Fractures of small bones.

(d) Minor burns.

(4) Expectant Treatment (Group T4) (BLUE Tag). This group comprises patients who have received serious and often multiple injuries, and whose treatment would be time-consuming and complicated with a low chance of survival. If fully treated, they may make heavy demands on medical manpower and supplies. Until the mass casualty situation is under control, they will receive appropriate supportive treatment. The extent of treatment depends on available supplies and manpower, and may involve the use of large doses of narcotic analgesics. These patients should not be abandoned, and every effort should be devoted to their comfort. The possibility of their survival, despite alarming injuries, must always be kept in mind. Examples are:

(a) Severe multiple injuries.

(b) Severe head or spinal injuries.

(c) Large doses of radiation.

(d) Widespread severe burns.

3. **TREATMENT CONCEPTS**.

a. The triage officer must be familiar with all aspects of combat casualty care (airway management, hemorrhage control, use of anesthesia, surgical procedures, fluid resuscitation and post-operative care, to name a few), but ever-present in his mind should be the following concepts:

(1) Asphyxia and hemorrhage are initially the primary causes of death in any mass casualty situation. (FMST.07.16c)

(2) Once at a definitive care facility, uncontrolled hemorrhage is the leading cause of death.

(3) Life has priority over limb, and preserving function overrides the correction of superficial anatomic defects.

(4) A casualty is in a constant state of change until the wound or injury has been repaired.

(5) Systemic disturbances caused by the wound continue until healing is complete.

(6) Combat conditions will dictate the modifications necessary in triage and care of mass casualties.

b. Priorities of Treatment. With the six concepts of treatment in mind, we can now address three basic priorities for treatment.

(1) First Priority:

(a) Asphyxia:

1. Mechanical obstruction of the airway.

2. Sucking chest wound.

3. Tension pneumothorax.

4. Maxillofacial trauma.

5. Massive hemo-pneumothorax.

(b) Shock:

1. Exsanguinating hemorrhage.

2. Major internal hemorrhage (thoracic, abdominal, and pelvic).

3. Visceral injuries/evisceration.

4. Cardio-pericardial injuries.

5. Massive muscle injury leading to Fat Embolization to the lung.

6. Full-thickness or deep-partial burns of 10 % or greater.

(2) Second Priority:

(a) Visceral injuries:

1. Perforation of G.I. Tract.

2. Pancreatic and Biliary system.

3. Genitourinary tract.

4. Thoracic wounds w/o asphyxia.

(b) Vascular injuries requiring repair:

1. Tourniquet cases.

(c) Closed head injuries:

1. Increasing loss of consciousness.

(d) Major burns:

1. Partial-thickness burns of greater than 25% TBSA in the low risk group, 20% TBSA in the high risk group.

2. Burns involving the poor risk group.

3. Burns involving the face, hands, feet, genitalia/perineum.

4. Circumferential limb burns.

(3) Third Priority:

(a) Brain and spinal injuries - requiring burr hole decompression.

(b) Soft tissue injuries.

(c) Debridement necessary with less than major muscle damage.

(d) Lesser fractures and dislocations.

(e) Eye injuries.

(f) Maxillofacial injuries w/o asphyxia.

(g) Burns under 20% in other locations other than priority 2.

4. **TRIAGE TOOLS**.

a. Trauma Score/Glasgow Coma Scale. In recent years, many methods have been proposed to speed up the classification and sorting of casualties. One method that has gained wide acceptance is the Trauma Score/Glasgow Coma Scale. Several studies have been performed using various parameters, i.e., respiratory rate, chest wall expansion, BP, capillary refill, etc. By assigning a number scale to each specific response, the likelihood of survival as a percentage can be calculated.

1. Glasgow Coma Scale.

a. Eye opening:

1) Spontaneous = 4.
2) To voice = 3.
3) To pain = 2.
4) None = 1.

b. Verbal Response.

1) Oriented = 5.
2) Confused = 4.
3) Inappropriate words = 3.
4) Incomprehensive words = 2.
5) None = 1.

c. Motor Response.

   1) Obeys Command = 6.

   2) Localized Pain = 5.
   3) Withdraw Pain = 4.
   4) Flexion (pain) = 3.Extension (pain) = 2.
   5) None = 1.

d. Total the points up and they equal:

   1) 13-15= 4.
   2) 9-12 = 3.
   3) 6-8 = 2.
   4) 4-5 = 1.
   5) <3 = 0

2. Trauma Scale.

   a. Respiratory Rate:

      1) 12-29 min = 4.
      2) 10-12 min = 3.
      3) 6-9 min = 2.
      4) 1-5 min = 1.
      5) None = 0.

   b. Systolic BP:

      1) >89 mmHg = 4.
      2) 76 mmHg = 3.
      3) 50-75 mmHg = 2.
      4) 1-49 mmHg = 1.
      5) None = 0.

   c. Total the points.

3. Total trauma Scale + GCS = % survival (1-12)

   12 = 99
   10 = 88
   8 = 67
   6 = 63
   4 = 33
   2 = 29

1 = 25
0 = 4

UNITED STATES MARINE CORPS
Mountain Warfare Training Center
Bridgeport, California 93517-5001

FMST.07.08
04/03/02

STUDENT HANDOUT

PREVENTIVE MEDICINE AND WATER PURIFICATION

TERMINAL LEARNING OBJECTIVE. Given a unit in a mountainous environment and the necessary equipment and supplies, perform preventive medicine, in accordance with the references. (FMST.07.08)

ENABLING LEARNING OBJECTIVES.

1) Without the aid of references, from a given list select the five important areas of personal hygiene, in accordance with the references. (FMST.07.08a)

2) Without the aid of references, from a given list select the correct steps taken to ensure proper immunization prior to deployment, in accordance with the references. (FMST.07.08b)

3) Without the aid of references, from a given list select the three methods of water purification, in accordance with the references. (FMST.07.08c)

4) Without the aid of references, from a given list select the four forms of halogens used for water purification, in accordance with the references. (FMST.07.08d)

5) Without the aid of references, from a given list select the definition of giardia, in accordance with the references. (FMST.07.08e)

6) Without the aid of references, from a given list select the most common drug used to treat giardia infection in the United States, in accordance with the references. (FMST.07.08f)

7) Without the aid of references, from a given list select the three important reasons for proper waste disposal, in accordance with the references. (FMST.07.08g)

8) Without the aid of references, from a given list select the correct methods of waste disposal, in accordance with the references. (FMST.07.08h)

16

1. **PERSONAL HYGIENE** Your body's defense mechanisms break down due to the physical demands of the arduous terrain, environment, stress and metabolic changes. Personal

hygiene becomes the key to prevent the spread of communicable diseases. Five important areas are: (WMC.6.17a)

a. Body.

(1) The body should be washed frequently in order to minimize the chances of small cuts and scratches developing into full blown infections and as a defense against parasitic infections.

(2) A daily bath or shower consisting of soap and hot water is ideal. However, when this is not possible, you should:

(a) Give yourself a "sponge bath" using soap and water, making sure particular attention is given to body creases, i.e., armpits, groin area, face ears, hands and feet.

(b) If water is in extremely short supply, you should take an "air bath" by:

(1) Remove all clothing and hang it up to air.

(2) Expose the body for two hours to the sunlight, which is ideal, but the effects will basically be the same if done indoors or during an overcast day. BE CAREFUL NOT TO SUNBURN!

(3) Shaving. Shaving at night before bedtime is preferable because shaving will remove facial oils, which aid in protecting the face from a harsh cold environment during the day.

b. Hair.

(1) Clean frequently.

(2) Inspect at least once a week for parasites. Have your mountain buddy inspect the areas not within your view. This is particularly important during the summer season. Any time the area you are operating in an area that contains wild life there is the possibility of parasites like fleas and ticks.

c. Hands. Hand washing with soap and water will cut down in the transmission of viral and parasitic infections via the person-person and fecal-oral route.

(1) Keep fingernails trimmed and clean to prevent accidentally scratching yourself and to prevent harborage for bacteria.

d. Feet. Are your prime sources of transportation, so take care of them!

(1) Inspect frequently for:

(a) Blisters.

(b) Infections, bacterial and fungal.

(2) Preventive measures:

(a) Keep your feet dry, change socks frequently.

(b) Use antifungal foot powder.

(c) Antiperspirants. Pretreatment for 2 weeks prior to field with an antiperspirant containing Aluminum Chlorhydrate. This will aid in the prevention of athlete's foot.

e. Oral Hygiene. The mouth and teeth should be cleaned at least daily to prevent tooth decay and gum disease.

(1) Ideally with:

(a) Tooth brush.

(b) Tooth paste.

(c) Dental floss.

(2) If these items are not available:

(a) A "chew stick" made from a clean twig about 8" long and finger width. Chew one end until it becomes frayed and brush-like, then use as a toothbrush.

(b) Field expedient dental floss can be pulled from the center lines of paracord.

(c) The gums should be stimulated at least once daily by rubbing them vigorously with a clean index finger.

2. **IMMUNIZATIONS (NAVMEDCOMINST 6230.3)**. In addition to being required, immunizations are your best front line defense against communicable disease.

a. Eight weeks prior to deployment: (WMC.6.17b)

(1) Ensure Basic Series Completed (BSC) on required immunizations.

(2) Determine if additional immunizations/chemophylaxis will be required in that geographical area of operations.

(3) Identify and flag health risk personnel: i.e. TB converters (NAVMEDCOMINST 6224.1), hepatitis (NAVMEDCOMINST 6230.1A), mononucleosis, parasitic infections and anemic personnel.

(4) Review medical intelligence reports:

(a) EPMU, Navy Environmental Preventive Medicine Unit.

(b) G-2, USMC.

(c) NAVDISVECTOECOLCONCEN, Navy Disease Vector Ecology and Control Centers.

(d) CDC - Center For Disease Control.

(e) WHO - World Health Organization.

3. **WATER.** A satisfactory source is one where quantity and quality is enough to supply the needs of all the troops. Water or ice taken from the environment should be considered contaminated and must be purified with available means. Be aware that cold weather slows the chemical reaction time of purification measures.

a. Sources.

(1) Rivers

(2) Streams

(3) Ponds

(4) Lakes

(5) Ice

(6) Snow

b. Water Source Selection. The choice of a water source is influenced by:

(1) Not contaminated by: (Check all to one mile upstream).

(a) Sewage

(b) Enemy pollution

(c) Chemicals

(d) Biologicals

(e) Radiologicals

NOTE: Check condition of vegetation or for dead animals.

(2) Quantity.

(3) Ease of procurement.

(4) Ease of purification.

(5) Freedom from turbidity.

(6) Excessive organic or non-organic contamination.

(7) Should be easily protectable.

4. **PURIFICATION.** Simply consists of removing or destroying enough impurities to make water safe and pleasant to drink.

a. The first step in purifying is to select a treatment method. (All figures are based on one quart of water.) The three different purifying methods are: (WMC.6.17c)

(1) Boiling. The standard recommendation is to bring the water to a rolling boil for 1 minute at sea level for complete sterilization of the water. Recent research showed that additional boiling time for sterilization was not necessary and that boiling for two minutes at any altitude rendered water safe for consumption.

(2) Halogens. The number of tablets and contact time must be considered. All charts based on one-quart canteen. (WMC.6.17d)

(a) Iodine tablets.

TYPE WATER	TABLETS	CONTACT TIME
CLEAR	1	25 MIN
CLOUDY	2	25 MIN
COLD	2	25 MIN

(b) Chlorine bleach. The strength of the solution, number of drops and contact time must be considered.

PERCENT	CLEAR WATER	CLOUDY WATER	COLD WATER
1%	10GTTS/30 MIN	20GTS/30 MIN	10GTTS/60MIN
4-6%	2GTTS/30 MIN	4GTTS/30 MIN	2GTTS/60 MIN
7-10%	1 GTTS/30 MIN	2GTTS/30 MIN	1 GTTS/60 MIN

(c) Iodine solution. The strength, number of drops and contact time must be considered.

PERCENT	CLEAR WATER	CLOUDY WATER	COLD WATER
2%	5GTTS/30 MIN	10GTTS/30 MIN	8GTTS/60 MIN

(d) Betadine solution.

CLEAR WATER	CLOUDY WATER	COLD WATER
8GTTS/30 MIN	16GTTS/30 MIN	8GTTS/60 MIN

(1) Filtration Most filters are not small enough to completely disinfect water. They remove particulate debris, which allows lower halogen dose and improved appearance and taste of "dirty" water. Filters clog quickly if the water is dirty or has a lot of suspended particles. NEVER USE FILTERS AS YOUR SOLE METHOD OF WATER PURIFICATION.

(a) Maximum Filter Pore Size:

1) Parasitic eggs and larvae: 20 micrometers (um).

2) Giardia, E. histolytica: 5 um.

3) Cryptosporidium: 1 um

4) Enteric bacteria: 0.2 um.

5) Viruses: 0.01 um (too small for field water filtration).

5. **GIARDIA**

a. Definition. (WMC.6.17e) An intestinal flagellate protozoa that can cause clinical illness in humans.

(1) Found worldwide in tropical and temperate areas, the infectious cysts are passed by humans and animals and can be transmitted by direct fecal-oral route or waterborne contamination (cysts maintain viability for months in cold water). Due to these factors, Giardia is both a common cause of diarrhea in travelers to foreign countries, and campers in the United States where it has become endemic and is now considered the most common human parasite.

(2) It is found with increasing frequency in formerly "pristine" mountain streams, rivers and lakes and outbreaks have occurred in remote recreational areas and resorts throughout the western mountain states.

(3) Giardiasis is currently a risk for anyone drinking untreated surface water in the western mountains and for communities with water treatment techniques inadequate for eliminating the cysts.

b. Pathophysiology. The organism has two forms: Trophozoite and cyst.

(1) These may be passed in the stool of an infected person. The trophozoites die within one hour outside the body, however the cysts are very hardy, retaining infectivity more than two months in cold water. If the cysts are ingested, excystation is encouraged by stomach acid, and the motile trophozoite attaches by suction disc to the wall of the duodenum and proximal jejunum, where they may cause symptoms of illness.

(2) Water, ice, or food prepared with contaminated water, are the major sources of infection, but direct person-to-person transmission can occur from sexual contact or hand-to-mouth contamination.

(3) Incubation time: 5 - 25 days (7 - 10 medium). Most infections are thought to be asymptomatic carrier states, however, a wide array of clinical syndromes may occur.

c. Symptoms.

(1) A number of people have abrupt onset of explosive diarrhea accompanied by abdominal cramps, foul gas, vomiting, fever and malaise. This commonly lasts three to four days before developing the more common subacute syndrome. In most individuals, the onset is more insidious, and symptoms are persistent or recurrent.

(2) Stools become mushy, greasy and malodorous. Over a period of weeks, watery diarrhea may alternate with soft stools and even constipation. Upper GI symptoms accompany stool changes, mid and upper abdominal cramping, substernal burning, acid indigestion, sulfurous belching, nausea and foul flatus, distention with an early "full" feeling are all typically increased after a meal.

(3) Symptoms of anorexia, fatigue and weight loss are common, but fever and vomiting are infrequent except during an acute onset of illness.

d. Laboratory Parameters.

(1) The "typical" victim should suggest a diagnosis of giardiasis. This individual may have a history of camping, vacationing at a mountain resort, or foreign travel. Symptoms of upper abdominal discomfort, foul gas and soft stools arose one or two weeks after a short trip or during a lengthy one and have fluctuated for weeks. Sometimes improving, the individual held off visiting the doctor, only to have the symptoms recur. Work or school is not interrupted, however, physical discomfort and embarrassment from the gas are concerns.

(2) Laboratory confirmation of giardiasis can be difficult. Stool samples are the first means of diagnosis. Trophozoites may be found in fresh, watery stools, but disintegrate rapidly. Cysts remain in fresh stools for at least 24 hours. Cysts passage, however, is extremely variable and not related to clinical symptoms. Three samples, taken every other day, should be examined for a definite diagnosis.

(3) Average duration of symptoms has varied from three to ten weeks. Acquired resistance has been shown. Immunity is suspected, but not proven and human reinfection definitely occurs.

e. Treatment. (WMC.6.17f) Complete cure of most Giardia infections can be achieved 90% of the time with *Metronidazole* (flagyl) 250 mg three time s a day for 5 days.

f. Prevention There is no safe or effective chemoprophylactic drug and prevention should be practiced by travelers and campers. The purity of American wild waters can no longer be assumed and in developing countries, even tap water should be suspected. In addition to choosing a technique for individual field water disinfection, individuals must become more responsible for prevention of water contamination. This requires attention to personal latrines and camper trailer waste disposal, a subject that most people do not want to touch.

6. **WASTE**. The importance of proper waste disposal cannot be over emphasized.

a. It serves to: (WMC.6.17g)

(1) Eliminate harborage for rodents and vermin.

(2) Prevents attraction of rodents and vermin.

(3) Prevents a pathogenic contamination source.

b. Two basic types of waste are organic and non-organic. The basic ways to dispose of them are burning, burying, or hauling it away. (WMC.6.17h)

(1) Organic wastes:

(a) Human waste - Burn or haul away to a designated waste pit area.

(b) Urine - Use only assigned marked areas.

(c) Edible garbage - Burn. It should not be left exposed for animals, vermin or the enemy.

(d) Paper -Burn.

(e) Contaminated bandages - Sterilize/burn.

(2) Non-organic waste:

(a) Metals - Haul away or bury.

(b) Liquids - Burn or bury.

UNITED STATES MARINE CORPS
Mountain Warfare Training Center
Bridgeport, California 93517-5001

FMST.07.19
04/03/02

STUDENT OUTLINE

REQUIREMENTS FOR SURVIVAL

TERMINAL LEARNING OBJECTIVE Given a survival situation in any type of environmental condition and minimal equipment and resources, apply the requirements for survival to facilitate individual or group survival and recovery per the references. (FMST.07.19)

ENABLING LEARNING OBJECTIVES

1. Without the aid of references and from a given list, choose one of the survival stressors, in accordance with the references. (WMC.A1.a)

2. Without the aid of references and from a given list, choose one of the priorities of work in a survival situation, in accordance with the references. (WMC.A1.b)

1. REQUIREMENTS FOR SURVIVAL

a. This mental "mind-set" is important in many ways. We usually call it the "will to survive" although you might call it "attitude" just as well. This basically means that, if you do not have the right attitude, you may still not survive.

A1

b. A guideline that can assist you is the acronym " SURVIVAL".

(1) <u>S</u>ize up.

(a) Size up the situation.

1. Conceal yourself from the enemy.

2. Use your senses to hear, smell, and see to determine and consider what is developing on the battlefield before you make your survival plan.

(b) Size up your surroundings.

1. Determine the rhythm or pattern of the area.

2. Note animal and bird noises and their movement.

3. Note enemy traffic and civilian movement.

(c) Size up your physical condition.

1. Check your wounds and give yourself first aid.

2. Take care to prevent further bodily harm.

3. Evaluate condition of self and unit prior to developing survival plan.

(d) Size up your equipment.

1. Consider how available equipment may affect survival senses; tailor accordingly.

(2) <u>U</u>ndue haste makes waste.

(a) Plan your moves so that you can move out quickly without endangering yourself if the enemy is near.

(3) <u>**R**</u>emember where you are.

(a) If you have a map, spot your location and relate it to the surrounding terrain.

(b) Pay close attention to where you are and where you are going. Constantly orient yourself.

(c) Try to determine, at a minimum, how your location relates to the following:

1. The location of enemy units and controlled areas.

2. The location of friendly units and controlled areas.

3. The location of local water sources.

4. Areas that will provide good cover and concealment.

(4) Vanquish fear and panic.

(a) Realistic and challenging training builds self-confidence and confidence for a unit's leadership.

(b) The feeling of fear and panic will be present. The survivor must control these feelings.

(5) Improvise and improve.

(a) Use tools designed for one purpose for other applications.

(b) Use objects around you for different needs. (i.e. use a rock for a hammer)

(6) Value living.

(a) Place a high value on living.

(b) Refuse to give into the problem and obstacles that face you.

(c) Draw strength from individuals that rise to the occasion.

(7) Act like the natives.

(a) Observe the people in the area to determine their daily eating, sleeping, and drinking routines.

(b) Observe animal life in the area to help you find sources of food and water.

NOTE: Remember that animal reactions can reveal your presence to the enemy. Animals cannot serve as an absolute guide to what you can eat and drink.

(8) Live by your wits, **but for now**, learn basic skills.

(a) Practice basic survival skills during all training programs and exercises.

2. **STRESS**. Stress has many positive benefits. Stress provides us with challenges: it gives us chances to learn about our values and strengths. Too much stress leads to distress. While many of these signs may not be self-identified, it remains critical that all survivors remain attentive to each other's signs of distress. Listed are a few common signs of distress found when faced with too much stress:

   a. Difficulty in making decisions (do not confuse this sign for a symptom of hypothermia).

   b. Angry outbursts.

   c. Forgetfulness.

   d. Low energy level.

   e. Constant worrying.

   f. Propensity for mistakes.

   g. Thoughts about death or suicide.

   h. Trouble getting along with others.

   i. Withdrawing from others.

   j. Hiding from responsibilities.

   k. Carelessness.

3. **SURVIVAL STRESSORS**. (FMST.07.19a) **Any** event can lead to stress. Often, stressful events occur simultaneously. These events are not stress, but they produce it and are called "stressors". In response to a stressor, the body prepares either to "fight or flee". Stressors add up. Anticipating stressors and developing strategies to cope with them are the two ingredients in the effective management of stress. It is essential that the survivor be aware of the types of stressors they will encounter.

   a. Injury, Illness, or Death: Injury, illness, and death are real possibilities a survivor has to face. Perhaps nothing is more stressful than being alone in an unfamiliar environment where you could die from hostile action, an accident, or from eating something lethal. Illness and injury can also add to stress by limiting your ability to maneuver, get food and drink, find shelter, and defend yourself.

   b. Uncertainty and Lack of Control: Some people have trouble operating in settings where everything is not clear-cut. The only guarantee in a survival situation is that nothing is guaranteed. This uncertainty and lack of control also add to the stress of being ill, injured, or killed.

c. Environment: A survivor will have to contend with the stressors of weather, terrain, and the variety of creatures inhabiting an area. Heat, cold, rain, winds, snow, mountains, insects, and animals are just a few of the challenges awaiting the Marine working to survive. Depending on how a survivor handles the stress of environment, his surroundings can be either a source of food and protection or a cause of extreme discomfort leading to injury, illness, or death.

d. Hunger and Thirst: Without food and water a person will weaken and eventually die. Getting and preserving food and water takes on increasing importance as the length of time in a survival setting increases. With the increased likelihood of diarrhea, replenishing electrolytes becomes critical. For a Marine used to having his provisions issued, foraging can be a significant source of stress.

e. Fatigue: It is essential that survivors employ all available means to preserve mental and physical strength. While food, water, and other energy builders may be in short supply, maximizing sleep to avoid deprivation is a very controllable factor. Training data collected at MCCDC demonstrates that individuals who lack sleep for 24 hours or more suffer a 25% decrease in effective performance. Further, sleep deprivation directly correlates with increased fear. Effective survival is highly unlikely when fatigue builds to a level where staying awake becomes a stressful evolution.

f. Isolation: Although Marines complain about higher headquarters, we become used to the information and guidance it provides, especially during times of confusion. Being in contact with others provides a greater sense of security and a feeling someone is available to help if problems occur. A significant stress in survival situations is that often a person or team has to rely solely on its own resources.

4. **NATURAL REACTIONS**. Man has been able to survive many shifts in his environment throughout the centuries. His ability to adapt physically and mentally to a changing world kept him alive. The average person will have some psychological reactions in a survival situation. These are some of the major internal reactions you might experience with the survival stresses.

   a. Fear: Fear is our emotional response to dangerous circumstances that we believe have the potential to cause death, injury, or illness. Fear can have a positive function if it encourages us to be cautious in situations where recklessness could result in injury. Unfortunately, fear can also immobilize a person. It can cause us to become so frightened that we fail to perform activities essential for survival.

   b. Anxiety: Anxiety can be an uneasy, apprehensive feeling we get when faced with dangerous situations. When used in a healthy way, anxiety urges us to act to end, or at least master, the dangers that threaten our existence. A survivor reduces his anxiety by performing those tasks that will ensure his coming through the ordeal alive. Anxiety can overwhelm a Marine to the point where he becomes easily confused and has difficulty thinking.

c. Anger and Frustration: Frustration arises when a person is continually thwarted in his attempts to reach a goal. The goal of survival is to stay alive until you can reach help or until help can reach you. To achieve this goal, Marines must complete some tasks with minimal resources. One outgrowth of frustration is anger. Getting lost, damaged or forgotten equipment, the weather, inhospitable terrain, enemy patrols, and physical limitations are just a few sources of frustration and anger. Frustration and anger encourage impulsive reactions, irrational behavior, poorly thought-out decisions, and, in some instances, an "I quit" attitude. If the Marine does not properly focus his angry feelings, he can waste much energy in activities that do little to further either his chances of survival or the chances of those around him.

d. Depression: Depression is closely linked with frustration and anger when faced with the privations of survival. A destructive cycle between anger and frustration continues until the person becomes worn down-physically, emotionally, and mentally. When a person reaches this point, he starts to give up, and his focus shifts from "What can I do" to "There is nothing I can do." If you allow yourself to sink into a depressed state, then it can sap all your energy and, more important, your will to survive.

e. Loneliness and Boredom: Man is a social animal and enjoys the company of others. Loneliness and boredom can be another source of depression. Marines must find ways to keep their minds productively occupied.

f. Guilt: The circumstances leading to your survival setting are sometimes dramatic and tragic. It may be the result of an accident or military mission where there was a loss of life. Perhaps you were the only, or one of a few, survivors. While naturally relieved to be alive, you simultaneously may be mourning the deaths of others that were less fortunate. Do not let guilt feelings prevent you from living.

5. **PRIORITIES OF WORK IN A SURVIVAL SITUATION.** (FMST.07.19b) Each survival situation will have unique aspects that alter the order in which tasks need to be accomplished. A general guideline is to think in blocks of time.

a. First 24 hours: The first 24 hours are critical in a survival situation. You must make an initial estimate of the situation. Enemy, weather, terrain, time of day and available resources will determine which tasks need to be accomplished first. They should be the following:

(1) Shelter.

(2) Fire.

(3) Water.

(4) Signaling.

b. Second 24 hours: After the first 24 hours have passed, you will now know you can survive. This time period needs to be spent on expanding your knowledge of the area. By completing the following tasks, you will be able to gain valuable knowledge.

(1) Tools and weapons: By traveling a short distance from your shelter to locate the necessary resources, you will notice edible food sources and game trails.

(2) Traps and snares: Moving further away from your shelter to employ traps and snares, you will be able to locate your shelter area from various vantagepoints. This will enable you to identify likely avenues of approach into your shelter area.

(3) Pathguards: Knowing the likely avenues of approaches, you can effectively place noise and casualty producing pathguards to ensure the security of your shelter area.

c. Remainder of your survival situation. This time is spent on continuously improving your survival situation until your rescue.

6. **GROUP SURVIVAL.** In-group survival, the group's survival depends largely on its ability to organize activity. An emergency situation does not bring people together for a common goal; rather, the more difficult and disordered the situation, the greater are the disorganized group's problems.

a. Groups Morale: High morale must come from internal cohesiveness and not merely through external pressure. The moods and attitudes can become wildly contagious. Conscious, well-planned organization and leadership on the basis of delegated or shared responsibility often can prevent panic. High group morale has many advantages.

(1) Individual feels strengthened and protected since he realizes that his survival depends on others whom he trusts.

(2) The group can meet failure with greater persistency.

(3) The group can formulate goals to help each other face the future.

b. Factors that Influence Group Survival: There are numerous factors that will influence whether a group can successfully survive.

(1) Organization of Manpower - Organized action is important to keep all members of the group informed; this way the members of the group will know what to do and when to do it, both under ordinary circumstances and in emergencies.

(2) Selective Use of Personnel - In well-organized groups, the person often does the job that most closely fits his personal qualifications.

(3) Acceptance of Suggestion and Criticisms - The senior man must accept responsibility for the final decision, but must be able to take suggestion and criticisms from others.

(4) Consideration of Time - On-the-spot decisions that must be acted upon immediately usually determines survival success.

(5) Check Equipment - Failure to check equipment can result in failure to survive.

(6) Acquiring survival knowledge and skills increases survival Knowledge and Skills - Confidence in one's ability to survive.

UNITED STATES MARINE CORPS
Mountain Warfare Training Center
Bridgeport, California 93517-5001

FMST.
07.20
04/03/02

STUDENT OUTLINE

SURVIVAL KIT

TERMINAL LEARNING OBJECTIVE. In a summer mountainous environment, construct a survival kit, in accordance with the references. (FMST.07.20)

ENABLING LEARNING OBJECTIVES

(1) Without the aid of references, choose from a given list in the criteria for a survival kit, in accordance with the references. (FMST.07.20a)

(2) Without the aid of references, choose from a given list one example of each criteria for a survival kit, in accordance with the references. (FMST.07.20b)

1. COMPONENTS FOR A SURVIVAL KIT

a. The environment is the key to the types of items you will need in your survival kit. How much equipment you put in your kit depends on how you will carry the kit. A kit on your body will have to be much smaller than one carried in a vehicle.

b. Always layer your survival kit, keeping the most important items on your body.

c. In preparing your survival kit, select items that can be used for more than one purpose.

A2

d. Your survival kit does not need to be elaborate. You only need functional items that will meet your needs and a case to hold them. The case might be a first aid case, an ammunition pouch, or another suitable case. This case should be:

(1) Water repellent or waterproof.

(2) Easy to carry or attach to your body.

(3) Suitable to accept various sized items.

(4) Durable.

e. When constructing a survival kit, you should have the following components: (FMST.07.20a)

(1) Fire starting items.

(2) Water procurement items.

(3) Food procurement items.

(4) Signaling items.

(5) First aid items.

(6) Shelter items.

2. **ITEMS CONTAINED WITHIN EACH COMPONENT** (FMST.07.20b)

a. Fire Starting Items.

(1) Matches.

(2) Magnifying glass.

(3) Flint and Steel.

(4) Lighter.

(5) Potassium Permanganate, with a container of sugar or anti-freeze.

(6) Prepackaged Tinder.

(a) Commercially Manufactured

(b) Cotton Balls and Petroleum Jelly

b. Water Procurement Items

(1) Water Disinfecting Chemicals.

(a) Iodine Tablets

(b) Betadine Solution

(c) Iodine Solution

(2) Metal Container. (Serves for boiling water)

(a) Canteen Cup

(b) Survival Kit Container

(c) Any suitable can that contained no petroleum products.

(3) Water Carrying Items.

(a) Canteen

(b) Plastic Bag

(c) Plastic/Metal/Glass Container which contained no petroleum products

c. Food Procurement Items

(1) Fish.

(a) Various sized hooks

(b) Various sized sinkers/weights

(c) Metal leaders and swivels

(d) Small weighted jigs

(e) Fishing line

i. Think about the size of fish for that environment when selecting weights and sizes.

(2) Game.

(a) Snares

i. Commercially Manufactured

ii. Aircraft Cable

iii. Tie Wire

(b) Bait

i. MRE Cheese Spread or Peanut Butter Package

(c) 550 Cord for Gill Net and Trap Construction

i. Engineer/Marking Tape

ii. Sling shot rubber and pouch

<u>d.</u> <u>Signaling Items</u>

(1) Day.

(a) Mirror

(b) Whistle

(c) Pyrotechnics (Smoke, Pen Flares)

(d) Air Panels

(2) Night.

(a) Pyrotechnics (Pen Flares, Star Clusters)

(b) Lights (Flashlight, Strobe, Chemlight)

(c) Whistle

<u>e.</u> <u>Shelter Items</u>

(1) Cordage.

(a) 550 Cord

(b) Wire

(c) Communication wire

(d) Tie wire

(2) Finger Saw.

(3) Sewing Kit with Needles for construction/repair of clothing.

(4) Tentage.

(a) Poncho

(b) Tarp

(c) Space blanket

(d) Plastic trash bags

f. First Aid Items

(1) Band-Aids.

(a) Steristrips

(b) Adhesive Tape

(c) Non-stick pads, 4x4's, Gauze, Battle Dressings

(d) Muslin Bandage

(2) Ointments.

(a) Burn

(b) Anti-septic

(3) Miscellaneous.

(a) Salt

(b) Sugar

(c) Eye Wash

(d) Alcohol prep pads

(e) Suture Kit

(f) Scalpel

(g) Vile of Yarrow

g. Miscellaneous items

(1) Fingernail clippers

(2) Compass

(3) Notebook with pen or pencil

(4) Wood eye screws and nails

(5) Surgical tubing

NOTE: It is assumed that the Marine is always carrying a high quality fixed bladed knife, a multi-tool knife, and a sharpening stone.

UNITED STATES MARINE CORPS
Mountain Warfare Training Center
Bridgeport, California 93517-5001

FMST.07.23
04/05/02

STUDENT OUTLINE

SIGNALLING AND RECOVERY

TERMINAL LEARNING OBJECTIVE Given a survival situation in any type of environmental condition and minimal equipment and resources, signal for aid to facilitate individual or group recovery per the references. (FMST 07.23)

ENABLING LEARNING OBJECTIVES

(1) Without the aid of references and from a given list, choose the audio international distress Signal, in accordance with the references. (FMST.07.23a)

(2) Without the aid of references and from a given list, choose the visual international distress signal, in accordance with the references. (FMST.07.23b)

(3) Without the aid of references and given a list of messages, match in writing the international code symbols to each corresponding message, in accordance with the references. (FMST.07.23c)

OUTLINE

1. **SIGNALING DEVICES**. The equipment listed below are items that may be on your body or items inside an aircraft. Generally, these items are used as signaling devices while on the move. They must be accessible for use at any moment's notice. Additionally, in a summer mountainous environment, Marines may experience areas that are snow covered and must be familiar with the effects that snow will have on specific signaling devices.

   a. Pyrotechnics. Pyrotechnics include star clusters and smoke grenades. When using smoke grenades in snow pack, a platform must be built. Without a platform, the smoke grenade will sink into the snow pack and the snow will absorb all smoke. A rocket parachute flare or a hand flare has been sighted as far away as 35 miles, with an average of 10 miles. Pyrotechnic flares are effective at night, but during daylight their detectability ranges are reduced by 90 percent.

A3

b. M-186 Pen Flare: The M-186 Pen Flare is a signaling device carried in the vest of all crew chiefs and pilots. Remember to cock the gun prior to screwing in the flare.

c. Strobe Light: A strobe light is generally carried in the flight vests of all crew chiefs and pilots. It can be used at night for signaling. Care must be taken because a pilot using goggles may not be able to distinguish a flashing strobe from hostile fire.

d. Flashlight: By using flashlights, a Morse code message can be sent. A SOS distress call consists of sending three dots, three dashes, and three dots. Keep repeating this signal.

e. Whistle: The whistle is used in conjunction with the audio international distress signal. It is used to communicate with forces on the ground.

f. AN/PRC-90 & AN/PRC-112: The AN/ PRC 90 survival radio is a part of the aviator's survival vest. The AN/PRC-112 will eventually replace the AN/PRC-90. Both radios can transmit either tone (beacon) or voice. Frequency for both is **282.8 for voice**, and **243.0 for beacon**. Both of these frequencies are on the UHF Band.

g. Day/Night Flare: The day/night flare is a good peacetime survival signal. The flare is for night signaling while the smoke is for day. A red cap identifies the older version flare with three nubbins while the new generation has three rings around the body for identification during darkness. The flare burns for approximately 20 second while the smoke burns for approximately 60 seconds.

NOTE: Once one end is used up, douse in water to cool and save the other end for future use.

h. Signal Mirror: A mirror or any shiny object can be used as a signaling device. It can be used as many times as needed. Mirror signals have been detected as far away as 45 miles and from as high as 16,000', although the average detection distance is 5 miles. It can be concentrated in one area, making it secure from enemy observation. A mirror is the best signaling device for a survivor; however, it is only as effective as its user. Learn how to use one now, before you find yourself in a survival situation.

(1) Military signal mirrors have instructions on the back showing how to use it. It should be kept covered to prevent accidental flashing that may be seen by the enemy.

(2) Any shiny metallic object can be substituted for a signal mirror.

(3) Haze, ground fog, or a mirage may make it hard for a pilot to spot signals from a flashing object. So, if possible, get to the highest point in your area when flashing. If you can't determine the aircraft's location, flash your signal in the direction of the aircraft noise.

AIMING THE SIGNAL MIRROR

2. METHODS OF COMMUNICATION

a. Audio: Signaling by means of sound may be good, but it does have some disadvantages.

(1) It has limited range unless you use a device that will significantly project the sound.

(2) It may be hard to pinpoint one's location due to echoes or wind.

(3) International Distress Signal. (FMST.07.23a) The survivor will make six blasts in one minute, returned by three blasts in one minute by the rescuer.

b. Visual: Visual signals are generally better than audio signals. They will pinpoint your location and can been seen at a greater distances under good weather conditions.

(1) The visual international distress symbol is recognized by a series of three evenly spaced improvised signaling devices. (FMST.07.23b)

3. **IMPROVISED SIGNALING DEVICES**. Improvised signaling devices are generally static in nature. They must be placed in a position to be seen by rescuers. They are made from any resources available, whether natural or man-made.

a. Smoke Generator: The smoke generator is an excellent improvised visual signaling device. It gives the survivor the flexibility to signal in either day or night conditions. This type of signal has been sighted as far away as 12 miles, with an average distance of 8 miles. Smoke signals are most effective in calm wind conditions or open terrain, but effectiveness is reduced with wind speeds above 10 knots. Build them as soon as time and the situation permits, and protect them until needed.

(1) Construct your fire in a natural clearing or along the edge of streams (or make a clearing). Signal fires under dense foliage will not be seen from the air.

(2) Find two logs, 6 - 10 inches in diameter, and approximately five feet long. Place the two logs parallel to each other with 3 - 4 feet spacing.

(3) Gather enough sticks, approximately two inches in diameter and four feet long, to lie across the first two logs. This serves as a platform for the fire.

(4) Gather enough completely dry branches to build a pyramid fire. The pyramid fire should be 4 feet by 4 feet by 2 feet high.

(5) Place your tinder under the platform.

(6) Gather enough pine boughs to lay on top of the pyramid fire. This serves to protect the fire and the tinder.

(7) To light, remove the pine bough and ignite the tinder. If available, construct a torch to speed up the lighting process, especially for multiple fires.

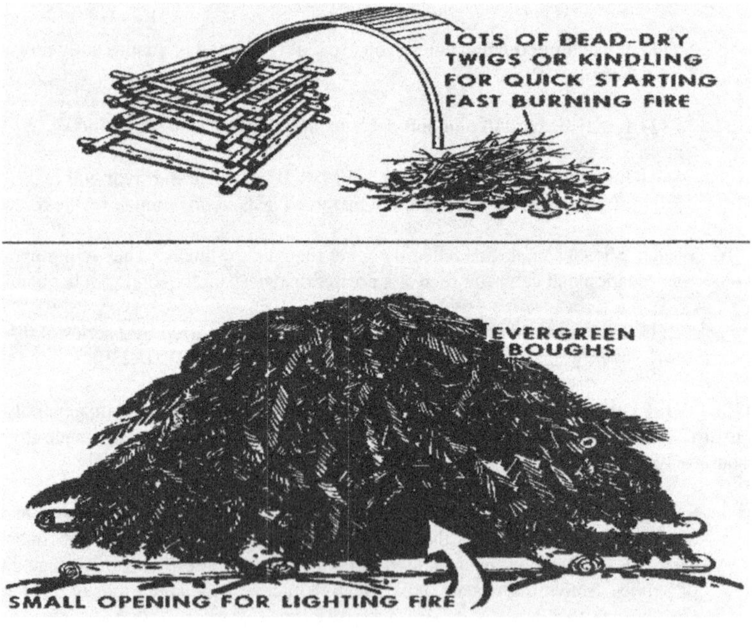

SMOKE GENERATOR

(8) To create a smoke effect during the day light hours, place the pine bough on the ignited fire.

(9) Placing a smoke grenade or colored flare under the platform will change the color of the smoke generated. Remember that you want the fire to draw in the colored smoke, which will create a smoke color that contrasts with the background will increase the chances of success.

b. Arrangement or alteration of natural materials: Such things as twigs or branches, can be tramped into letters or symbols in the snow and filled in with contrasting materials. To attract more attention, ground signals should be arranged in big geometric patterns.

(1) International symbols. The following symbols are internationally known. (FMST.07.23c)

Number	Message	Code symbol
1	REQUIRE ASSISTANCE	V
2	REQUIRE MEDICAL ASSISTANCE	X
3	NO OR NEGATIVE	N
4	YES OR AFFIRMATIVE	Y
5	PROCEED IN THIS DIRECTION	↑

INTERNATIONAL SYMBOLS

(2) Shadows: If no other means are available, you may have to construct mounds that will use the sun to cast shadows. These mounds should be constructed in one of the International Distress Patterns. To be effective, these shadow signals must be oriented to the sun to produce the best shadows. In areas close to the equator, a North—South line gives a shadow anytime except noon. Areas further north or south of the equator require the use of East—West line or some point of the compass in between to give the best result.

(3) Size: The letters should be large as possible for a pilot or crew to spot. Use the diagram below to incorporate the size to ratio for all letter symbols.

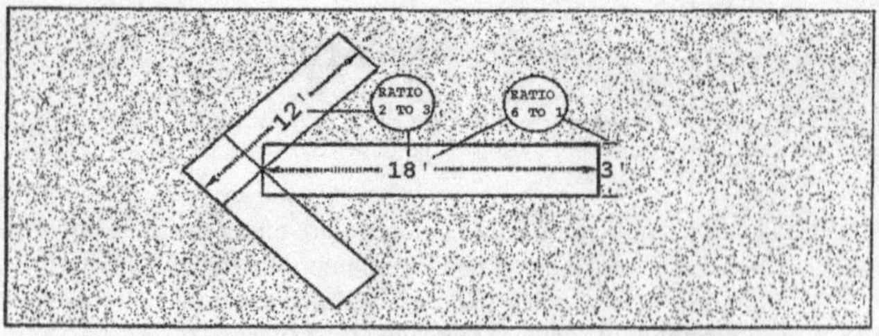

SIZE AND RATIO

(4) Contrast: When constructing letter symbols, contrast the letter from the surrounding vegetation and terrain. Ideally, bring material from another location to build the letter. This could be clothing, air panels, space blanket, etc.

(a) On snow, pile bough or use sea dye from a LPP (Life preserver, personal). Fluorescent sea dye markers have been sighted as far away as 10 miles, although the average detection distance is 3 miles.

4. **AIR TO GROUND COMMUNICATIONS**. Air to ground communications can be accomplished by standard aircraft acknowledgments.

a. Aircraft will indicate that ground signals have been seen and understood by:

(1) Rocking wings from side to side: This can be done during the day or in bright moonlight.

b. Aircraft will indicate that ground signals have been seen but not understood by:

(1) Making a complete, clockwise circle during the day or in bright moonlight.

5. **RECOVERY**. Marines trapped behind enemy lines in future conflicts may not experience quick recovery. Marines may have to move to a place that minimizes risk to the recovery force. No matter what signaling device a Marine uses, he must take responsibility for minimizing the recovery force's safety.

a. Placement Considerations: Improvised signaling devices, in a hostile situation, should not be placed near the following areas due to the possibility of compromise:

(1) Obstacles and barriers.

(2) Roads and trails.

(3) Inhabited areas.

(4) Waterways and bridges.

(5) Natural lines of drift.

(6) Man-made structures.

(7) All civilian and military personnel.

b. Tactical Consideration: The following tactical considerations should be adhered to prior to employing an improvised signaling device.

(1) Use the signals in a manner that will not jeopardize the safety of the recovery force or you.

(2) Locate a position, which affords observation of the signaling device and facilitates concealed avenues of escape (if detected by enemy forces). Position should be located relatively close to extract site in order to minimize "time spent on ground" by the recovery force.

(3) Maintain continuous security through visual scanning and listening while signaling devices is employed. If weapon systems are fire and/or observation should cover available, signaling devices.

(4) If enemy movement is detected in the area, attempt to recover the signaling device, if possible.

(5) Employ improvised signaling devices only during the prescribed times, if briefed in the mission order.

c. Recovery Devices: In mountainous terrain, a helicopter landing may be impossible due to ground slope, snow pack, or vegetation. The survivor must be familiar with recovery devices that may be aboard the aircraft.

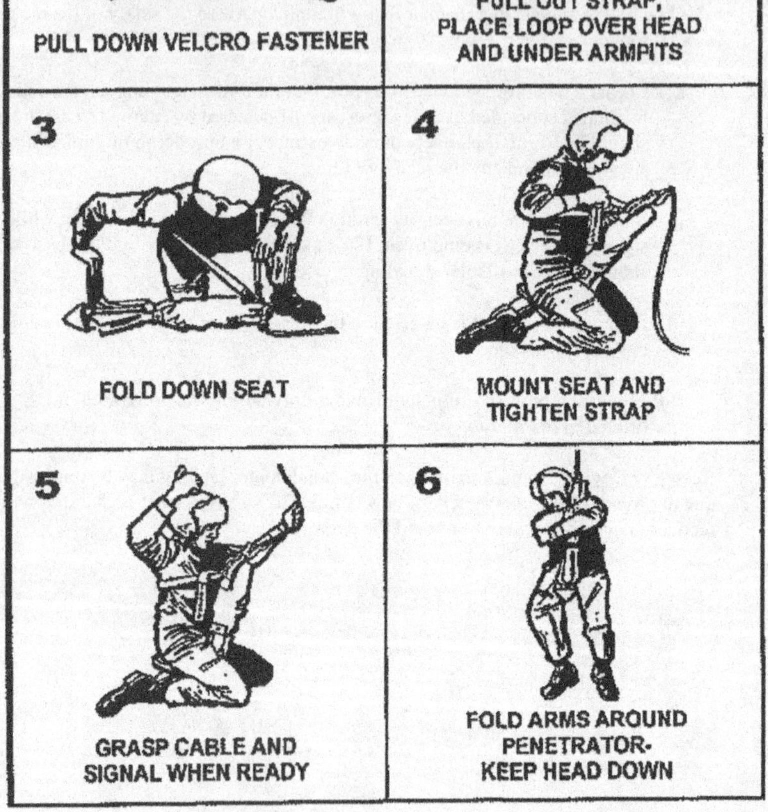

JUNGLE PENETRATOR

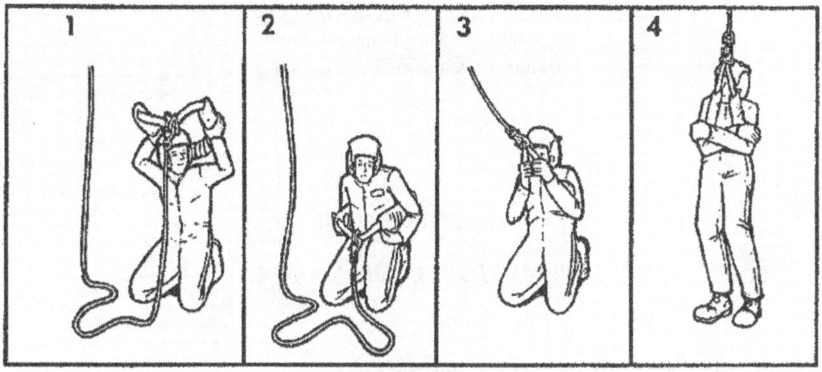

SLING HOIST

d. Recovery by other than aircraft. Recovery by means other than aircraft may occur. Unit Sop's should include signaling and link-up with forces at the following locations:

(1) Border Crossings. The evader who crosses into a neutral country is subject to detention by that country for the duration of the war.

(2) FEBA/FLOT.

(a) Static: Recovery along a static FEBA is always difficult. Under these conditions, enemy and friendly forces can be expected to be densely arrayed and well camouflaged, with good fields of fire. Attempts to penetrate the FEBA should be avoided.

(b) Advancing: Individuals isolated in front of advancing friendly units should immediately take cover and wait for the friendly units to overrun their position.

(c) Retreating: Individuals between opposing forces should immediately take cover and wait for enemy units to pass over their position. After most enemy units have moved on, evaders should try to link up with other isolated friendly elements and return to friendly forces.

(3) Link-up with friendly patrols. Unit authentication numbers and/or locally developed codes may assist the evader to safely make contact in or around the FEBA and when approached by friendly forces.

UNITED STATES MARINE CORPS
Mountain Warfare Training Center
Bridgeport, California 93517-5001

FMST.07.21
FMST.07.04
04/07/02

STUDENT OUTLINE

EXPEDIENT SHELTERS AND FIRES

TERMINAL LEARNING OBJECTIVES

(1) In a summer mountainous environment, construct expedient shelters, in accordance with the references. (FMST.07.21)

(2) In a summer mountainous environment, construct fires, in accordance with the references. (FMST.07.04)

b. **ENABLING LEARNING OBJECTIVES**

(1) Without the aid of references and from a given list, choose the four characteristics of a safe expedient shelter, in accordance with the references. (FMST.07.21a)

(2) Without the aid of references and from a given list, choose the four hazards avoid when using natural shelters, in accordance with the references. (FMST.07.21b)

(3) Without the aid of references and from a given list, choose one of man-made expedient shelters, in accordance with the references. (FMST.07.21c)

(4) Without the aid of references, start a fire using a primitive method, in accordance with the references. (FMST.07.04a)

A4

OUTLINE

1. **BASIC CHARACTERISTICS FOR SHELTER**. (FMST.07.21a) Any type of shelter, whether it is a permanent building, tentage, or an expedient shelter must meet six basic characteristics to be safe and effective. The characteristics are:

    a. Protection from the Elements: The shelter must provide protection from rain, snow, wind, sun, etc.

    b. Heat Retention: It must have some type of insulation to retain heat; thus preventing the waste of fuel.

    c. Ventilation: Ventilation must be constructed, especially if burning fuel for heat. This prevents the accumulation of carbon monoxide. Ventilation is also needed for carbon dioxide given off when breathing.

    d. Drying Facility: A drying facility must be constructed to dry wet clothes.

    e. Free from Natural Hazards: Shelters should not be built in areas of avalanche hazards, under rock fall or "standing dead" trees have the potential to fall on your shelter.

    f. Stable: Shelters must be constructed to withstand the pressures exerted by severe weather.

2. **NATURAL SHELTERS**. Natural shelters are usually the preferred types because they take less time and materials construct. The following may be made into natural shelters with some modification.

    a. Caves or Rock Overhangs: Can be modified by laying walls of rocks, logs or branches across the open sides.

    b. Hollow Logs: Can be cleaned or dug out, then enhanced with ponchos, tarps or parachutes hung across the openings.

    c. Hazards of Natural Shelters: (FMST.04.21b)

        (1) Animals: Natural shelters may already be inhabited (i.e. bears, coyotes, lions, rats, snakes, etc.). Other concerns from animals may be disease from scat or decaying carcasses.

        (2) Lack of Ventilation: Natural shelters may not have adequate ventilation. Fires may be built inside for heating or cooking but may be uncomfortable or even dangerous because of the smoke build up.

        (3) Gas Pockets: Many caves in a mountainous region may have natural gas pockets in them.

(4) Instability: Natural shelters may appear stable, but in reality may be a trap waiting to collapse.

3. **MAN-MADE SHELTERS**. (FMST.04.21c) Many configurations of man-made shelters may be used. Over-looked man-made structures found in urban or rural environments may also provide shelter (i.e. houses, sheds, or barns). Limited by imagination and materials available, the following man-made shelters can be used in any situation.

a. Poncho Shelter.

b. Sapling Shelter.

c. Lean-to.

d. Double lean-to.

e. A-frame Shelter.

f. Fallen Tree Bivouac.

4. **CONSTRUCTION OF MAN-MADE SHELTERS**. To maximize the shelter's effectiveness, Marines should take into consideration the following prior to construction.

a. Considerations.

(1) Group size.

(2) Low silhouette and reduced living area dimensions for improved heat conservation.

(3) Avoid exposed hill tops, valley floors, moist ground, and avalanche paths.

(4) Create a thermal shelter by applying snow, if available, to roof and sides of shelter.

(5) Location of site to fire wood, water, and signaling, if necessary.

(6) How much time and effort needed to build the shelter.

(7) Can the shelter adequately protect you from the elements (sun, wind, rain, and snow). Plan on worst case scenario.

(8) Are the tools available to build it. If not, can you make improvised tools?

(9) Type and amount of materials available to build it.

(10) When in a tactical environment, you must consider the following:

(a) Provide concealment from enemy observation.

(b) Maintain camouflaged escape routes.

(c) Use the acronym BLISS as a guide.

B – Blend in with the surroundings.
L –Low silhouette.
I – Irregular shape.
S – Small.
S – Secluded located.

b. Poncho Shelter: This is one of the easiest shelters to construct. Materials needed for construction are cord and any water-repellent material (i.e. poncho, parachute, and tarp). It should be one of the first types of shelter considered if planning a short stay in any one place.

(1) Find the center of the water-repellent material by folding it in half along its long axis.

(2) Suspend the center points of the two ends using cordage.

(3) Stake the Four Corners down, with sticks or rocks.

c. Sapling Shelter: This type of shelter is constructed in an area where an abundance of saplings are growing. It is an excellent evasion shelter.

(1) Find or clear an area so that you have two parallel rows of saplings at least 4' long and approximately 1 ½' to 2' apart.

(2) Bend the saplings together and tie them to form several hoops, which will form the framework of the shelter.

(3) Cover the hoop with a water-repellent covering.

(4) The shelter then may be insulated with leaves, brush, snow, or boughs.

(5) Close one end permanently. Hang material over the other end to form a door.

d. Lean-to: A lean-to is built in heavily forested areas. It requires a limited amount of cordage to construct. The lean-to is an effective shelter but does not offer a great degree of protection from the elements.

(1) Select a site with two trees (4-12" in diameter) spaced far enough apart that a man can lie down between them. Two sturdy poles can be substituted by inserting them into the ground the proper distance apart.

(2) Cut a pole to support the roof. It should be at least 3-4" in diameter and long enough to extend 4-6" past both trees. Tie the pole horizontally between the two trees, approximately 1 meter off the deck.

(3) Cut several long poles to be used as stringers. They are placed along the horizontal support bar approximately every 1 ½' and lay on the ground. All stringers may be tied to or laid on the horizontal support bar. A short wall or rocks or logs may be constructed on the ground to lift the stringers off the ground, creating additional height and living room dimensions.

(4) Cut several saplings and weave them horizontally between the stringers. Cover the roof with water-repellent and insulating material.

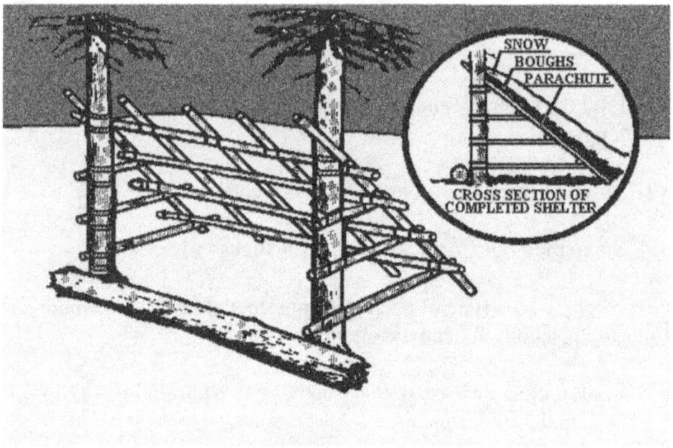

LEAN-TO

e. Double lean-to: The double lean-to shelter is constructed for 2-5 individuals. It is constructed by making two lean-tos and placing them together.

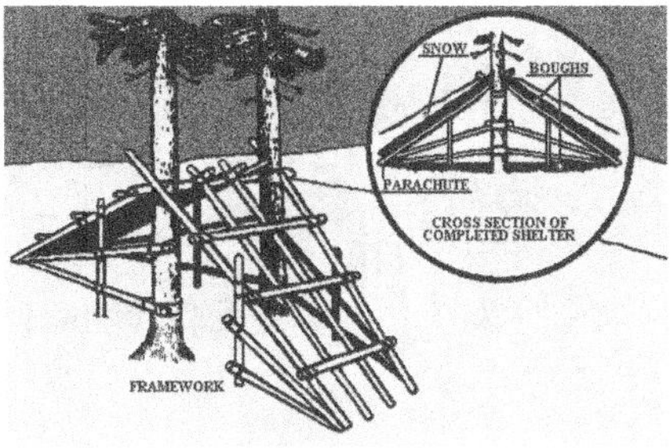

DOUBLE LEAN-TO

f. A-Frame Shelter: An A-Frame shelter is constructed for 1-3 individuals. After the framework is constructed, bough/tentage is interwoven onto the frame and snow, if available, is packed onto the outside for insulation.

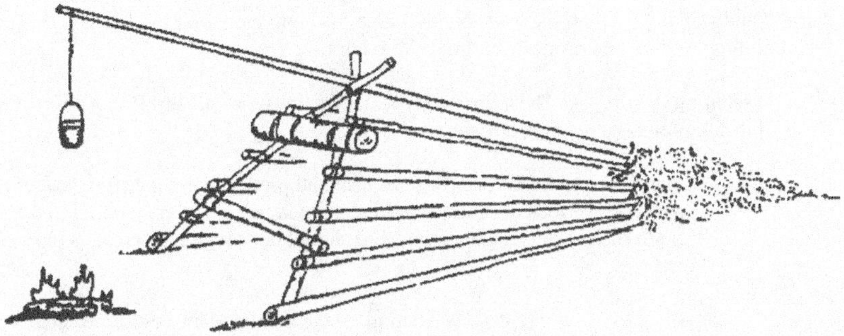

A-FRAME

g. Fallen Tree Bivouac: The fallen tree bivouac is an excellent shelter because most of the work has already been done.

(1) Ensure the tree is stable prior to constructing.

(2) Branches on the underside are cut away making a hollow underneath.

(3) Place additional insulating material to the top and sides of the tree.

(4) A small fire is built outside of the shelter.

FALLEN TREE BIVOUAC

5. **REFLECTOR WALLS.** Heating a shelter requires a slow fire that produces lots of steady heat over a long period of time. A reflector wall should be constructed for all open-ended shelters. A reflector wall is constructed with a flat rock or a stack of green logs propped behind the fire. A surprising amount of heat will bounce back from the fire into the shelter.

6. **FIRES**. Fires fall into two main categories: those built for cooking and those built for warmth and signaling. The basic steps are the same for both: preparing the fire lay, gathering fuel, building the fire, and properly extinguishing the fire.

   a. Preparing the fire lay: There are two types of fire lays: fire pit and Dakota hole. Fire pits are probably the most common.

      (1) Create a windbreak to confine the heat and prevent the wind from scattering sparks. Place rocks or logs used in constructing the fire lay parallel to the wind. The prevailing downwind end should be narrower to create a chimney effect.

      (2) Avoid using wet rocks. Heat acting on the dampness in sandstone, shale, and stones from streams may cause them to explode.

      (3) Dakota Hole: The Dakota Hole is a tactical fire lay. Although no fire is 100% tactical, this fire lay will accomplish certain things:

         (a) Reduces the signature of the fire by placing it below ground.

         (b) Provides more of a concentrated heat source to boil and cook, thus preserving fuel and lessening the amount of burning time.

         (c) By creating a large air draft, the fire will burn with less smoke than the fire pit.

         (d) It is easier to light in high winds.

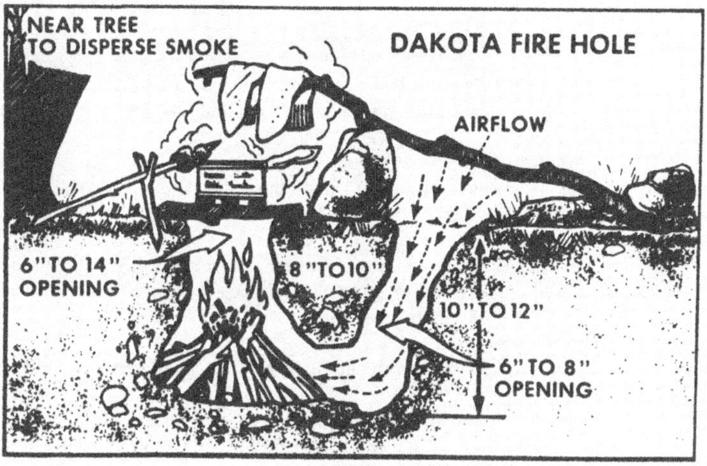

DAKOTA HOLE

b. Gather Fuel: Many Marines take shortcuts when gathering firewood. Taking a few extra minutes can mean the difference between ease and frustration when building a fire.

(1) Tinder: Tinder is the initial fuel. It should be fine and dry. Gather a double handful of tinder for the fire to be built and an extra double handful to be stored in a dry place for the following morning. Dew can moisten tinder enough to make lighting the fire difficult. Some examples are:

(a) Shredded cedar/juniper bark, pine needles.

(b) Dry grass.

(c) Slivers shaved from a dry stick.

(d) Hornet's nest.

(e) Natural fibers from equipment supplemented with pine pitch (i.e., cotton battle dressing).

(f) Cotton balls and petroleum jelly or Char-cloth.

NOTE: Sticks used for tinder should be dry and not larger than the diameter of a toothpick.

(2) Kindling: This is the material that is ignited by the tinder that will burn long enough to ignite the fuel.

(a) Small sticks/twigs pencil-thick up to the thickness of the thumb. Ensure that they are dry.

(b) Due to a typically large resin content, evergreen limbs often make the best kindling. They burn hot and fast, but typically do not last long.

(3) Fuel Wood: Fuel Wood is used to keep the blaze going long enough to fulfill its purpose. Ideally, it should burn slow enough to conserve the woodpile, make plenty of heat, and leave an ample supply of long-lasting coals.

(a) Firewood broken from the dead limbs of standing trees or windfalls held off the ground will have absorbed less moisture and therefore should burn easily.

(b) Refrain from cutting down live, green trees.

(c) Softwoods (evergreens and conifers) will burn hot and fast with lots of smoke and spark, leaving little in the way of coals. Hardwoods (broad leaf trees) will burn slower with less smoke and leave a good bed of coals.

(d) Learn the woods indigenous to the area. Birch, dogwood, and maple are excellent fuels. Osage orange, ironwood, and manzanita, though difficult to break up, make terrific coals. Aspen and cottonwood burn clean but leave little coals.

(e) Stack your wood supply close enough to be handy, but far enough from the flames to be safe. Protect your supply from additional precipitation.

(f) If you happen to go down in an aircraft that has not burned a mixture of gas and oil may be used. Use caution when igniting this mixture.

c. Building the Fire. The type of fire built will be dependent upon its intended use, either cooking or heating and signaling.

(1) Cooking Fires. The following listed fires are best used for cooking:

(a) Teepee Fire: The teepee fire is used to produce a concentrated heat source, primarily for cooking. Once a good supply of coals can be seen, collapse the teepee and push embers into a compact bed.

TEEPEE FIRE

(2) Heating and Signaling Fires.

(a) Pyramid Fire: Pyramid fires are used to produce large amounts of light and heat. They will dry out wet wood or clothing.

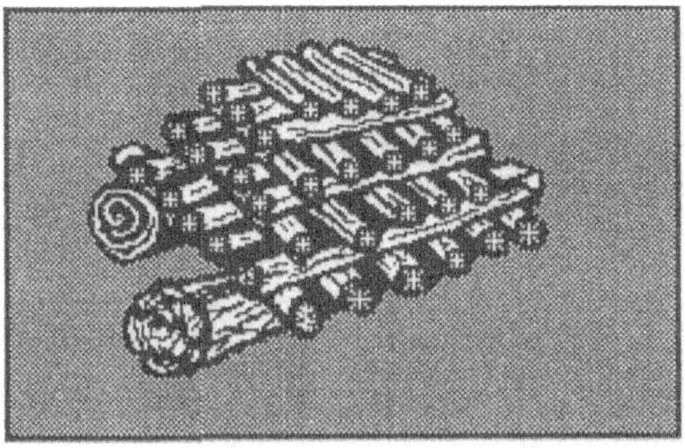

PYRAMID FIRE

d. Starting Fires. Starting a fire is done by a source of ignition and falls into two categories, modern igniters and primitive methods.

(1) Modern Methods: Modern igniters use modern devices we normally think of to start a fire. Reliance upon these methods may result in failure during a survival situation. These items may fail when required to serve their purpose.

(a) Matches and Lighters. Ensure you waterproof these items.

(b) Convex Lens. Binocular, camera, telescopic sights, or magnifying lens are used on bright, sunny days to ignite tender.

(c) Flint and Steel: Sometimes known as metal matches or "Mag Block". Scrape your knife or carbon steel against the flint to produce a spark onto the tinder. Some types of flint & steel designs will have a block of magnesium attached to the device, which can be shaved onto the tinder prior to igniting. Other designs may have magnesium mixed into the flint to produce a higher quality of spark.

(2) Primitive Methods. Primitive fire methods are those developed by early man. There are numerous techniques that fall into this category. The only method that will be taught at MCMWTC is the Bow & Drill.

(3) Bow & Drill. The technique of starting a fire with a bow & drill is a true field expedient fire starting method, which requires a piece of cord and knife from your survival kit to construct. The components of the bow & drill are bow, drill, socket, and fireboard, ember patch, and bird's nest.

(a) Bow: The bow is a resilient, green stick about 3/4 of an inch in diameter and 30-36 inches in length. The bowstring can be any type of cord, however, 550 cord works best. Tie the string from one end of the bow to the other, without any slack.

(b) Drill: The drill should be a straight, seasoned hardwood stick about 1/2 to 3/4 of an inch in diameter and 8 to 12 inches in length. The top end is tapered to a blunt point to reduce friction generated in the socket. The bottom end is slightly rounded to fit snugly into the depression on the fireboard.

(c) Socket: The socket is an easily grasped stone or piece of hardwood or bone with a slight depression on one side. Use it to hold the drill in place and to apply downward pressure.

(d) Fire board: The fire board is a seasoned softwood board which should ideally be 3/4 of an inch thick, 2-4 inches wide, and 8-10 inches long. Cut a depression 3/4 of an inch from the edge on one side of the fireboard. Cut a U-shape notch from the edge of the fireboard into the depression. This notch is designed to collect and form an ember, which will be used to ignite the tinder.

(e) Ember Patch: The ember patch is made from any type of suitable material (i.e., leather, aluminum foil, and bark). It is used to catch and transfer the ember from the fireboard to the bird's nest. Ideally, it should be 4 inches by 4 inches in size.

(f) Bird's Nest: The bird's nest is a double handful of tinder, which will be made into the shape of a nest. Tinder must be dry and finely shredded material (i.e., outer bark from juniper/cedar/sage brush or inner bark from cottonwood/aspen or dry grass/moss). Lay your tinder out in two equal rows about 4 inches wide and 8-12 inches long. Loosely roll the first row into a ball and knead the tinder to further break down the fibers. Place this ball perpendicular onto the second row of tinder and wrap. Knead the tinder until all fibers of the ball are interwoven. Insert the drill half way into the ball to form a partial cylinder. This is where the ember will be placed.

(4) Producing a fire using the bow & drill.

(a) Place the ember patch under the U-shaped notch.

(b) Assume the kneeling position, with the left foot on the fireboard near the depression.

(c) Load the bow with the drill. Ensure the drill is between the wood of the bow and bow string. Place the drill into the depression on the fireboard. Place the socket on the tapered end of the drill.

(d) Use the left hand to hold the socket while applying downward pressure.

(e) Use the right hand to grasp the bow. With a smooth sawing motion, move the bow back and forth to twirl the drill.

(f) Once you have established a smooth motion, smoke will appear. Once smoke appears, apply more downward pressure and saw the bow faster.

(g) When a thick layer of smoke has accumulated around the depression, stop all movement. Remove the bow, drill, and socket from the fire board, without moving the fireboard. Carefully remove your left foot off the fireboard.

(h) Gently tap the fireboard to ensure all of the ember has fallen out of the U-shaped notch and is lying on the ember patch. Remove the fireboard.

(i) Slowly fan the black powder to solidify it into a glowing ember. Grasping the ember patch, carefully drop the ember into the cylinder of the bird's nest.

(j) Grasp the bird's nest with the cylinder facing towards you and parallel to the ground. Gently blow air into the cylinder. As smoke from the nest becomes thicker, continue to blow air into the cylinder until fire appears.

(5) Trouble Shooting the Bow & Drill

(a) Drill will not stay in depression- Apply more downward pressure and/or increase width/depth of depression.

(b) Drill will not twirl- lessens the amount of downward pressure and/or tighten bowstring.

(c) Socket smoking- Lessen the amount of downward pressure. Wood too soft when compared to hardness of drill. Add some lubrication: animal fat, oil, or grease.

(d) No smoke- Drill and fireboard are the same wood. Wood may not be seasoned. Check drill to ensure that it is straight. Keep left hand locked against left shin while sawing.

(e) Smoke but no ember- U-shaped notch not cut into center of the depression.

(f) Bowstring runs up and down drill- use a locked right arm when sawing. Check drill to ensure that it is straight. Ensure bowstring runs over the top of the left boot.

(g) Birds nest will not ignite- Tinder not dry. Nest woven too tight. Tinder not kneaded enough. Blowing too hard (ember will fracture).

e. Extinguishing the Fire. The fire must be properly extinguished. This is accomplished by using the drown, stir, and feel method.

(1) Drown the fire by pouring at water in the fire lay.

(2) Stir the ember bed to ensure that the fire is completely out.

(3) Check the bed of your fire by feeling for any hot spots.

(4) If any hot spots are found, start the process all over again.

UNITED STATES MARINE CORPS
Mountain Warfare Training Center
Bridgeport, California 93517-5001

FMST.07.22
04/07/02

STUDENT OUTLINE

SURVIVAL NAVIGATION

LEARNING OBJECTIVES

TERMINAL LEARNING OBJECTIVE. In a summer mountainous environment, navigate in a survival situation, in accordance with the references. (FMST.07.22)

ENABLING LEARNING OBJECTIVES

(1)Without the aid of references and from a given list, choose one of the considerations for travel, in accordance with the references. (FMST.07.22a)

(2)Without the aid of references and from a given list, choose two methods for locating the North Star, in accordance with the references. (FMST.07.22b)

OUTLINE

1. **CONSIDERATIONS FOR STAYING OR TRAVELLING**. (FMST.07.22a)

   a. Stay with the aircraft or vehicle if possible. More than likely somebody knows where it was going. It is also a ready-made shelter.

   b. Leave only when:

      (1) Certain of present location; have known destination and the ability to get there.

      (2) Water, food, shelter, and/or help can be reached.

      (3) Convinced that rescue is not coming.

A5

c. If the decision is to travel, the following must also be considered:

(1) Which direction to travel and why.

(2) What plan is to be followed.

(3) What equipment should be taken

(4) How to mark the trail.

(5) Predicted weather.

d. If the tactical situation permits leave the following information at the departure point:

(1) Departure time.

(2) Destination.

(3) Route of travel/direction.

(4) Personal condition.

(5) Available supplies.

2. DAYTIME SURVIVAL NAVIGATION

a. Sun Movement. It is generally taken for granted that the sun rises in the east and sets in the west. This rule of thumb, however, is quite misleading. In fact, depending on an observer's latitude and the season, the sun could rise and set up to 50 degrees off of true east and west.

b. The following diagram and terms are essential to understanding how the sun and stars can help to determine direction:

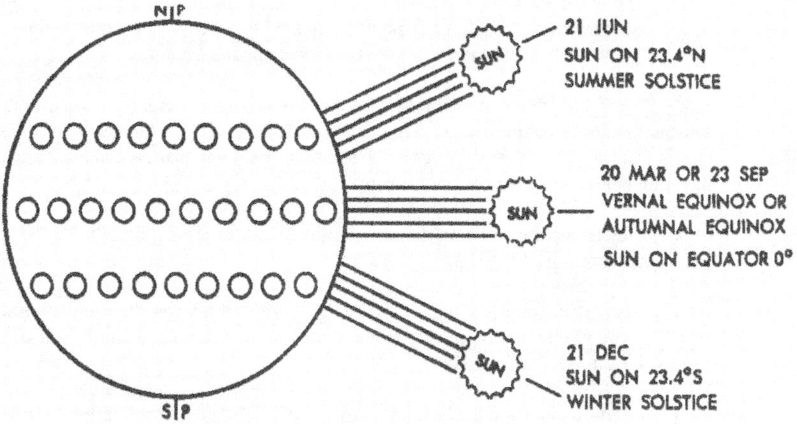

Position of the Sun at Equinox and Solstice

(1) Summer/Winter Solstice: (21 June/21 December) Two times during the year when the sun has no apparent northward or southward motion.

(2) Vernal/Autumnal Equinox: (20 March/23 September) Two times during the year when the sun crosses the celestial equator and the length of day and night are approximately equal.

c. Sun's Movement. As reflected in the diagram above, the sun continuously moves in a cycle from solstice to equinox, throughout each day, however, the sun travels a uniform arc in the sky from sunrise to sunset. Exactly halfway along its daily journey, the sun will be directly south of an observer (or north if the observer is in the Southern Hemisphere). This rule may not apply to observers in the tropics (between 23.5 degrees north and south latitude) or in the Polar Regions (60 degrees latitude). It is at this point that shadows will appear their shortest. The time at which this occurs is referred to as "local apparent noon."

d. Local Apparent Noon. Whenever using any type of shadow casting device to determine direction, "local apparent noon" (or the sun's highest point during the day) must be known. Local apparent noon can be determined by the following methods.

(1) Knowing sunrise and sunset from mission orders, i.e., sunrise 0630 and sunset 1930. Take the total amount of daylight (13 hours), divide by 2 (6 hours 30 minutes), and add to sunrise (0630 plus 6 hours 30 minutes). Based on this example, local apparent noon would be 1300.

(2) Using the string method. The string method is used to find two equidistant marks before and after estimated local apparent noon. The center point between these two marks represents local apparent noon.

e. Sun's Bearing. With an understanding of the sun's daily movement, as well as its seasonal paths, a technique is derived that will determine the true bearing of the sun at sunrise and sunset. With the aid of a circular navigational chart, we can accurately navigate based on the sun's true bearing:

(1) Determine the sun's maximum amplitude at your operating latitude using the top portion of the chart.

(2) Scale the center baseline of the chart where 0 appears as the middle number; write in the maximum amplitude at the extreme north and south ends of the baseline.

(3) Continue to scale the baseline; you should divide the baseline into 6 to 10 tick marks that represent equal divisions of the maximum amplitude.

(4) From today's date along the circumference, draw a straight line down until it intersects the baseline.

(5) The number this line intersects is today's solar amplitude. If the number is left of 0, it is"north" amplitude; if the number is right of 0, it is a "south" amplitude. Use the formula at the bottom of the chart to determine the sun's bearing at sunrise or sunset.

LATITUDE (N OR S)	5	10	15	20	25	30	35	40	45	50	55	60
MAXIMUM AMPLITUDE	24°	24°	24°	25°	26°	27°	29°	31°	34°	38°	44°	53°

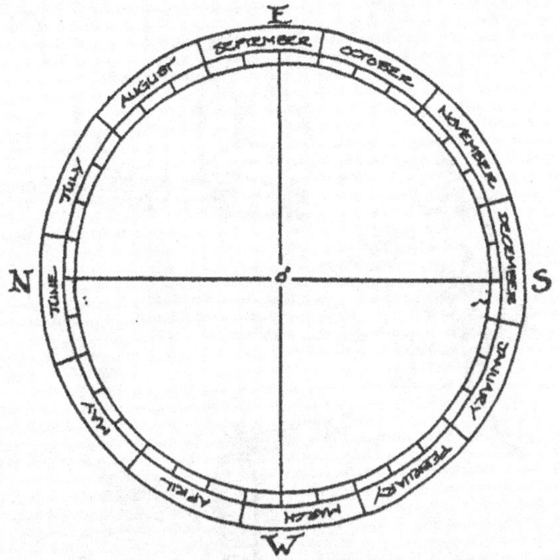

NORTH AMPLITUDES	SOUTH AMPLITUDES
SUNRISE BEARING = 90° − AMPLITUDE	SUNRISE BEARING = 90° + AMPLITUDE
SUNSET BEARING = 270° + AMPLITUDE	SUNSET BEARING = 270° − AMPLITUDE

Circular Navigational Chart

f. Shadow Stick Construction. This technique will achieve a cardinal direction within 10 degrees of accuracy if done within two hours of local apparent noon. Once again, this technique may be impractical near the Polar Regions, as shadows tend to be very long; similarly, in the tropics shadows are generally very small.

(1) Get a straight, 3-6 foot stick free of branches and pointed at the ends and 3-5 small markers: i.e., sticks, rocks, or nails.

(2) Place the stick upright in the ground and mark the shadow tip with a marker.

(3) Wait 10-15 minutes and mark shadow tip again with a marker.

(4) Repeat this until all of the markers are used.

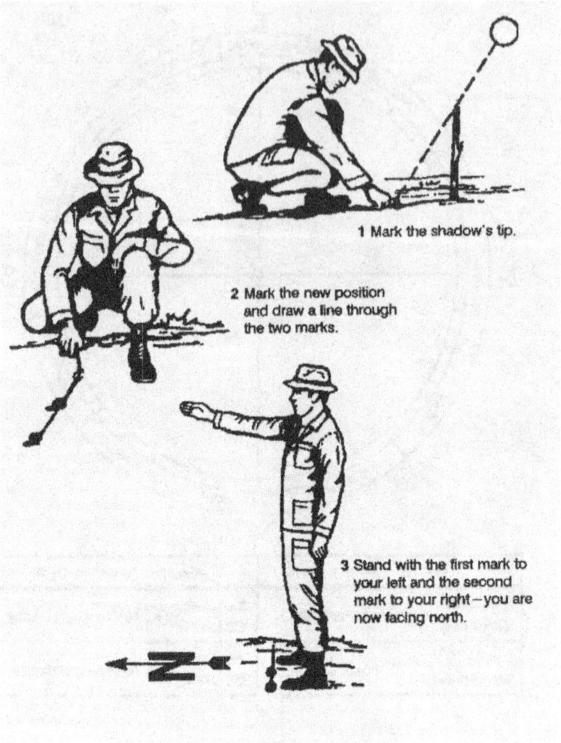

SHADOW STICK METHOD

(5) The markers will form a West—East line.

(6) Put your left foot on the first marker and your right foot on the last marker, you will then be facing north.

3. **POCKET NAVIGATOR** The only material required is a small piece of paper or other flat-surface material upon which to draw the trace of shadow tips and a 1 to 2 inch pin, nail, twig, wooden matchstick, or other such device to serve as a shadow-casting rod.

a. Set this tiny rod upright on your flat piece of material so that the sun will cause it to cast a shadow. Mark the position where the base of the rod sits so it can be returned to the same spot for later readings. Secure the material so that it will not move and mark the position of the material with string, pebbles, or twigs, so that if you have to move the paper you can put it back exactly as it was. Now, mark the tip of the rod shadow.

b. As the sun moves, the shadow-tip moves. Make repeated shadow-tip markings every 15 minutes. As you make the marks of the shadow tip, ensure that you write down the times of the points.

c. At the end of the day, connect the shadow-tip markings. The result will normally be a curved line. The closer to the vernal or autumnal equinoxes (March 21 and September 23) the less pronounced the curvature would be. If it is not convenient or the tactical situation does not permit to take a full day's shadow-tip markings, your observation can be continued on subsequent days by orienting the pocket navigator on the ground so that the shadow-tip is aligned with a previously plotted point.

d. The markings made at the sun's highest point during the day, or solar noon, is the north—south line. The direction of north should be indicated with an arrow on the navigator as soon as it is determined. This north-south line is drawn from the base of the rod to the mark made at solar noon. This line is the shortest line that can be drawn from the base of the pin to the shadow-tip curve.

e. To use your pocket navigator, hold it so that with the shadow-tip is aligned with a plotted point at the specified point. I.e.; if it is now 0900 the shadow-tip must be aligned with that point. This will ensure that your pocket navigator is level. The drawn arrow is now oriented to true north, from which you can orient yourself to any desired direction of travel.

f. The pocket navigator will work all day and will not be out of date for approximately one week.

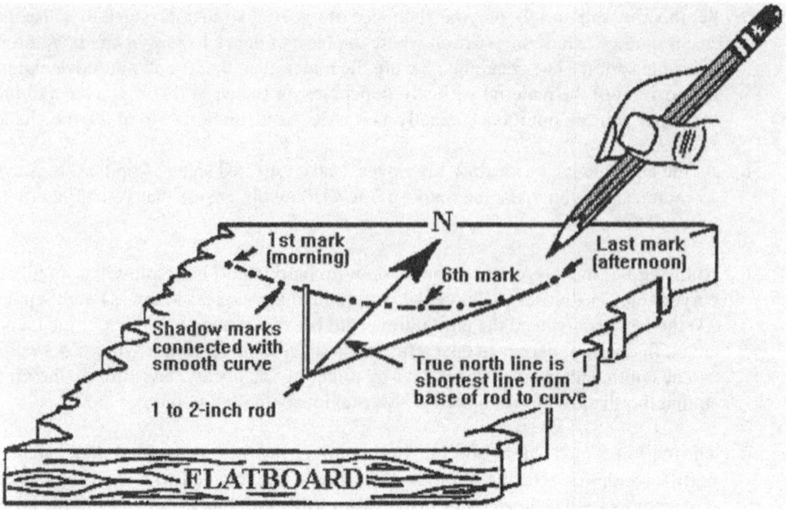

POCKET NAVIGATOR

4. NIGHTTIME SURVIVAL NAVIGATION

a. Mark North. To aid you in navigating at night, it is beneficial to watch where the sun goes down. If you're going to start moving after dark mark the northerly direction.

b. Locating the North Star. There are two methods used in locating the North Star. (FMST.07.22b)

(1) Using the Big Dipper (*Ursa Major*). The best indictors are the two "dippers ". The North Star is the last star in the handle of the Little Dipper, which is not the easiest constellation to find. However, the Big Dipper is one of the most prominent constellations in the Northern Hemisphere. The two lowest stars of the Big Dipper's cup act as pointers to the North Star. If you line up these two stars, they make a straight line that runs directly to the North Star. The distance to the North Star along this line is 5 times that between the two pointer stars.

(2) Using Cassiopeia (*Big M or W*). Draw a line straight out from the center star, approximately half the distance to the Big Dipper. The North Star will be located there.

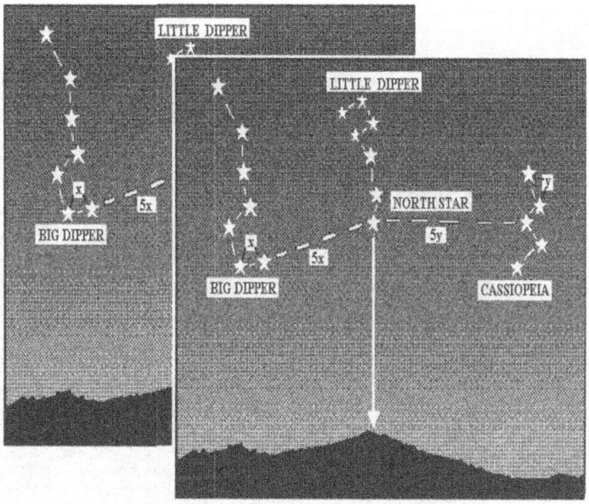

LOCATING THE NORTH STAR

NOTE: Because the Big Dipper and Cassiopeia rotate around the North Star, they will not always appear in the same position in the sky. In the higher latitudes, the North Star is less effective for the purpose of orienting because it appears higher in the sky. At the center of the Arctic Circle, it would be directly overhead and all directions lead south.

c. Southern Cross. In the Southern Hemisphere, Polaris is not visible. There, the Southern Cross is the most distinctive constellation. An imaginary line through the long axis of the Southern Cross, or True Cross, points towards a dark spot devoid of stars approximately three degrees offset from the South Pole. The True Cross should not be confused with the larger cross nearby know as the False Cross, which is less bright, more widely spaced, and has five stars. Two closely spaced, very bright stars that trail behind the crosspiece can confirm the True Cross. These two stars are often easier to pick out than the cross itself. Look for them. Two of the stars in the True Cross are among the brightest stars in the heavens; they are the stars on the southern and eastern arms. The stars on the northern and western arms are not as conspicuous, but are bright.

NOTE: The imaginary point depicted in the picture is the dark spot devoid of stars.

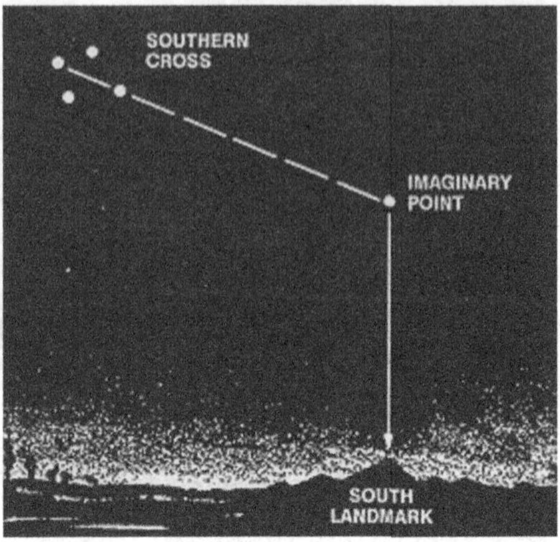

SOUTHERN CROSS

d. Moon Navigator. Like the sun, the moon rises in the east and sets in the west. Use the same method of the shadow stick as you did during the day.

5. **IMPROVISED COMPASSES**. There are three improvised techniques to construct a compass.

a. Synthetic technique. The required items are a piece of synthetic material, (i.e., parachute cloth), and a small piece of iron or steel that is long, thin, and light. Aluminum or yellow metals won't work (only things that rust will do). A pin or needle is perfect, but a straightened paper clip, piece of steel baling wire or barbed wire could also work.

(1) Stroke the needle repeatedly in one direction against the synthetic material. Ensure that you lift the material a few inches up into the air at the end of each stroke, returning to the beginning of the needle before descending for another stroke in the same direction. Do this approximately 30 strokes. This will magnetize the needle.

(2) Float the metal on still water using balled up paper, wood chip, or leaf. Gather some water in a non-magnetic container or a scooped out recess in the

ground, such as a puddle. Do not use a "tin can" which is made of steel. (Aluminum can would be fine.) Place the float on the water, then the metal on it. It will slowly turn to orient itself.

b. Magnet technique. You will achieve the same results by using a magnet. Follow the same steps as you did with the synthetic material. The magnets you are most likely to have available to you are those in a speaker or headphones of a radio.

c. Magnetization through a battery. A power source of 2 volts or more from a battery can be used with a short length of insulated wire to magnetize metal. Coil the wire around a needle. If the wire is non-insulated, wrap the needle with paper or cardboard. Attach the ends to the battery terminals for 5 minutes.

d. Associated problems with improvised compasses. The following are common problems with all improvised compasses.

(1) Soft steel tends to lose its magnetism fairly quickly, so you will have to demagnetize your needle occasionally, though you should not have to do this more than two or three times a day.

(2) Test your compass by disturbing it after it settles. Do this several times. If it returns to the same alignment, you're OK. It will be lined up north and south, though you will have to determine by other means which end is north. Use the sun, stars, or any other natural signs in the area.

(3) Remember that this will give magnetic north. In extreme northern latitudes, the declination angle can be extreme.

6. SURVIVAL NAVIGATION TECHNIQUES

a. Navigator.

(1) Employ a navigation method.

(2) Find the cardinal direction.

(3) Pick a steering mark in the desired direction of travel.

b. Maintain a Log: The possibility may arise when you will not have a map of the area. A log will decrease the chance of walking in circles.

(1) Construction.

(a) Use any material available to you i.e., paper, clothing, MRE box, etc.

(b) Draw a field sketch annotating North, prominent terrain features, and Friendly/enemy position.

(2) Maintenance.

(a) Annotate distance traveled, elevation gained and lost, and cardinal directions.

(b) Maintain and update field sketch as movement progresses.

(c) Ensure readability of your field sketch. (I.e., don't clutter the sketch so much that it can't be read.)

c. <u>During Movement Constantly Refer To</u>.

(1) Log.

(2) Steering marks.

d. <u>Actions If You Become Lost</u>.

(1) Immediate action

(a) Orient your sketch. This will probably make your mistake obvious.

(2) Corrective action

(a) Backtrack using steering marks until you have determined the location of your error.

(b) Re-orient your sketch.

(c) Select direction of travel and continue to march.

UNITED STATES MARINE CORPS
Mountain Warfare Training Center
Bridgeport, California 93517-5001

FMST.07.08
04/07/02

STUDENT OUTLINE

WATER PROCUREMENT

LEARNING OBJECTIVES

TERMINAL LEARNING OBJECTIVE Given a unit in a cold weather environment and the necessary equipment and supplies, perform cold weather preventive medicine, in accordance with the references. (FMST.07.08)

ENABLING LEARNING OBJECTIVES

(1) Without the aid of references from a given list, choose one of the techniques for finding incidental water, in accordance with the references. (FMST.07.08i)

(2) Without the aid of references from a given list, choose one of the four hazardous fluids to avoid substituting for potable water, in accordance with the references. (FMST.07.08j)

(3) Without the aid of references, construct a water-producing device, in accordance with the references. (FMST.07.08k)

OUTLINE

1. WATER INTAKE

a. Thirst is not a strong enough sensation to determine how much water you need.

b. The best plan is to drink, utilizing the OVER DRINK method. Drink plenty of water anytime it is available and particularly when eating.

A6

c. Dehydration is a major threat. A loss of only 5 % of your body fluids causes thirst, irritability, nausea, and weakness; a 10% loss causes dizziness, headache, inability to walk, and a tingling sensation in limbs; a 15% loss causes dim vision, painful urination, swollen tongue, deafness, and a feeling of numbness in the skin; also a loss of more than 15% body fluids could result in death.

d. Your water requirements will be increased if:

(1) You have a fever.

(2) You are experiencing fear or anxiety.

(3) You evaporate more body fluid than necessary. (i.e., not using the proper shelter to your advantage)

(4) You have improper clothing.

(5) You ration water.

(6) You overwork.

2. **INCIDENTAL WATER**. (FMST.07.08.i)

a. During movement, you may have to ration water until you reach a reliable water source. Incidental water may sometimes provide opportunities to acquire water. Although not a reliable or replenished source, it may serve to stretch your water supply or keep you going in an emergency. The following are sources for incidental water:

(1) Dew: In areas with moderate to heavy dew, tying rags or tuffs of fine grass around your ankles can collect dew. While walking through dewy grass before sunrise, the rags or grass will saturate and can be rung out into a container. The rags or grass can be replaced and the process is repeated.

(2) Rainfall: Rainwater collected directly in clean container or in plants that contain no harmful toxins is generally safe to drink without disinfecting. The survivor should always be prepared to collect rainfall at a moment's notice. An inverted poncho works well to collect rainfall.

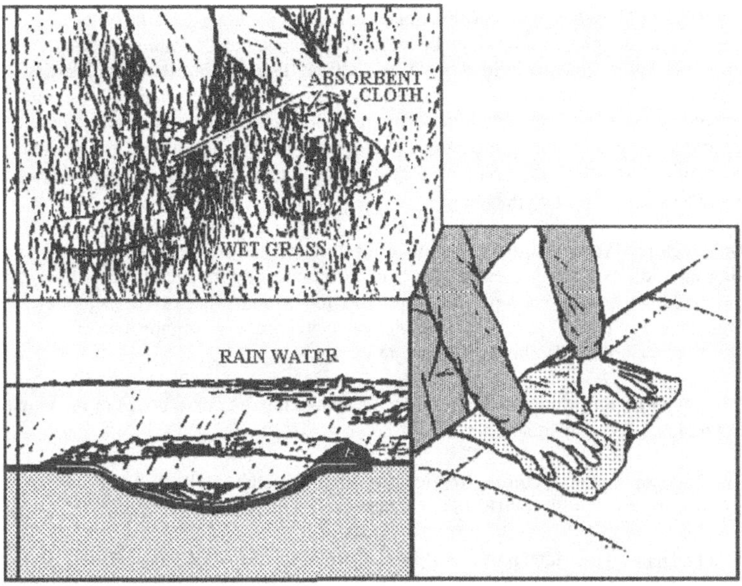

3. **HAZARDOUS FLUIDS** (FMST.07.08j)

a. Survivors have occasionally attempted to augment their water supply with other fluids, such as alcoholic beverages, urine, blood, or seawater. While it is true that each of these fluids has a high water content, the impurities they contain may require the body to expend more fluid to purify them. Some hazardous fluids are:

(1) Seawater: Seawater in more than minimal quantities are actually toxic. The concentration of sodium and magnesium salts is so high that fluid must be drawn from the body to eliminate the salts and eventually the kidneys cease to function.

(2) Alcohol: Alcohol dehydrates the body and clouds judgment. Super-cooled liquid, if ingested, can cause immediate frostbite of the throat, and potential death.

(3) Blood: Blood, besides being salty, is a food. Drinking it will require the body to expend additional fluid to digest it.

(4) Urine: Drinking urine is not only foolish, but also dangerous. Urine is nothing more than the body's waste. Drinking it only places this waste back into the body, which requires more fluid to process it again.

4. **WATER QUALITY**. Water contains minerals, toxins, and pathogens. Some of these, consumed in large enough quantities may be harmful to human health. Pathogens are our primary concern. Pathogens are divided into Virus, Cysts, Bacteria, and Parasites. Certain pathogens are more resistant to chemicals and small enough to move through microscopic holes in equipment (i.e., T-shirt, parachute). Certain pathogens also have the ability to survive in extremely cold water temperatures. Pathogens generally do not live in snow and ice. Water quality is divided into three levels of safety with disinfecting as the most desired level, then purified, followed by potable.

   a. Disinfecting. Water disinfecting removes or destroys harmful microorganisms. Giardia cysts are an ever-present danger in clear appearing mountain water throughout the world. By drinking non-potable water you may contract diseases or swallow organisms that could harm you. Examples of such diseases or organisms are Dysentery, Cholera, Typhoid, Flukes, and Leeches.

   b. Remember impure water, no matter how overpowering the thirst, is one of the worst hazards in a survival situation.

   c. The first step in disinfecting is to select a treatment method. The two methods we will discuss are as follows: (FMST.07.08j)

      (1) Heat. The Manual of Naval Preventive Medicine (P-5010) states that you must bring the water to a rolling boil before it is considered safe for human consumption. This is the most preferred method.

         (a) Bringing water to the boiling point will kill 99.9% of all Giardia cysts. The Giardia cyst dies at 60°C and Cryptosporidium dies at 65C. Water will boil at 14,000' at 86°C and at 10,000' at 90C. With this in mind you should note that altitude does not make a difference unless you are extremely high.

      (2) Chemicals. There are numerous types of chemicals that can disinfect water. Below are a few of the most common. In a survival situation, you will use whatever you have available.

         (a) Iodine Tablets.

         (b) Chlorine Bleach.

         (c) Iodine Solution.

         (d) Betadine Solution.

         (e) Military water purification tablets. These tablets are standard issue for all DOD agencies. These tablets have a shelf life of four years from the date of manufacture, unless opened. Once the seal is broken, they have a shelf life of one year, not to exceed the initial expiration date of four years.

49703
Month / Year / Batch Number

(3) Water Disinfecting Techniques and Halogen Doses.

Iodination techniques Added to 1 liter or quart of water	Amount for 4 PPM	Amount for 8 PPM
Iodine tablets Tetraglcine hydroperiodide EDWGT Potable Aqua Globaline	½ tablet	1 tablet
2% iodine solution (tincture)	0.2 ml 5 gtts	0.4 ml 10 gtts
10% povidone-iodine solution*	0.35 ml 8 gtts	0.70 ml 16 gtts
Chlorination techniques	Amount for 5 PPM	Amount for 10 PPM
Household bleach 5% Sodium hypochlorie	0.1 Ml 2 gtts	0.2 ml 4 gtts
AquaClear Sodium dichloroisocyanurate		1 tablet
AquaCure, AquaPure, Chlor-floc Chlorine plus flocculating agent		8 PPM 1 tablet

*Providone-iodine solutions release free iodine in levels adequate for disinfecting, but scant data is available.

Measure with dropper (1 drop=0.05 ml) or tuberculin syringe

PPM-part per million gtts-drops ml-milliliter

Concentration of **Contact time in minutes at various water temperatures**

Halogen

	5 C / 40 F	15 C / 60 F	30 C / 85 F
2 PPM	240	180	60
4 PPM	180	50	45
8 PPM	60	30	15

NOTE: These contact times have been extended from the usual recommendations to account for recent data that prolonged contact time is needed in very cold water to kill *Giardia* cysts.

NOTE: Chemicals may not destroy Cryptosporidium.

d. Purification. Water purification is the removal of organic and inorganic chemicals and particulate matter, including radioactive particles. While purification can eliminate offensive color, taste, and odor, it may not remove or kill microorganisms.

(1) Filtration. Filtration purifying is a process by which commercial manufacturers build water filters. The water filter is a three-tier system. The first layer, or grass layer, removes the larger impurities. The second layer, or sand layer, removes the smaller impurities. The final layer, or charcoal layer (**not the ash but charcoal from a fire**), bonds and holds the toxins. All layers are placed on some type of straining device and the charcoal layer should be at least 5-6 inches thick. Layers should be changed frequently and straining material should be boiled. Remember that this is not a disinfecting method, cysts can possibly move through this system.

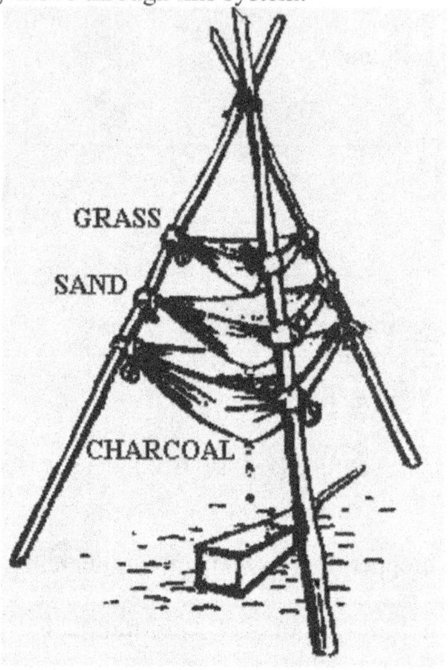

WATER FILTER

(2) Commercial Water Filters. Commercial water filters are generally available in most retail stores and may be with you. Understanding what the filter can do is the first step in safeguarding against future illnesses.

(a) A filter that has a .3 micron opening or larger will not stop Cryptosporidium.

(b) A filter system that does not release a chemical (i.e., iodine) may not kill all pathogens.

(c) A filter that has been overused may be clogged. Usage may result in excessive pumping pressure that can move harmful pathogens through the opening.

e. Potable. Potable indicates only that a water source, on average over a period of time, contains a "minimal microbial hazard," so the statistical likelihood of illness is acceptable.

(1) Sedimentation. Sedimentation is the separation of suspended particles large enough to settle rapidly by gravity. The time required depends on the size of the particle. Generally, 1 hour is adequate if the water is allowed to sit without agitation. After sediment has formed on the bottom of the container, the clear water is decanted or filtered from the top. Microorganisms, especially cysts, eventually settle, but this takes longer and the organisms are easily disturbed during pouring or filtering. Sedimentation should not be considered a means of disinfecting and should be used only as a last resort or in an extreme tactical situation.

5. **SOLAR STILLS.** (FMST.07.08k)

a. Solar stills are designed to supplement water reserves. Contrary to belief, they will not provide enough water to meet the daily requirement for water.

b. AboveGround Solar Still. This device allows the survivor to make water from vegetation. To make the aboveground solar still, locate a sunny slope on which to place the still, a clear plastic bag, green leafy vegetation, and a small rock.

(1) Construction.

(a) Fill the bag with air by turning the opening into the breeze or by "scooping" air into the bag.

(b) Fill the bag half to three-quarters full of green leafy vegetation. Be sure to remove all hard sticks or sharp spines that might puncture the bag.

CAUTION
Do not use poisonous vegetation. It will provide poisonous liquid.

(c) Place a small rock or similar item in the bag.

(d) Close the bag and tie the mouth securely as close to the end of the bag as possible to keep the maximum amount of air space. If you have a small piece of tubing, small straw, or hollow reed, inserts one end in the mouth of the bag before tying it securely. Tie off or plug the tubing so that air will not escape. This tubing will allow you to drain out condensed water without untying the bag.

(e) Place the bag, mouth downhill, on a slope in full sunlight. Position the mouth of the bag slightly higher than the low point in the bag.

(f) Settle the bag in place so that the rock works itself into the low point in the bag.

(g) To get the condensed water from the still, loosen the tie and tip the bag so that the collected water will drain out. Retie the mouth and reposition the still to allow further condensation.

(h) Change vegetation in the bag after extracting most of the water from it.

(i) Using 1-gallon zip-loc. bag instead of trash bags is a more efficient means of construction.

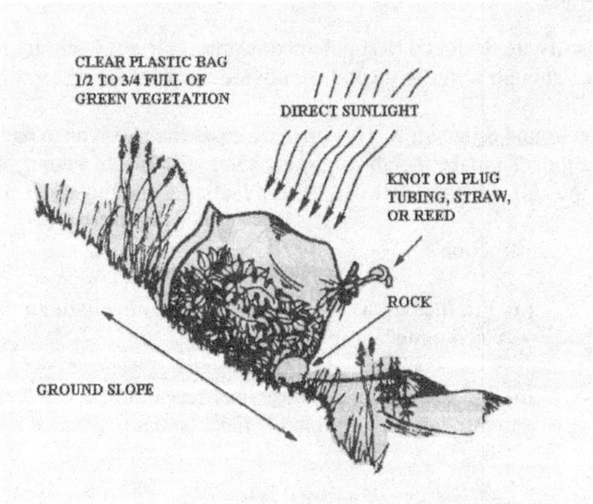

<u>ABOVE GROUND SOLAR STILL</u>

c. <u>BelowGround Solar Still</u>. Materials consist of a digging stick, clear plastic sheet, container, rock, and a drinking tube.

(1) Construction.

(a) Select a site where you believe the soil will contain moisture (such as a dry streambed or a low spot where rainwater has collected). The soil should be easy to dig, and will be exposed to sunlight.

(b) Dig a bowl-shaped hole about 1 meter across and 24 inches deep.

(c) Dig a sump in the center of the hole. The sump depth and perimeter will depend on the size of the container you have to place in it. The bottom of the sump should allow the container to stand upright.

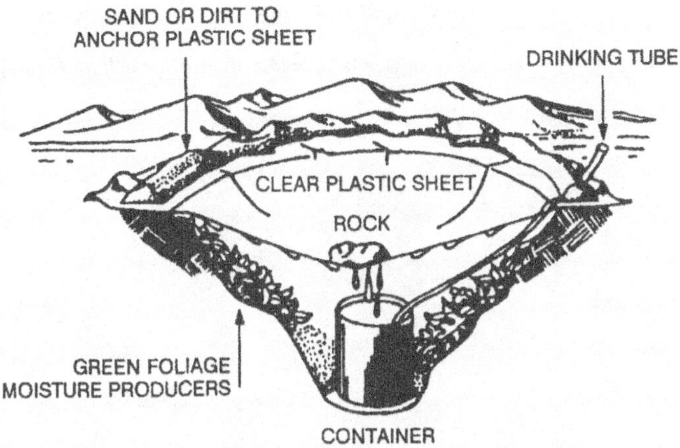

<u>**BELOW GROUND SOLAR STILL**</u>

(d) Anchor the tubing to the container's bottom by forming a loose overhand knot in the tubing. Extend the unanchored end of the tubing up, over, and beyond the lip of the hole.

(e) Place the plastic sheet over the hole, covering its edges with soil to hold in place. Place a rock in the center of the plastic sheet.

(f) Lower the plastic sheet into the hole until it is about 18 inches below ground level. Make sure the cone's apex is directly over the container. Ensure the plastic does not touch the sides of the hole because the earth will absorb the moisture.

(g) Put more soil on the edges of the plastic to hold it securely and prevent the loss of moisture.

(h) Plug the tube when not in use so that moisture will not evaporate.

(i) Plants can be placed in the hole as a moisture source. If so, dig out additional soil from the sides.

(j) If polluted water is the only moisture source, dig a small trough outside the hole about 10 inches from the still's lip. Dig the trough about 10 inches deep and 3 inches wide. Pour the polluted water in the trough. Ensure you do not spill any polluted water around the rim of the hole where the plastic touches the soil. The trough holds the polluted water and the soil filters it as the still draws it. This process works well when the only water source is salt water.

(k) Three stills will be needed to meet the individual daily water intake needs.

UNITED STATES MARINE CORPS
Mountain Warfare Training Center
Bridgeport, California 93517-5001

WMC
04/07/02

STUDENT HANDOUT

FORAGING PLANTS AND INSECTS FOR SURVIVAL USES

PURPOSE The purpose of this period of instruction is to familiarize the Wilderness Medicine Student with the potential resources for survival sustenance.

OUTLINE

1. GENERAL CONSIDERATIONS. There are few places without some type of edible vegetation. Plants contain vitamins, minerals, protein, carbohydrates, and dietary fiber. Some plants also contain fats. The following are general considerations:

a. Do not assume that because birds or animals have eaten a plant, it is edible by humans.

b. Poor plant recognition skills will seriously limit your ability to survive.

c. Plant dormancy and snowfall make foraging plants difficult during the winter months.

d. Plants generally poison by:

(1) Ingestion: When a person eats a part of a poisonous plant.

(2) Contact: When a person makes contact with a poisonous plant that cause any type of skin irritation or dermatitis.

(3) Absorption: When a person absorbs the poison through the skin, which can interrupt a bodily function.

(4) Inhalation: Poisoning can occur through the inhalation of smoke that contains poisonous plant residue.

A7

e. Plant properties can change throughout the growing season. Plants can be edible during certain periods while poisonous in others.

2. PLANT RECOGNITION. Plant recognition can take years of study to learn. The following is used as a general guideline.

a. Types of plants that should be avoided are:

(1) All mushrooms.

(2) Plants with a milky sap.

(3) White and yellow berries should be avoided, as they are almost always poisonous.

(4) Plants with shiny leaves should be considered poisonous.

(5) Plants that are irritants to the skin should not be eaten.

(6) All beans and peas should be avoided.

b. Edible Plants: There are several edible plants found at the Marine Corps Mountain Warfare Training Center. All are easy to recognize and many of the same plants are located throughout the United States and abroad.

(1) Wild Onions - Complete plant.

(2) Dandelion - Roots and leaves.

(3) Mallow - Leaves, stem and flowers.

(4) Watercress -Complete plant.

(5) Cattail - Root, stalk and stem.

(6) Bull Thistle - The flower.

(7) Juniper Tree - Berries and cambium layer.

(8) All Conifers - Cambium layer, needles, and nuts within the cones.

(9) Currants - Berries.

c. Preparing an edible plant. All edible plants should be boiled twice with a change of water to remove any bitterness, acids, or bacteria that the plant may contain.

3. PLANT TESTING PROCEDURE. Always adopt the following procedure when trying unknown plants as food.

a. Wait 8 hours without eating.

b. Select a plant that grows in sufficient quantity in the local area. Separate the part of the plant you wish to test root, stem, leaf, or flower. Certain parts of plants are poisonous while the, other parts may be edible.

c. Rub a portion of the plant you have selected on your inner forearm. Wait 15 minutes and look for any swelling, rash, or irritation.

d. Boil the plant or plant part in changes of water. The toxic properties of many plants are water-soluble or can be destroyed by heat. Cooking and discarding two changes of water can lessen the amount of poisonous material or remove it completely. These boiling periods should last at least 5 minutes each.

e. Place 1 teaspoon of the prepared plant food in the mouth and chew for 5 minutes, but do not swallow. If unpleasant effects occur (burning, bitter, or nauseating taste), remove the plant from the mouth at once and discard it as a food source. If no unpleasant effects occur, swallow the plant material and wait 8 hours.

f. If after 8 hours no unpleasant effects have occurred (nausea, cramps, and diarrhea), eat two tablespoonfuls and wait 8 hours.

g. In no unpleasant effects have occurred at the end of this 8-hour period, the plant may be considered edible.

h. Completely document and sketch the plant in a logbook to refer to for future use. This will aid in future procurement of this plant. If plant properties have changed, you will have to repeat the plant testing procedure.

NOTE: Have charcoal ready to eat. If any side effects occur, attempt to induce vomiting and immediately eat charcoal. Charcoal will absorb toxins and reduce the chance of Death.

4. MEDICINAL PLANTS One of the many things that hinder a survivor's ability to survive is medical problems. Injuries incurred will reduce survival expectancy and the ability to evade.

a. Coyote Willow: Coyote Willow is a thick forming shrub with clustered stems and very narrow leaves. It habits wet soils, especially riverbanks, sandbars, and silt flats.

(1) Drinking tea made from the inner bark can reduce fever.

(2) The dried and powdered inner bark can be used to stop bleeding.

(3) Administering a half-cup of willow charcoal dissolved in water can treat diarrhea.

(4) The Coyote Willow roots can be mashed and applied to tooth aches.

b. Yarrow: Yarrow is a plant (tall, usually not branched, with many white, slat topped group flowers). Do not confuse Yarrow with Water Hemlock.

(1) Rubbing a handful of crushed Yarrow flowers and leaves on any exposed can make insect repellent skin.

(2) Placing a Yarrow leaf poultice on the wound can stop bleeding.

(3) Relief from many rashes, including poison oak and ivy, can be achieved by applying a Yarrow leaf compress to the effected area.

(4) The Yarrow root can be chewed to relieve the pain of a toothache or break a fever.

(5) A potent anesthetic can be made by scrubbing fresh Yarrow roots in water to clean them. Once the roots are clean, crush them into a spongy mass and apply gently to the wounded area.

c. Tea To make a tea, place the plants in a container and pour in two cups of boiling water. Let the mixture steep for 20 minutes and drink.

d. Poultice: Place plant (Coyote Willow, Yarrow, etc.) in gauze or other similar material and fold it so the gauze will hold the plant in place. Put gauze in a larger cloth, about 6"x8", and roll the sides inward. Fold the cloth over without losing the plant. While boiling water, dip the bottom portion of the cloth containing the plant into the hot water by holding the edges. Keep the plant submerged in the boiling water until it becomes saturated. Bring the cloth straight up and with a twisting motion, wring the excess water back into the pot. Apply the poultice to the affected area as soon as it has cooled enough to place on the wound. To be effective, the poultice should be as hot as you can tolerate it.

e. Compress: A compress is made just like a poultice; except it is cold when applied to the wound. NOTE: Poultice or Compress should be applied for 1 to 24 hours, as needed. When applying a poultice, you may experience a throbbing pain as it draws out the infection and neutralizes toxins. When the pain subsides, the poultice has accomplished its task and should be removed. Apply a fresh poultice as needed until the desired level of healing has been reached.

5. POISONOUS PLANTS. Rarely will a survivor have an ideal means of killing large game, however there are certain plants that can aid the survivor. The two plants that we will talk about are the Water Hemlock and Monkshood.

a. Preparing Poisonous Plants: Once a poisonous plant has been located, dig up the root of the plant. The roots of the Water Hemlock and Monkshood generally grow 8 to 10 inches deep. Extreme caution must be used when handling the root. Do not handle the root without a barrier between your hands and the root. A barrier can be gloves, socks, T-shirt, or even moss. Split the root lengthwise to expose the inside of the root where the toxin is located. With the root split, rub the tip of your spear/arrow inside of the root opening. In a slow controlled manner, work from the bottom of the tip to the top. Once the tip is thoroughly coated, allow the toxin to dry and apply another coat to the tip. Continue to apply coat after coat of toxin, until the root is completely drained of its' toxin. You are now finished and the tip is ready for use.

NOTE: Do not eat the meat in the vicinity of the poisoned wound.

CAUTION: Do not apply poison to the actual point and edges of tip. This avoids accidental puncture to you. Wash and boil all gloves and clothing used as a barrier before wearing

6. INSECTS. Insects are the most abundant life form on earth and are an excellent survival food. They are easy to catch and provide 65-80% protein; compared to 20% for beef. They aren't too appetizing, but personal bias has no place in a survival situation. The focus must remain on maintaining your health.

a. Insects to avoid.

(1) All adults that sting or bite.

(2) Hairy or brightly colored insects.

(3) All Caterpillars.

(4) Insects that have a pungent odor.

(5) All spiders.

(6) Disease carriers like ticks, flies, or mosquitoes.

b. Edible Insects.

(1) Insect Larvae.

(2) Grasshoppers.

(3) Beetles.

(4) Grubs.

(5) Ants.

(6) Termites.

(7) Worms.

c. Foraging for Insects. One must be careful not to expend more energy harvesting food than can be replaced. For example, catching insects such as grasshoppers can become exasperating and tiring.

(1) At night grasshoppers climb tall plants and clings to the stalks near the top. They can be picked from the plants in the early morning while they are chilled and dormant.

(2) Dig for worms in damp humus soil, under rocks/logs or look for them on the ground after it has rained.

(3) Carpenter ants are found in dead trees and stumps, which can be gathered by hand.

(4) Most other insects can be found in rotten logs, under rocks, and in open grassy areas.

d. Preparing Edible Insects.

(1) Insects with a hard outer shell have parasites. Remove the wings and barbed legs before cooking.

(2) Drop worms into potable water for at least a half-hour. They will naturally purge themselves. You can either cook or eat them raw.

(3) Most other insects can be eaten raw. Cooking insects will improve their taste. If the thought of eating insects is unbearable, grind them into a paste and mix with other foods.

UNITED STATES MARINE CORPS
Mountain Warfare Training Center
Bridgeport, California 93517-5001

WMC.A8.0
05/02/01

STUDENT HANDOUT

SURVIVAL TRAPS AND SNARES

PURPOSE: The purpose of this class is to familiarize the student with some of the techniques used to take game. Game can be useful for both food and clothing.

OUTLINE

1. GENERAL TIPS AND TECHNIQUES FOR TRAPPING

a. General Tips. Knowing a few general hints and tips will make the trapping of animals much easier and considerably more effective. The eight general tips for trapping are:

(1) Know your game: Knowing the habits of the animal you want to trap will help lure it into your trap. Such things as when and where they move, feed, and water will help you determine where the trap can be most effectively placed.

A8

(2) Keep things simple: You don't have time in a survival situation to construct elaborate traps and they do not necessarily do a better job.

(3) Set traps in the right place: Animals will travel and stop in certain locations. That is where to best employ traps.

(4) Cover up your scent: Animals will avoid an area, which smells strange to them. Smoke from your fire is the best cover to use.

(5) Use the right type of trap: Some traps work better than others do (i.e. a deadfall works better on squirrels than a spring pole).

(6) Use the right size trap: Adjust the size of your traps to the size of the animal.

(7) Check traps: Check your traps twice daily: morning and evening. Checking your traps less than twice a day can allow your game to escape, rot, or be taken by other predators.

(8) Bait your traps: Bait of any type will add to your chances of success.

b. General Techniques. A general technique is the method in which the trap is intended to kill or hold the animal.

(1) Strangulation: This method strangles the animal, such as a snare.

(2) Mangling: This method crushes the animal, such as a deadfall.

(3) Entanglement: This method entangles the animal, such as a net.

(4) Live: This method holds the animal, such as a box trap.

c. Types of Triggers. There are three basic triggers used for all traps and path guards. Depending on the situation, variations of these triggers can be used.

(1) Puite Figure 4.

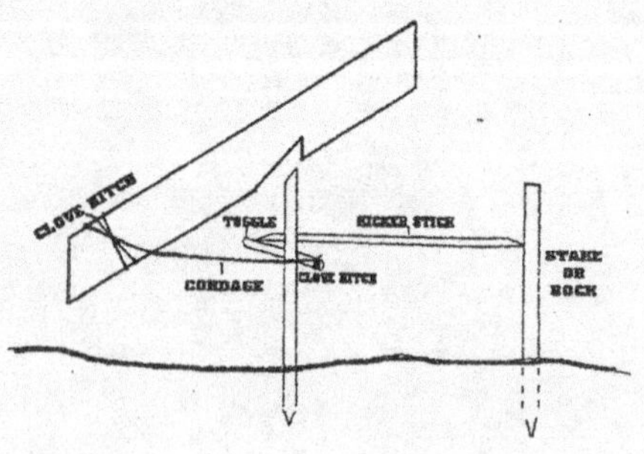

(2) Toggle.

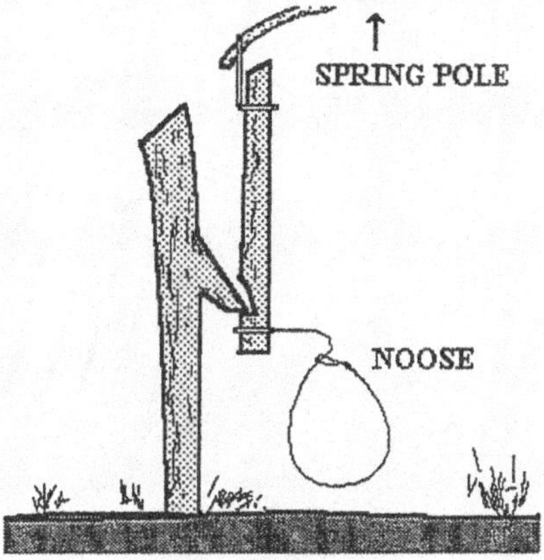

TOGGLE

(c) Universal.

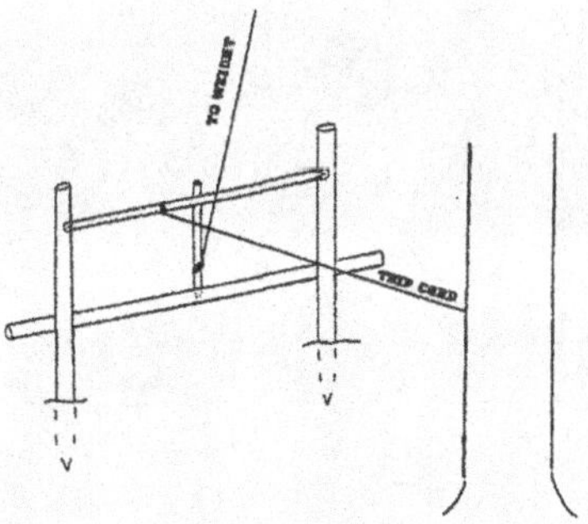

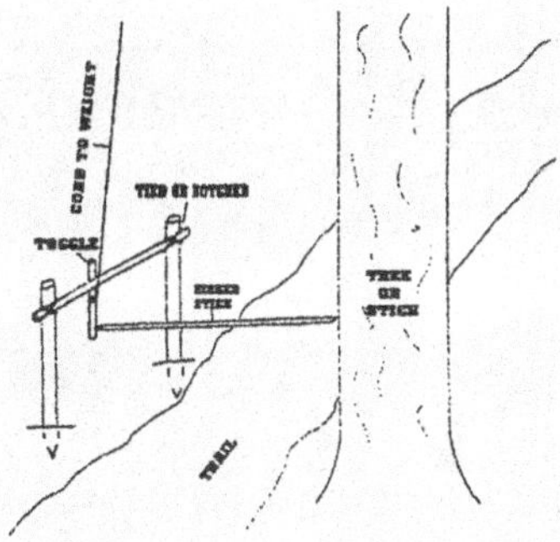

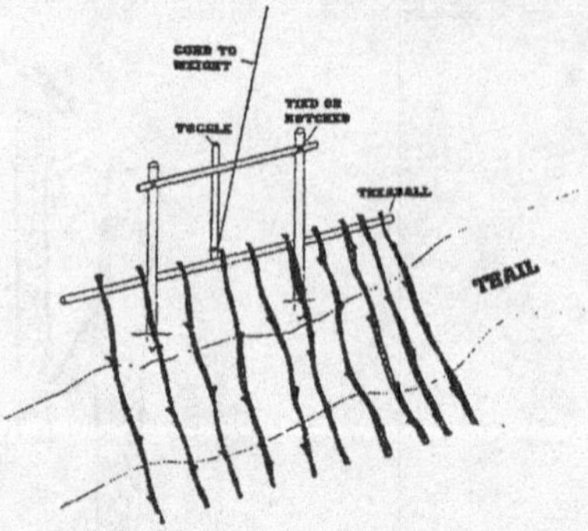

2. TRAPS. Traps are c
action is generally caused by either a weight or spring loaded device.

a. Puite Deadfall: The Puite deadfall requires a knife and piece of cord to construct. It is designed to mangle small rodents. The trigger for the deadfall is the Puite figure 4.

b. Spring Pole: The spring pole requires a small sapling and cordage to construct. The trigger for the spring pole is the toggle. It is designed to lift the animal off the ground; not allowing predatory animals to take your game. Remember, the trigger can not be so tight that the intended game can not set it off.

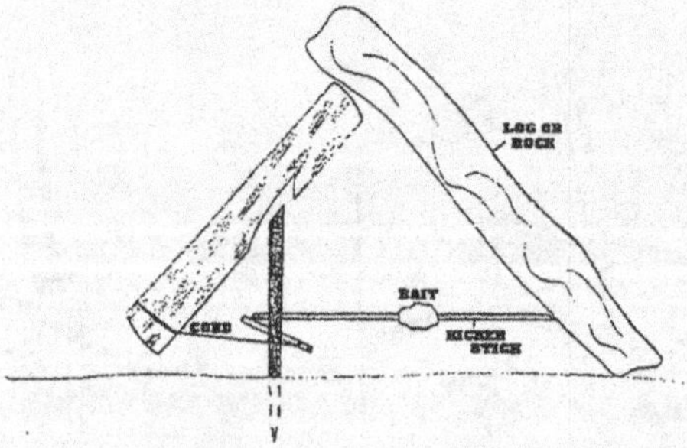

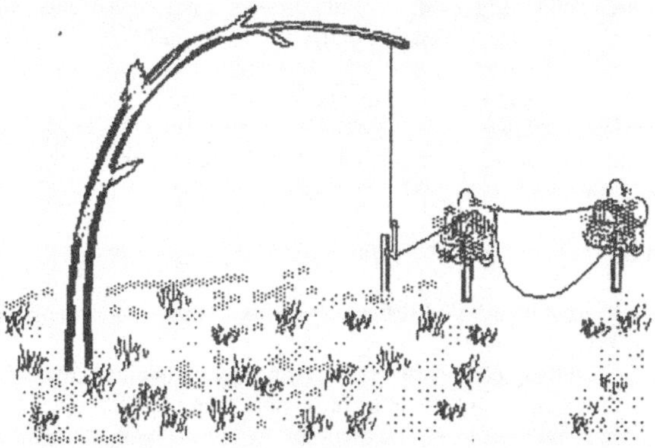

SPRING POLE

c. Box Trap: The box trap requires limited cordage to construct. It is designed to hold live small rodents and birds. The box trap trigger is the Puite figure 4.

3. PATH GUARDS Path guards is designed to protect and provide security for your shelter area against the enemy and predatory animals. They are classified into noise and casualty producing path guards.

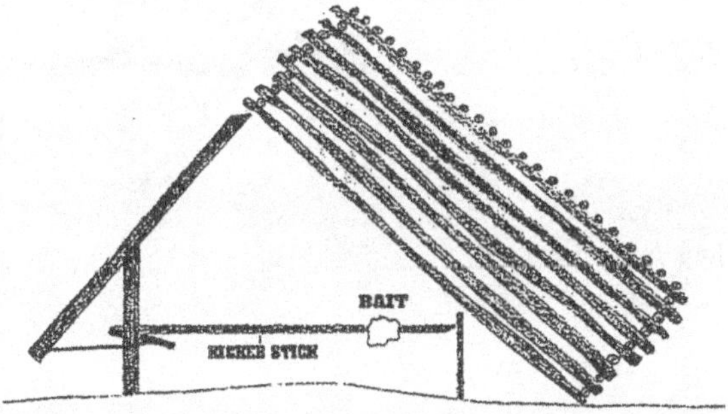

BOX TRAP

a. Noise producing path wards. Noise producing pathguards serve as an alarm for your shelter area. When triggered, it should produce some type of loud noise. Although construction can vary, depending on materials available, one example is as follows:

(1) Secure a young sapling to a universal trigger.

(2) At the end of the sapling, tie several pieces of metal to the sapling. Use what ever is available for metal.

(3) Camouflage the metal on the ground.

(4) When triggered, the sapling should swing back and forth, causing the metal rattle.

b. Casualty producing path wards. Casualty producing path guards, when triggered, should cause death or injury to the enemy or predatory animal. Tips should be poisoned as discussed in Survival Plant Uses class. Triggers for this type of path guard should be the universal triggers.

(1) Log Jerk

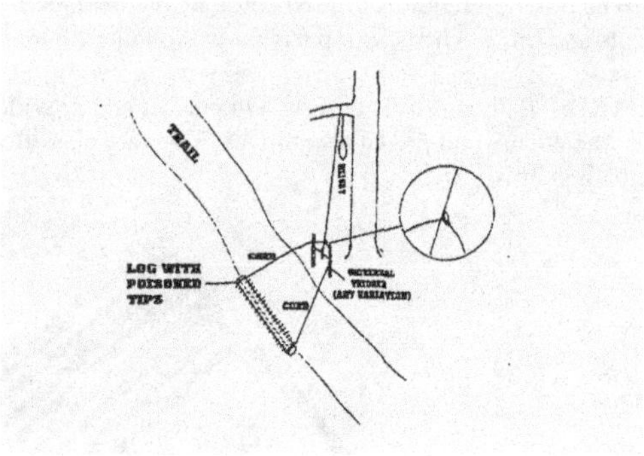

(2) Fish Hook Nightmare

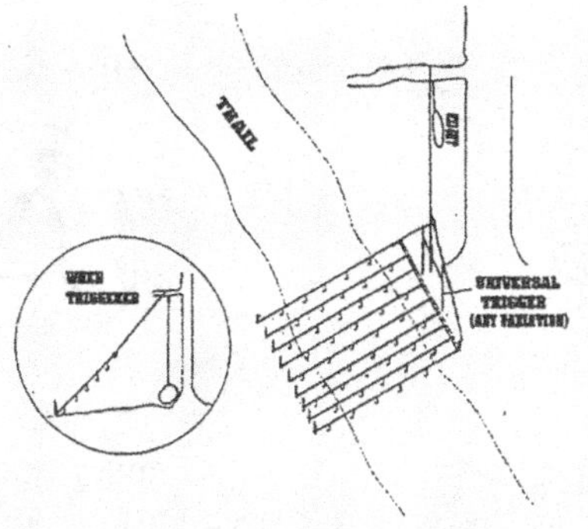

4. SNARING. A snare a loop at one end, which tightens, nd less time consuming to constru

a. Requirements for snaring there are five requirements too effectively employ snares. They are as follows:

(1) Wire: Although snares can be used with rope or cord, they are less effective than wire. Wire should have memory and resist kinking. Aircraft cable type in size 1/32 to 3/16 inch should be used. This type of wire prevents animal chew out and resists breakage. Remember, you want the smallest diameter cable capable of holding the animal.

(2) Locking device: A locking device is imperative for a snare to work properly. Locking devices secure the snare around the animal's neck. There are several methods available for a locking device.

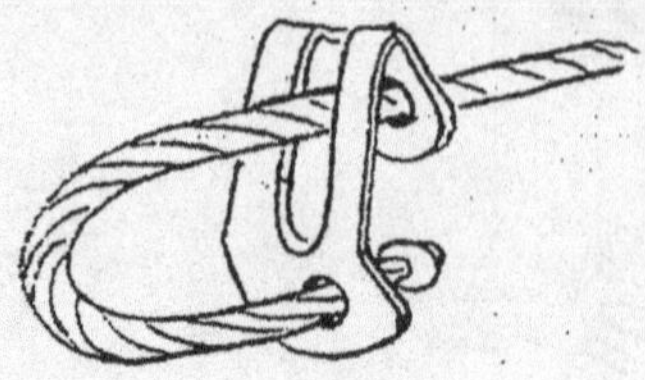

COMMERCIAL **BENT WASHER**

(3) Channelization: The objective of channelization is to force the animal to move through the snare. Remember, animals will take the path of least resistance.

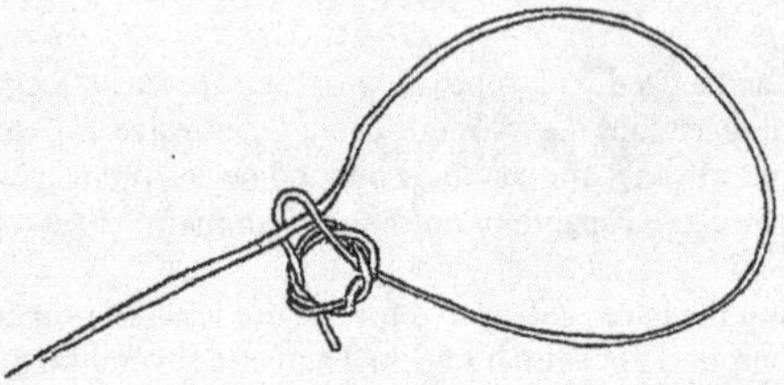

KNOT

(4) Loop size: Probably the single most important requirement in snaring. A correctly employed snare will have the snare holding the animal around the neck. Loop size is placed on the snare according to the intended animal. Too large will result in a body or leg catch, resulting in possible chew out or breakage. Too small will enable the animal to force the snare to the side, resulting in a miss. Additionally, the loop must be placed with specific ground clearance. The following chart will give a general guideline.

ANIMAL	NOOSE SIZE	GROUND CLEARANCE
SQUIRREL	2 1/2 TO 3 INCHES	1/2 TO 1 1/2 INCHES
RABBIT	4 TO 5 1/2 INCHES	1 1/2 TO 3 INCHES
RACCOON	6 INCHES	3 TO 4 INCHES
FOX	7 TO 10 INCHES	8 TO 10 INCHES
COYOTES	12 TO 14 INCHES	12 INCHES
BOBCAT	9 INCHES	8 INCHES

Note: (1) Noose size is the diameter of the snare loop.

(2) Ground clearance is measured from the bottom of the loop to the ground.

(5) Bait. In a survival situation, you will not be able to employ numerous snares. Baiting all snares increases your chances of success.

b. Snares although there are numerous ideas to employ snares, here are a few ideas.

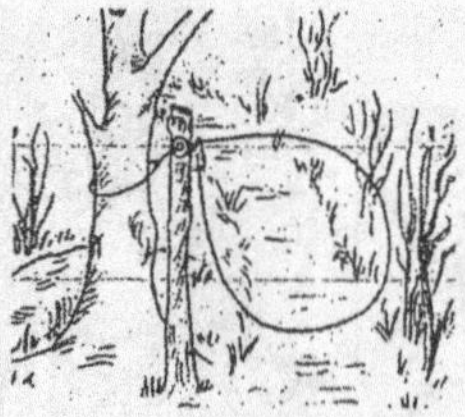

TRAIL SET

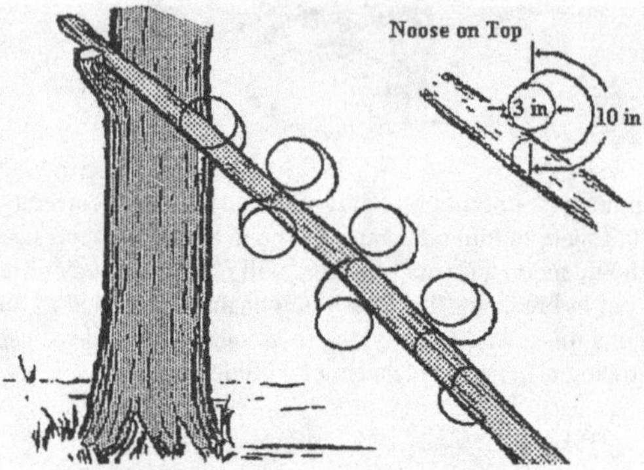

SQUIRREL POLE

DEN SET

<u>**CUBBIE SET**</u>

5. FISH. In a mountainous region, fish are normally an abundant resource. Not only are they a food source, the "left-overs" provide an excellent bait for traps and snares.

a. Fishing locations: Fishing in mountain streams is generally best done with a hand line. When fishing these streams, always look for these places to fish.

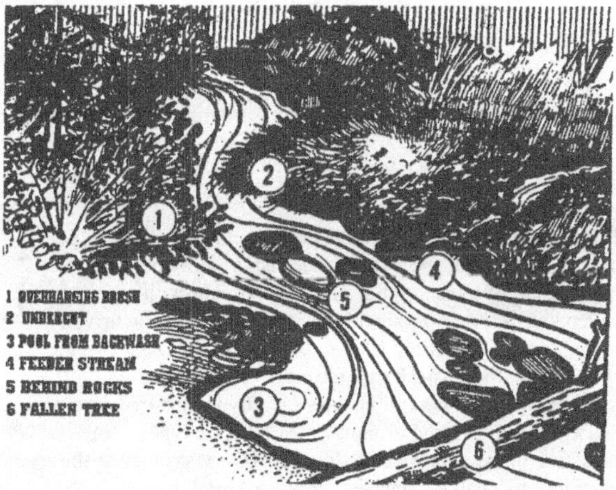

<u>**FISHING LOCATIONS**</u>

b. Expedient hooks: Although hooks should be carried in a survival kit, the survivor should be able to construct additional hooks if the situation arises. Expedient hooks are made to become lodged in the throat of the fish. Below are a few examples.

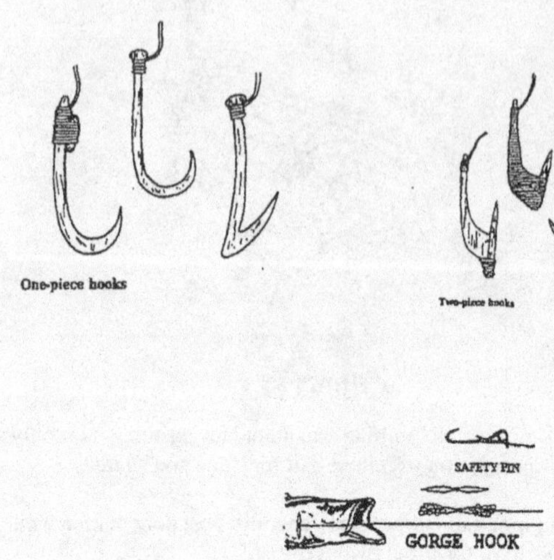

EXPEDIENT HOOKS

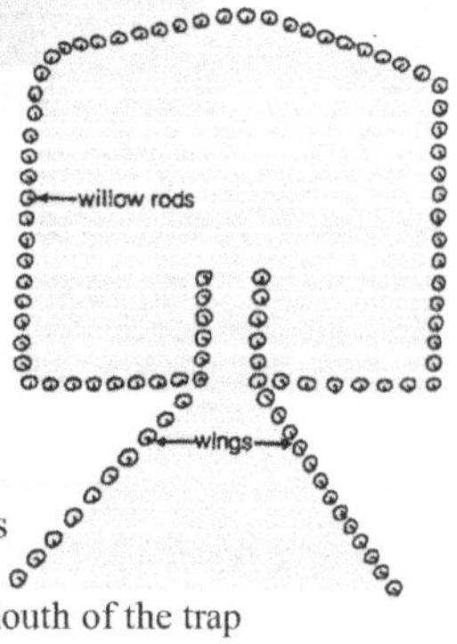

c. Fish traps: A fish trap can be effective if you have a shallow stream and time to construct it. A basic fish trap in nothing more than a barricade of rocks or sticks across a stream with another barricade using a funnel-type entrance which fish can be driven into but have a difficult time finding their way out. Once fish are trapped between these two barricades they may be speared, clubbed, or grabbed. This can be very effective when certain types of fish are moving in large groups to spawn. This type of trap is very effective in catching fish. A door can be constructed at the mouth of the trap so that excess fish can be kept live until needed.

FISH TRAP

d. Set Lines: Set lines are an effective method of fishing which conserves energy. Put them out over night with several boated hooks attached. Place them with the hooks either on the bottom or suspended off the bottom, until you have determined where the fish are feeding.

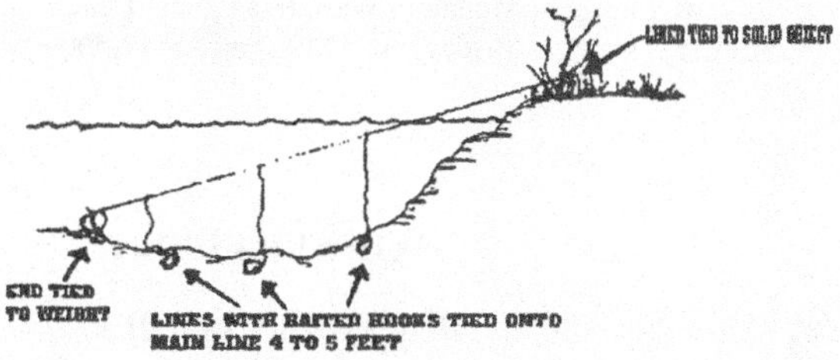

SET LINES

e. Netting: A gill net is most effective in still water, e.g., a lake (near the inlet and outlet are good locations) or back water in a large stream (for survival doesn't hesitate to block the stream). Nets can be constructed using the inner cords of parachute shroud lines. Place floats on top and weights on the bottom to keep the net vertical in the water. When the ice is on a lake, the fish are inclined to stay deeper. The smaller the mesh, the smaller the fish you can catch, but a small mest will still entangle a large fish. A mesh of 3 inches is a good standard. A gill net is very time consuming and requires allot of material to construct.

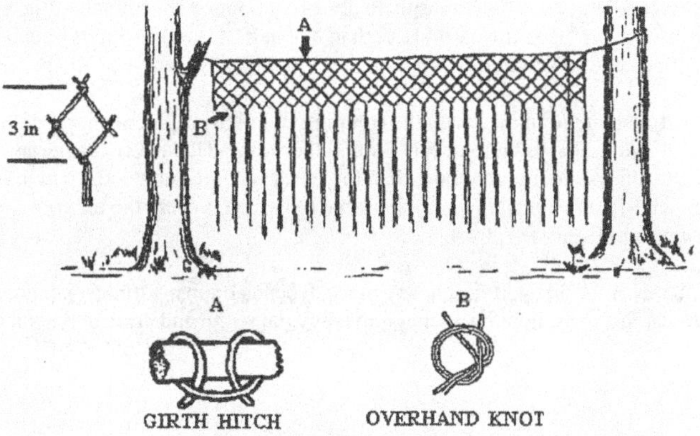

GILL NET

UNITED STATES MARINE CORPS
Mountain Warfare Training Center
Bridgeport, California 93517-5001

WMC
04/07/02

STUDENT HANDOUT

MOUNTAIN WEATHER

LESSON PURPOSE: The purpose of this period of instruction is to introduce you to weather patterns, the invisible aspects of weather, and some visible clues to help you forecast the weather. This lesson relates to all operations in the mountains at MWTC.

OUTLINE

1. GENERAL

a. The earth is surrounded by the atmosphere, which is divided, into several layers. The world's weather systems are in the troposphere, the lower of these layers. This layer reaches as high as 40,000 feet.

b. Dust and clouds in the atmosphere absorb or bounce back much of the energy that the sun beams down upon the earth. Less than one half of the sun's energy actually warms the earth's surface and lower atmosphere.

c. Warmed air, combined with the spinning (rotation) of the earth, produces winds that spread heat and moisture more evenly around the world. This is very important because the sun heats the Equator much more than the poles and without winds to help restore the balance, much of the earth would be impossible to live on. Where the air-cools, you can get clouds, rain, snow, hail, fog, frost, etc.

d. The weather that you find in any place depends on many things, i.e. how hot the air is, how moist the air is, how it is being moved by the wind, and especially, is it being lifted or not?

A10

2. PRESSURE

a. All of these factors are related to air pressure, which is the weight of the atmosphere at any given place. The lower the pressure, the more likely are rain and strong winds.

b. In order to understand this we can say that the air in our atmosphere acts very much like a liquid.

c. Areas with a high level of this liquid would exert more pressure on the Earth and be called a "high pressure area".

d. Areas with a lower level would be called a "low pressure area".

e. In order to equalize the areas of high pressure it would have to push out to the areas of low pressure.

f. The characteristics of these two pressure areas are as follows:

(1) High-pressure area. Flows out to equalize pressure.

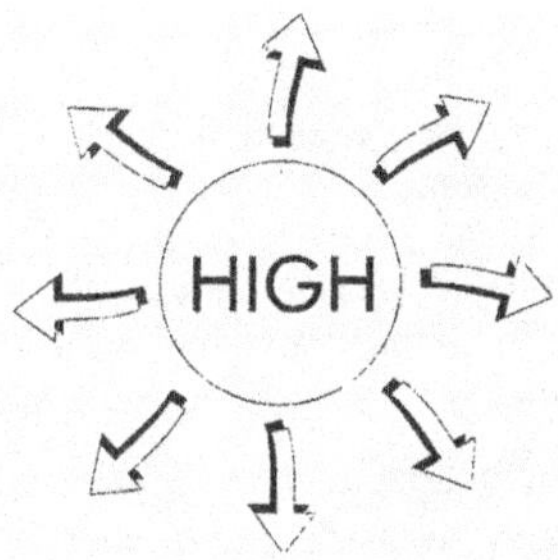

(2) Low pressure area. Flows in to equalize pressure.

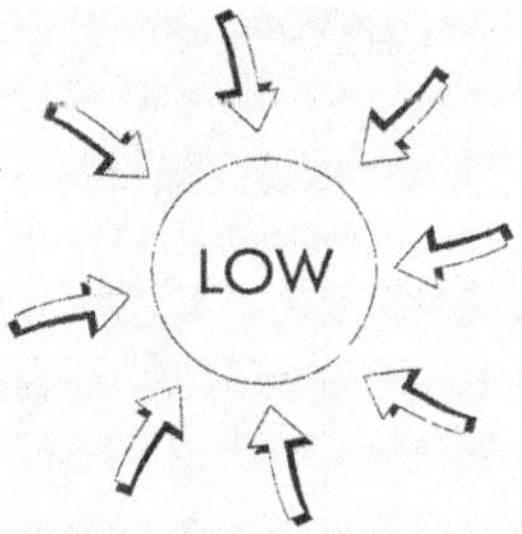

g. The air from the high-pressure area is basically just trying to gradually flow out to equalize its pressure with the surrounding air, while the low pressure is beginning to build vertically. Once the low has achieved equal pressure, it can't stop and continues to build vertically; causing turbulence, which results in bad weather.

NOTE: When looking on the weather map, you will notice that these differences in pressure resemble contour lines. These contour lines are called isobars and are translated to mean equal pressure area.

h. Isobars: Pressure is measured in millibars or another more common measurement is inch mercury.

i. Fitting enough, areas of high pressure are called ridges and areas of low pressure are called troughs.

NOTE: The average air pressure at sea level is:

29.92 inches mercury.
1,013 millibars.

j. As we go up in altitude, the pressure (or weight) of the atmosphere decreases.

EXAMPLE: At 18,000 feet in elevation it would be 500 millibars vice 1,013 millibars at sea level.

3. HUMIDITY Humidity is the amount of moisture in the air. All air holds water vapor, although it is quite invisible.

a. Air can hold only so much water vapors, but the warmer the air, the more moisture it can hold. When the air has all the water vapor that it can hold, the air is said to be saturated (100% relative humidity).

b. If the air is then cooled, any excess water vapor condenses; that is, it's molecules join to build the water droplets we can see.

c. The temperature at which this happens is called the "condensation point". The condensation point varies depending on the amount of water vapor and the temperature of the air.

d. If the air contains a great deal of water vapor, condensation will form at a temperature of 20°C (68°F). But if the air is rather dry and does not hold much moisture, condensation may not form until the temperature drops to 0°C (32°F) or even below freezing.

e. Adiabatic Lapse Rate: The adiabatic lapse rate is the rate that air will cool on ascent and warm on descent. The rate also varies depending on the moisture content of the air.

(1) Saturated Air = 2.2°F per 1,000 feet.

(2) Dry Air = 5.5°F per 1,000 feet.

4. WINDS. As we stated earlier, the uneven heating of the air by the sun and rotation of the earth causes winds. Much of the world's weather depends on a system of winds that blow in a set direction. This pattern depends on the different amounts of sun (heat) that the different regions get and also on the rotation of the earth.

a. Above hot surfaces, air expends (air molecules spread out), and move to colder areas where it cools and becomes denser, and sinks to the earth's surface. This forms a circulation of air from the poles along the surface or the earth to the Equator, where it rises and moves towards the poles again.

b. Once the rotation of the earth is added to this, the pattern of the circulation becomes confusing.

c. Because of the heating and cooling, along with the rotation of the earth, we have these surface winds. All winds are named from the direction they originated from:

(1) Polar Easterlies: These are winds from the polar region moving from the east. This is air that has cooled and settled at the poles.

(2) Prevailing Westerlies: These winds originate from approximately 30 degrees North Latitude from the west. This is an area where prematurely cooled air, due to the earth's rotation, has settled back to the surface.

(3) Northeast Tradewinds: These are winds that originate from approximately 30 degrees North from the Northeast. Also prematurely cooled air.

d. Jet Stream: A jet stream can be defined as a long, meandering current of high speed winds near the tropopause (transition zone between the troposphere and the stratosphere) blowing from generally a westerly direction and often exceeding 250 miles per hour. The jet stream results from:

(1) Circulation of air around the poles and Equator.

(2) The direction of air flow above the mid latitudes.

(3) The actual path of the jet stream comes from the west, dipping down and picking up air masses from the tropical regions and going north and bringing down air masses from the Polar Regions.

NOTE: The average number of long waves in the jet stream is between three and five depending on the season. Temperature differences between polar and tropical regions influence this. The long waves influence day to week changes in the weather; there are also short waves that influence hourly changes in the weather.

5. AIR MASSES. As we know, all -of these patterns move air. This air comes in parcels known as "air masses". These air masses can vary in size from as small as a town too as large as a country. These air masses are named for where they originate:

a. Maritime: Over water.

b. Continental: Over land.

c. Polar: Above 60 degrees North.

d. Tropical: Below 60 degrees North.

e. Combining these give us the names and description of the four types of air masses:

(1) Continental Polar: Cold, dry air mass.

(2) Maritime Polar: Cold wet air mass.

(3) Continental Tropical: Dry, warm air mass.

(4) Maritime Tropical: Wet, warm air mass.

ABBREVIATED MEANINGS
CP = Continental Polar.
MP = Maritime Polar.
CT = Continental
Tropical. MT =
Maritime Tropical.

f. The thing to understand about air masses; they will not mix with another air mass of a different temperature and moisture content. When two different air masses collide, we have a front, which will be covered in more detail later in this period of instruction.

g. Here are some other types of winds that are peculiar to mountain environments but don't necessarily affect the weather:

(1) Anabolic wind: These are winds that blow up mountain valleys to replace warm rising air and are usually light winds.

(2) Catabolic wind: These are winds that blow down mountain valley slopes caused by the cooling of air and are occasionally strong winds.

6. LIFTING/COOLING. As we know, air can only hold so much moisture depending on its temperature. If we cool this air beyond its saturation point, it must release this moisture in one form or another, i.e. rain, snow, fog, dews, etc. There are three ways that air can be lifted and cooled beyond its saturation point.

a. Orographic uplift: This happens when an air mass is pushed up and over a mass of higher ground such as a mountain. Due to the adiabatic lapse rate, the air is cooled with altitude and if it reaches its saturation point we will receive precipitation.

OROGRAPHIC UPLIFT

b. Convection effects: This is normally a summer effect due to the sun's heat radiating off of the surface and causing the air currents to push straight up and lift air to a point of saturation.

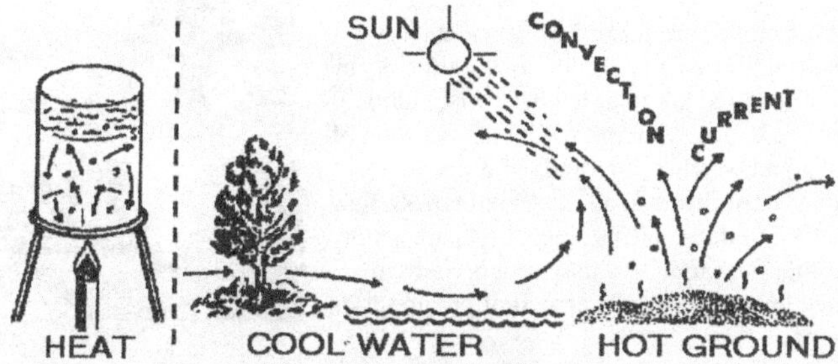

CONVECTION EFFECTS

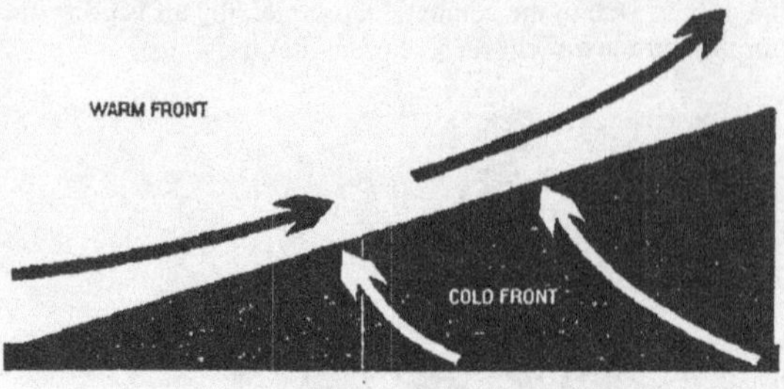

FRONTAL LIFTING

c. Frontal ... rature content ... is forced aloft; fi ... where we receive ... ifting is not uncomı

7. CLOUDS. Anytime air is lifted or cooled beyond its saturation point (100% relative humidity), clouds are formed. Clouds are one of our signposts to what is happening. Clouds can be described in many different ways; they can also be classified by height or appearance, or even by the amount of area covered, vertically or horizontally.

a. Cirrus: These clouds are formed of ice crystals at very high altitudes (usually 20,000 to 35,000 feet) in the mid-latitudes and are thin, feathery type clouds. These clouds can give you up to 24 hours warning of approaching bad weather, hundreds of miles in advance of a 'warm front. Frail, scattered types, such as "mare-tails" or dense cirrus layers, tufts are a sign of fair weather but predictive may be a prelude to approaching lower clouds, the arrival of precipitation and the front.

b. Cumulus: These clouds are formed due to rising air currents and are prevalent in unstable air that favors vertical development. These currents of air create cumiliform clouds that give them a piled or bunched up appearance, looking similar to cotton balls. Within the cumulus family there are three different types to help us to forecast the weather:

(1) Cotton puffs of cumulus are Fair Weather Clouds but should be observed for possible growth into towering cumulus and cumulonimbus.

(2) Towering cumulus are characterized by vertical development. Their vertical lifting is caused by some type of lifting action, such as connective currents found on hot summer afternoons or when wind is forced to rise up the slope of a mountain or possibly the lifting action that

may be present in a frontal system. The towering cumulus has a puffy and "cauliflower-shaped" appearance.

(3) Cumulonimbus clouds are characterized in the same manner as the towering cumulus, form the familiar "thunderhead" and produce thunderstorm activity. These clouds are characterized by violent updrafts, which carry the tops of the clouds to extreme elevations. Tornadoes, hail and severe rainstorms are all products of this type of cloud. At the top of the cloud, a flat anvil shaped form appears as the thunderstorm begins to dissipate.

c. Stratus: Stratus clouds are formed when a layer of moist air is cooled below its saturation point. Stratiform, clouds lie mostly in horizontal layers or sheets, resisting vertical development. The word "stratus" is derived from the Latin word "layer". The stratus cloud is quite uniform and resembles fog. It has a fairly uniform base and a dull, gray appearance. Stratus clouds make the sky appear heavy and will occasionally produce fine drizzle or Very light snow with fog. However, because there is little or no vertical movement in the stratus clouds, they usually do not produce precipitation in the form of heavy rain or snow.

d. As previously stated, clouds are formed when air is lifted to a point where it cools to its saturation point. We also know that frontal lifting affects our fronts, which produce the largest portion of our precipitation.

8. FRONTS. As we know, fronts often happen when two air masses of different moisture and temperature content interact. One of the ways we can identify that this is happening is by the progression of the clouds.

a. Warm Front: A warm front occurs when warm air moves into and over a slower (or stationary) cold air mass. Since warm air is less dense, it will rise naturally so that it will push the cooler air down and rise above it. The cloud you will see at this stage is cirrus. From the point where it actually starts rising, you will see stratus. As it continues to rise, this warm air-cools by the cold air and, this, receiving moisture at the same time. As it builds in moisture, it darkens becoming "nimbus-stratus", which means rain of thunderclouds. At that point some type of moisture will generally fall.

b. Cold Front. A cold front occurs when a cold air mass (colder than the ground that it is traveling over) overtakes a warm air mass that is stationary or moving slowly. This cold air, being denser, will go underneath the warm air, pushing it higher. Of course, no one can see this "air", but they can see clouds and the clouds themselves can tell us what is happening. The cloud progression to look for is cirrus to cirrocumulus to cumulus and, finally, to cumulonimbus.

c. Occluded Front. Cold fronts move faster than warm ones so that eventually a cold front overtakes a warm one and the warm air becomes progressively lifted from the surface. The zone of division between cold air ahead and cold air behind is called a "cold occlusion". If the air behind the front is warmer than ahead, it is a warm occlusion. Most land areas experience more occlusions than other types of fronts. In the progression of clouds leading to fronts, orographic uplift can play part in deceiving you of the actual type of front, i.e. progression of clouds leading to a warm front with orographic cumulus clouds added to these. The progression of clouds in an occlusion is a combination of both progressions from a warm and cold front.

9. USING PRESSURE AS AN INDICATOR. A very important factor of telling us what might happen is the pressure. As we know, low pressure or dropping pressure normally indicates deteriorating weather whereas high pressure usually gives us more good weather or clearing of bad weather. There are a couple of ways to monitor our pressure and are as follows:

a. Barometer: A barometer could be described as a pan of mercury with a tube leading out of the pan. Pressure from the atmosphere causes the mercury to rise in the tube.

(1) The tube is marked in millibars and the station that's reading these millibars will know how much it should rise for that location. Once again, if it rises more than normal, it would be considered a high-pressure reading.

b. Altimeter: Another means that is used to measure pressure is an altimeter, which is commonly used by mountaineers. It works like this:

(1) As you rise in elevation the pressure becomes less, thus allowing the needle in the altimeter to rise. If the needle rises without you rising with it, there is less pressure in the atmosphere than before and thus, a low-pressure area.

c. Contrail Lines: A basic way of identifying a low-pressure area is to note the contrail lines from jet aircraft. If they don't dissipate within two hours, that indicates a low-pressure area in your area. This usually occurs about 24 hours prior to an oncoming front.

d. Lenticulars: These are optical, lens-shaped cumulus clouds that have been sculpted by the winds. This indicates moisture in the air and high winds aloft. When preceding a cold front, winds and clouds will begin to lower.

10. USING SIGNS FROM NATURE. These signs will give you a general prediction of the incoming weather conditions. Try to utilize as many signs together as possible, which will improve your prediction. All of these signs have been tested with relative accuracy, but shouldn't be depended on 100%. But in any case you will be right more times than wrong in predicting the weather. From this we can gather as much information as needed and compile it along with our own experience of the area we are working in to help us form a prediction of incoming weather. The signs are as follows:

a. A spider's habits are very good indicators of what weather conditions will be within the next few hours. When the day is to be fair and relatively windless, they will spin long filaments over which they scout persistently. When precipitation is imminent, they shorten and tighten their snares and drowse dully in their centers.

b. Insects are especially annoying two to four hours before a storm.

c. If bees are swarming, fair weather will continue for at least the next half day.

d. Large game such as deer, elk, etc., will be feeding unusually heavy four to six hours before a storm.

e. When the smoke from a campfire, after lifting a short distance with the heated air, beats downward, a storm is approaching. Steadily rising smoke indicates fair weather.

f. A red sun or sky in the morning indicates nearing rain due to the excessive amount of moisture in the air.

g. A red evening sky indicates that the air contains so little moisture that rain within the next 24 hours is highly improbable.

h. A gray, overcast evening sky indicates that moisture carrying dust particles in the atmosphere have become overloaded with water; this condition favors rain.

i. A gray morning sky indicates dry air above the haze caused by the collecting of moisture on the dust in the lower atmosphere; you can reasonably a fair day.

j. When the setting sun shows a green tint at the top as it sinks behind clear horizon, fair weather is probable for most of the next 24 hours.

k. A rainbow in the late afternoon indicates fair weather ahead. However, a rainbow in the morning is a sign of prolonged bad weather.

l. A corona is the circle that appears around the sun or the moon. When this circle grows larger and larger, it indicates that the drops of water in the atmosphere are evaporating and that the weather will probably be clear. When this circle shrinks by the hour, it indicates that the water drops in the atmosphere are becoming larger, forming into clouds, rain is almost sure to fall.

m. In fair weather, air currents flow down streams and hillsides in the early morning and start drifting back up towards sunset. Any reversal of these directions warns of a nearing storm.

n. When the breeze is such that the leaves show their undersides, a storm is likely on the way.

o. It is so quiet before a storm, that distant noises can be heard more clearly. This is due to the inactivity of wildlife a couple of hours before a storm.

p. When in the mountains, the sight of morning mist rising from ravines is a good sign of clear weather the rest of the day.

q. A heavy dew or frost in the morning is a sign of fair weather for the rest of the day. This is due to the moisture in the atmosphere settling on the ground vice in the form of precipitation such as rain, snow, etc.

UNITED STATES MARINE CORPS
Mountain Warfare Training Center
Bridgeport, California 93517-5001

FMST.07.42
03/14/02

STUDENT HANDOUT

RESCUE PHILOSOPHY FOR SWIFTWATER RESCUE TEAMS

TERMIANL LEARNING OBJECTIVE In a swift water rescue scenario, organize a swift water rescue, in accordance with the reference. (FMST.07.42)

ENABLING LEARNING OBJECTIVES.

(1) With out the aid of reference, from a given list select the correct components of rescue preparation, in accordance with the reference. (FMST.07.42a)

(2) With out the aid of reference, from a given list select the correct role description of the rescue team members, in accordance with the reference. (FMST.07.42b)

(3) With out the aid of reference, from a given list select the correct methods of rescue, in accordance with the reference. (FMST.07.42c)

OUTLINE.

1. **COMPONENTS OF RESCUE PREPARATION**. (FMST.07.42a) These components comprise the backbone of both technical skills and the proper mental attitude required of a rescue professional. Remember that this course only provides you with the first component.

a. Training. Provides the students with the basic technical skills required to perform a safe rescue, these can be gained by attending an approved SWIFTWATER RESCUE COURSE.

b. Practice. Use of the skills learned in training.

c. Experience. Gained through advanced training, practical exercises and real-world situations.

B1

d. Judgment. The combination and application of skills for a given situation. Remember that every rescue situation is different. Don't get into a SOP mindset.

2. **PRIORITIES OF SAFETY**. Most of you have heard the saying that "It is better to be judged by twelve then carried by six". With that in mind you must realize that, at times, there is no way to affect a safe rescue. If you place yourself or your team in a hostile environment, from which you cannot escape, you are adding to the problem, not assisting in the solution. For this reason we set priorities of safety.

a. Self-aid. Be able to rescue yourself.

b. Buddy-aid. Ensure all safety measures are in force and that team members can rescue each other in an emergency.

c. Victim-aid. Only after self and buddy-aid are ensured should you attempt to rescue the victim. This will enable you to concentrate and focus your entire effort on the rescue.

3. **ROLES OF THE TEAM MEMBERS**. (FMST.07.24b) Often when dealing with rescues, you will find yourself working with outside agencies and organizations. In order to successfully coordinate the entire effort the Incident Command System (ICS) was developed. In swiftwater rescue situations there are four specific roles that must be filled.

a. Incident Commander. Is in overall command of the rescue efforts. Does not necessarily have the technical knowledge to conduct the rescue, but must be able to coordinate the rescue effort.

b. Rigger. Personnel responsible for equipment and rope systems, i.e. establishing the tension diagonal or the tyrolean high line.

c. Gofer. This group of individuals fills all other supporting roles, i.e. Communications, logistics, medevac, accounting, equipment accountability etc.

d. Rescuer. The individuals who are actually conducting the rescue.

4. **COMPONENTS OF SEARCH AND RESCUE**. In order to successfully complete a rescue there is four objectives that must be met. In order to remember them we use the acronym LAST.

a. Locate. This is the search in the search and rescue business.

b. Access. The determination of the assets, equipment, technical skills, and personnel required safely reaching and extracting the victim.

c. Stabilization. Providing the on scene medical attention required in order to extract the victim without causing further injury or aggravating an existing condition.

d. Transport. Removing the victim from the hostile environment to a safe one where further treatment can be provided.

5. **METHODS OF RESCUE**. (FMST.07.42c) In any rescue, we must access the situation to determine the best techniques to apply. Bearing in mind the self-aid, buddy-aid, victim-aid concept as well as time, weather, victim's condition, etc. In all cases we will attempt the lowest risk methods first, while preparing for a higher risk option. In all cases remember the KISS principle.

a. Reach. Use a ladder, fire hose, stick, paddle, rifle or any other sturdy object to extend your reach in order to reach the victim.

b. Throw. Get some type of floatation device to the victim. Or throw a line to the victim in order to pendulum or pull him/her to the shore.

c. Row. Putting a boat in the water to reach the victim.

d. Go and Tow. These are the contact rescues that, while look good on film, put you at the extreme risk of becoming a second victim.

e. Helo. Most civilian helicopter crews are not trained in SAR evolutions. Military aircraft and crews are a better bet. Several uses for helos are in the support role. Locating the victim, inserting and extracting teams and moving equipment are some of these roles. If using the helo in a direct role it is better to use a qualified SAR swimmer/diver to hook-up or pick-up victims.

UNITED STATES MARINE CORPS
Mountain Warfare Training Center
Bridgeport, California 93517-5001

FMST.07.25
03/14/02

STUDENT HANDOUT

PERSONAL AND TEAM EQUIPMENT

TERMINAL LEARNING OBJECTIVE As a member of a swiftwater rescue team, maintain swiftwater rescue equipment, in accordance with the reference. (FMST.07.25)

ENABLING LEARNING OBJECTIVES.

(1) Without the aid of references, from a given identify the correct items of personal equipment, in accordance with the reference. (FMST.07.25a)

(2) Without the aid of references, from a given identify the correct items of team equipment, in accordance with the reference. (FMST.07.25b)

OUTLINE.

1. **PERSONAL EQUIPMENT**. (FMST.07.25a) The following items of personal equipment should be in each man's possession for a swiftwater rescue.

   a. Dry Suit. Warmest, most versatile, full body protection available. Extra thermal layers can be added for varying water temperatures. Provides wind protection out of the water and provides protection against contaminated water.

   b. Wet Suit. Excellent protection from flotsam. Variety of thick nesses for thermal protection. Insulating quality is degraded in moving water due to the flushing effect.

   c. Footwear. Primary is full sole scuba booties. You can use tennis shoes or jungle boots with heavy wool socks.

   d. Gloves. Neoprene is recommended.

   e. Hoods. Wetsuit hood or pile hood for insulation. Remember 30-50% of your body heat losses can occur through the head and neck region of your body if left unprotected.

B2

f. Helmet. Protects you from flotsam and impact injuries. Should be padded and ventilated to allow water to flow through.

g. Knife. Fixed or locking blade attached to the vest, handle down.

h. Whistle. Required for communication. Best without "pea".

i. Eye Protection. Protects subject from debris.

j. Personal Floatation Device. This is the single most important piece of equipment. Jackets should provide at least 15 lbs. of floatation. Proper fit is a priority. Rescue vests with a "Blow-Out" belt should only be used by trained professionals.

2. **TEAM EQUIPMENT**. (FMST.07.25b) The following is a list of some of the team equipment for a typical four-man team. Of course the situation will dictate which equipment will be required for the mission on hand, but all hands should be familiar with this equipment. This is only a list. The uses and application of this gear, along with practical application exercises later in this course will provide you with the hands-on training and working knowledge to properly and safely use this equipment.

QTY	ITEM
1	300 ft, 1 Static Rope
4	150 ft, 11 mm Static Rope
25	85 Locking Stubai Carabiners
4	2" Pulleys
2	Rescue Figure 8
6	3 ft, 7mm Prussic
20	15 ft, 1" Tubular Nylon Runners
2	Ascenders
3	Throw Bags
2 pr	Fins
1	Rescue PFD
1	Self Bailing Raft (13 or 14 foot)
1	Oar Frame w/ Oars, or Seven Paddles
3	Radios (waterproofed)
1	Medical Kit
1	Litter

NOTE: This is not a complete list of rescue equipment. However, it is a good basic load.

UNITED STATES MARINE CORPS
Mountain Warfare Training Center
Bridgeport, California 93517-5001

FMST.07.41
03/14/02

STUDENT HANDOUT

SWIFTWATER TERMINOLOGY AND DYNAMICS

TERMINAL LEARNING OBJECTIVES In a swift water rescue scenario, evaluate swift water river conditions in accordance with the reference. (FMST.07.41)

ENABLING LEARNING OBJECTIVES.

1) Without the aid of reference, from a given list correctly identify the swiftwater terminology, in accordance with the reference. (FMST.07.41a)

2) Without the aid of reference, from a given list correctly identify the classifications of rivers, in accordance with the reference. (FMST.07.41b)

OUTLINE.

1. **SWIFTWATER TERMINOLOGY**. (FMST.07.41a)

a. Hydraulic or Hydraulic effect: Movement of water caused by pressure.

b. Eddy: Horizontal reversal. Current is blocked by an obstacle creating a vacuum which, when filled, causes water to flow upstream.

c. Eddy Fence: Reversal line between downstream current and the eddy. This fence can be very strong and the water flow may vary in height by several inches.

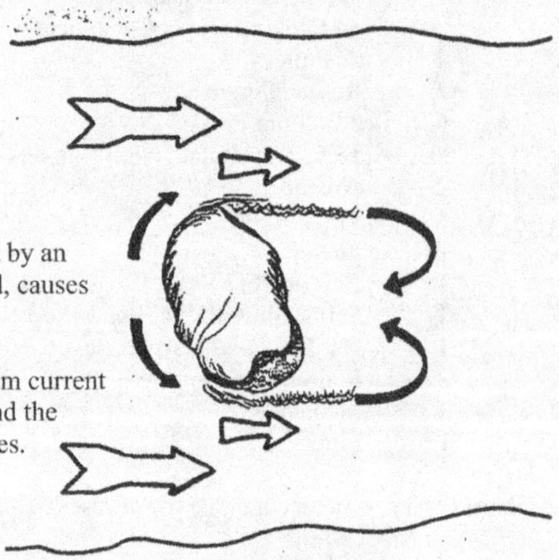

EDDY/EDDY FENCE

B3

d. Hole, Stopper, Keeper: Vertical reversal as water flows over an obstacle the main current follows the base of the river causing a surface vacuum. Downstream water flows back upstream to fill this void. This sets up a recirculating effect.

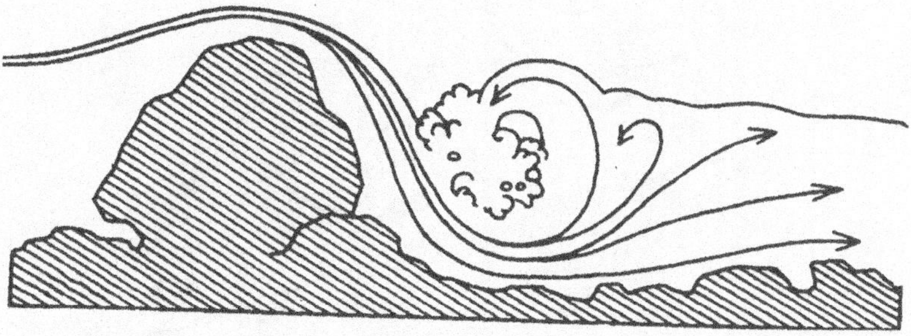

HOLE/STOPPER/KEEPER

e. Standing Wave. Haystack: Series of waves in the main channel caused by obstacles, ledges, or increases in the flow rate, gradient, or volume.

STANDING WAVES/HAYSTACKS

f. Downstream V. Tonal: Represent largest volume of water or main channel.

g. Upstream V: Caused by obstacles submerged just below the surface.

h. Strainer: Debris, logs, small trees on flooded banks, which allow water to flow through but will not allow a body to pass.

STRAINER / GUARD RAIL

STRAINER / TREE

i. Cushion, Pillow: Build-up of water on the upstream side of obstacles or the outside of curves.

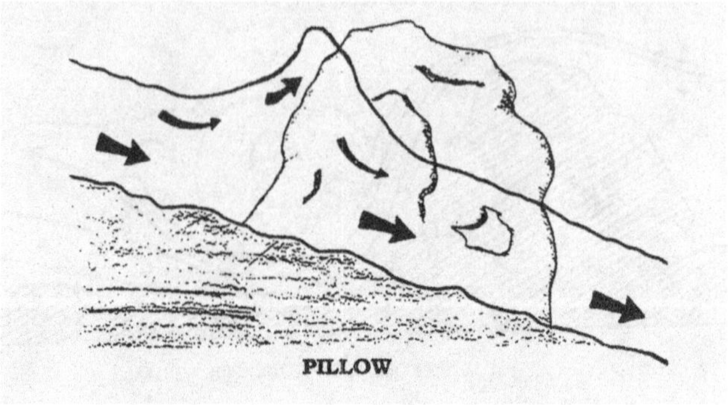

j. Confluence: Merging of streams.

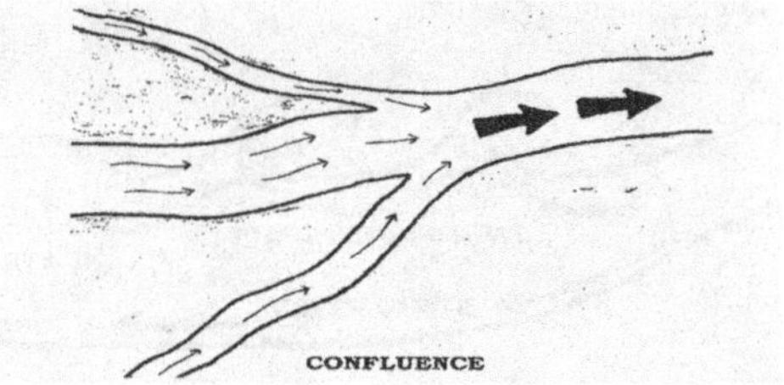

k. River Right. On the right side of the river looking downstream.

l. River Left. On the left side of the river looking downstream.

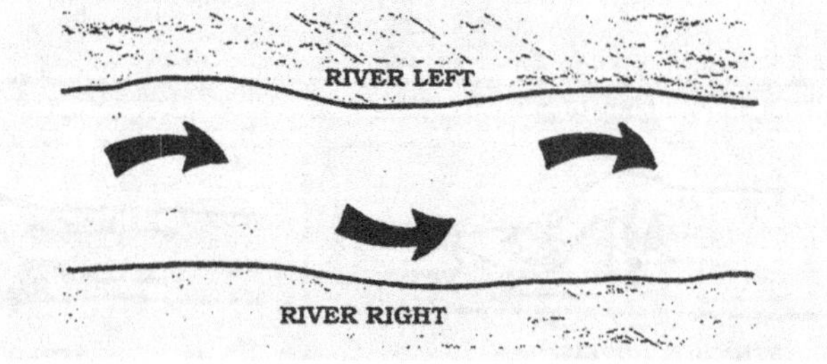

2. **CLASSIFICATIONS OF RIVERS**: (FMST.07.41b)

a. Class 1. Moving water, few obstacles, small ripples or waves.

b. Class 2. Moving water, clear channels, wide waves to 3', no scouting required.

c. Class 3. Rapids, narrow channels that require complex maneuvering, waves capable of flipping rafts or swamping canoes, scouting required.

d. Class 4. Long hazardous rapids, precise boat handling in rough water, scouting required, rescues hazardous.

e. Class 5. Severe, extended, extremely violent rapids, professional boat handlers and scouting required. Rescues extremely hazardous, situations are life threatening.

f. Class 6. Only Expert/Professional River Runners. Rescue may be impossible.

g. CLASSES U DON'T EVEN THINK ABOUT IT! Waters are not navigable.

3. **RIVER STRENGTH**.

a. Volume of Water. The volume of water in a given stretch of the river between two points is based on a simple formula and is measured in Cubic Feet Per Second (CFS). Simply take the river depth x river width x river velocity = volume of water in CFS.

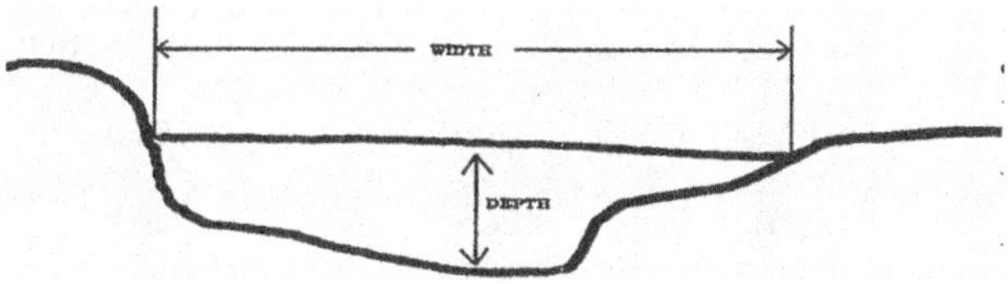

For example: 6 feet deep x 100 feet wide x 2 feet per second = 1,200 CFS

b. River Velocity. In order to determine the velocity or current speed measure a 100-foot section of the river and drop a float at the upstream end. Then, time its movement to the lower end and use the chart below.

TIME IN SECONDS	VELOCITY FT PER SEC	VELOCITY MPH
5	20	13
10	10	7
15	6.7	5
17	5.9	4
20	5	3.5
25	4	3
29	3.5	2.5
37	2.7	2
50	2	1.5
80	1.3	1
110	0.7	0.5

NOTE: Should the shape of the river change by becoming wider or narrower, once you have the flow in CFS, you can determine the new velocity by using the following formula:

$$\text{Velocity} = \frac{\text{Volume in CFS}}{\text{Width x Depth}}$$

c. Force Exerted. The velocity of the water will determine the force exerted X's the amount of exposed area under pressure. For simplicity we will use the following chart:

CURRENT VELOCITY	FORCE EXERTED ON OBJECT		
MPH	LEGS RAFT	BODY	WRAPPED
3	17	34	168
6	67	134	672
9	151	302	1,512
12	269	538	2,688

4. **LOW HEAD DAMS**. Low head dams are used throughout the world for irrigation, flood control, at hydroelectric plants and for many other uses. Basically they are pouring over obstacles that create hydraulics, keeper holes, and recirculating currents at their bases. Unlike natural obstacles, that rarely span the entire river and that allow you to be "flushed" out of the bottom or sides, low head dams create a hydraulic that extends from shore to shore.

BOIL LINE

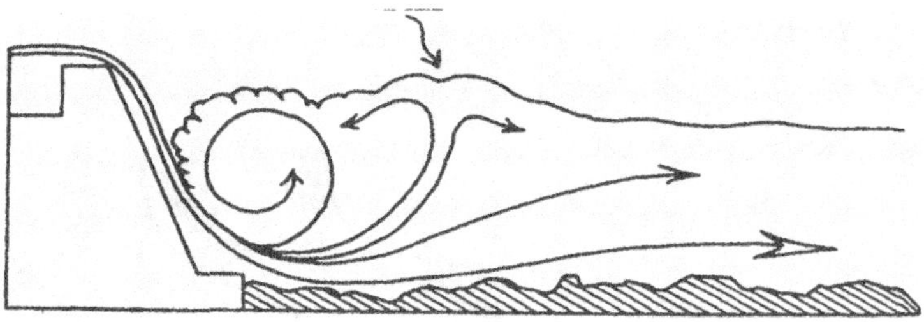

a. The size of the hydraulic created is based on several denominators:

(1) Height of drop.

(2) Angle of drop.

(3) Volume of water CFS.

(4) Velocity of water MPH.

(5) Bottom composition of runout.

b. The degree of difficulty for the rescuer is also compounded by:

(1) Water depth below dam.

(2) Debris build-up.

(3) Dam construction/conformation.

(4) Lack of access.

(5) Lack of natural anchors.

(6) Distance between riverbanks.

c. The basic hydrology is the same as any other pour over obstacle. As the river flows over the dam, the water follows the face of the dam to the river bottom. When sufficient speed diminishes the water rebounds to the surface and continues downstream. This flow has caused a vacuum or pocket directly below the dam that must be filled. Therefore, when the water rebounds to the surface, some of it flows back upstream to form a vertical reversal or pour-over eddy. The line of demarcation for the upstream/downstream flow is called the boil line. The shape and angle of the dam face and the volume and velocity of the water flowing over the dam create the distance and width of the boil line.

5. **REVIEW**.

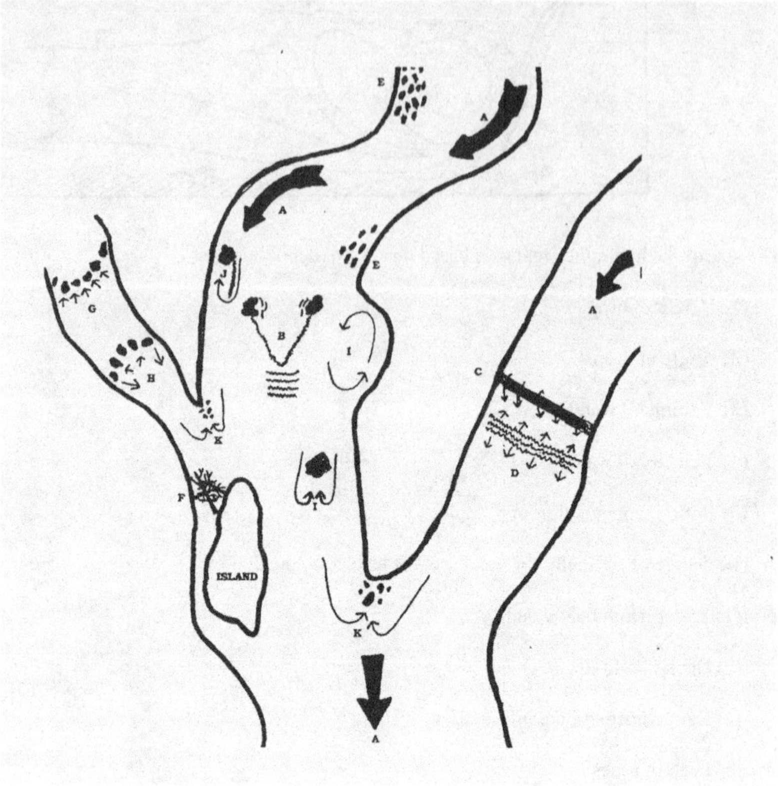

A. Main Current
B. Downstream V
C. Low head Dam
D. Boil Line
E. Rock Garden
F. Debris Pile/Strainer
G. Frowning Hole
H. Smiling Hole
I. Eddy
J. Eddy Fence
K. Confluence

UNITED STATES MARINE CORPS
Mountain Warfare Training Center
Bridgeport, California 93517-5001

FMST.07.24
04/14/02

STUDENT HANDOUT

COMMUNICATIONS

TERMINAL LEARNING OBJECTIVES Given a casualty in a swift water environment, assignment as a member of a rescue team, and a whistle, employ swift water rescue communication signals to support mission requirements per the reference. (FMST.07.24)

ENABLING LEARNING OBJECTIVES.

(1) In a summer mountainous environment, prior to swift water rescue operations, when the instructor verbally states a hand and arm signal, the student will demonstrate that signal, in accordance with the references. (FMST.07.24a)

(2) In a summer mountainous environment, standing at the edge of a swift water stream, when the instructor blasts a whistle a varied amount of times, the student will focus their attention in the proper direction, in accordance with the references. (FMST.07.24b)

OUTLINE.

1. **RELIABLE COMMUNICATION**. In many cases time is critical in a SWR scenario. Therefore immediate, understandable communication is a must. Radios, bull horns, and voice work well in some situations, but wet equipment, dead batteries, and ambient noise levels can render these methods ineffective. Therefore, rescue professionals have developed a series of simple hand and arm, and whistle signals. Any rescue team can develop their own signals. The only requirement is that all hands understand the signals and that they are briefed to any other agency you are working with.

2. **HAND AND ARM SIGNALS**. (FMST.07.24a)

ACTION	MEANING
1 hand extended over head	Require Assistance
Both hands extended forming an "O" with one hand touching The head	Okay
Extend both hands above head, then point left/right	Move in direction indicated
Extend both hands above head, wave, then point left/right	Eddy out in direction indicated
Cross both arms over chest	Require medical assistance or Corpsman

3. **WHISTLE BLASTS**. (FMST.07.24b)

ACTION	MEANING
One Blast	Stop/Attention
Two Blasts	Upstream
Three Blasts	Downstream
Three Blasts Repeated	Emergency

UNITED STATES MARINE CORPS
Mountain Warfare Training Center
Bridgeport, California 93517-5001

FMST.0
7.27
04/14/02

STUDENT HANDOUT

THROW BAGS AND FOOT ENTANGLEMENTS

TERMINAL LEARNING OBJECTIVES In a swiftwater rescue scenario employ a throw bag, in accordance with the reference. (FMST.07.27)

ENABLING LEARNING OBJECTIVES.

1) Without the aid of reference, provided with a throw bag and swimming victim, rescue the victim, in accordance with the reference. (FMST.07.27a)

OUTLINE.

1. **THE THROW BAG**: The throw bag is very simply a cloth bag with a small float insert, packed with 50 to 70 feet of 1/4" to 3/8" floating polypropylene or spectra/poly line. The concept is to hold the exposed end of the line and throw the packed bag to a swimmer. As the bag flies, the rope deploys enabling you to reach a victim 50 to 70 feet from shore. It is used either from a standing position, or a moving position on open shores.

    a. Standing Procedures. (FMST.07.27a)

1. Assume a stable belay position and ensure your footing is solid. Avoid loose, sloped or wet rock.

2. Attract victim by blowing your whistle (1 blast). Show him/her the throw bag and yell, "grab the rope". NOTE: If the victim grabs the bag rather than the rope, the rope will continue to deploy to its full length.

3. Grasping the exposed end of the rope, throw the bag underhand in a smooth motion directly at the victim. You may need to lead them in order to score a direct hit.

4. Tell the victim to place the rope over his/her upstream shoulder. This will establish a good ferry angle.

B5

5. Do not wrap the rope around yourself or tie it off at anytime. You may have to release the victim to avoid an obstacle or strainer. Use a dynamic belay and pull the victim to shore or pendulum the victim to other rescuers downstream.
6. If available have other rescuers downstream to recover the victim or repeat the procedure if you miss!

b. Moving Procedures. The moving procedures are the same as the standing procedures except you are on a smooth, open shoreline, so you move slightly downstream of the victim as you pull them gently ashore.

2. **FOOT ENTRAPMENTS AND VERTICAL PINNINGS**. These scenarios are caused by a variety of actions. Swimmers put their feet down or fishermen, in waders, get their feet stuck between rocks and are pushed over by the current. Swimmers are stuck in a strainer. Rafters get wrapped on a rock with victims pinned between the raft and rock or kayakers drop down a vertical shoot and get lodged in the rocks at the bottom. Whatever way it occurs you must respond quickly in order to save the victim. Initially take the following precautions:

   1. Position upstream spotters, to provide warning of boats or debris.

   2. Plan alternate methods if initial try doesn't work.

   3. Have back-up rescuers downstream in order to retrieve the victim.

   a. Once these actions are complete conduct the rescue using one of the five following methods.

   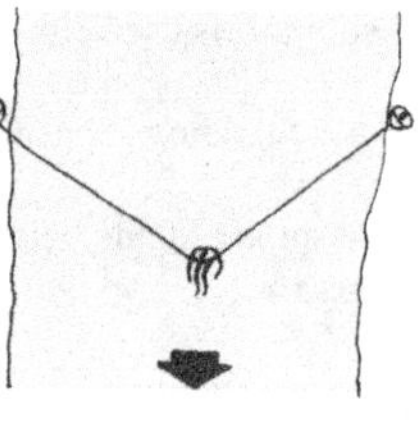

   1. Bechel-Ray One Line Method. The simplest and safest method. Get a rescuer on both sides of the river. Establish a rope between the two rescuers and downstream of the victim. Walk upstream until the rope hooks the victim. If you can position the line near the victims ankles you may be able to pop him/her loose and the downstream safety can pick him/her up. If not you can haul the victim upstream until he/she is clear of the obstacle, then by one rescuer playing out line and the other retrieving it you can bring the victim ashore. Remember to bring the victim to the shore with the medevac capability.

   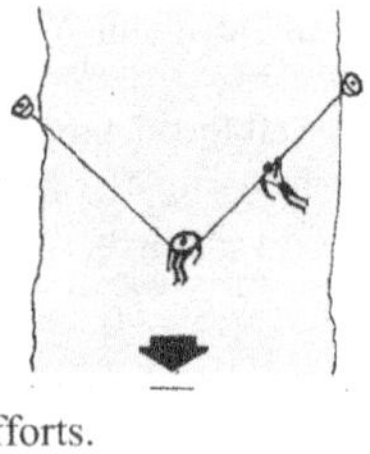

   2. The Strong Swimmer-variation. If the previous method doesn't work, leave the line in place and anchor it to stabilize the victim above the surface. Then have a "strong swimmer" go down one side of the rope, as would a tension diagonal. This rescuer is not tied in. Upon arrival the rescuer can readjust the line on the victim, attempt to pull them free, give them a PFD and helmet, or if a debris pile or rock is near, get on it and assist in further efforts.

3. Two Line Loop. There are two variations on this method both of which are applied using the initial rope used in the Bechel-Ray attempt.

(1) Deploy a second rope across the river and hook it onto the first line. Allow it to go one-half way down the rope to the victim. Pull at a 90° angle to rope #1. This will establish a mechanical advantage and pull the victim loose.

(2) Once the victim is free walk the line downstream trapping the victim in a loop. Then by releasing slack on line one, you can bring him/her ashore.

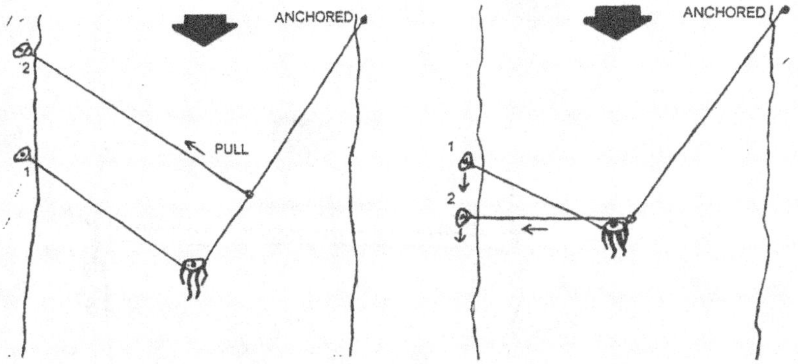

4. European Tethered Rescuer Lower. Using a rescue vest with a "blow-out" belt establish an upstream belay and lower the tethered rescuer to the victim. The rescuer's body will block the water forming an eddy around the victim enabling him to be pulled free. Ensure that the rescuer has a working rescue vest, as this will enable him to cut away if the situation dictates.

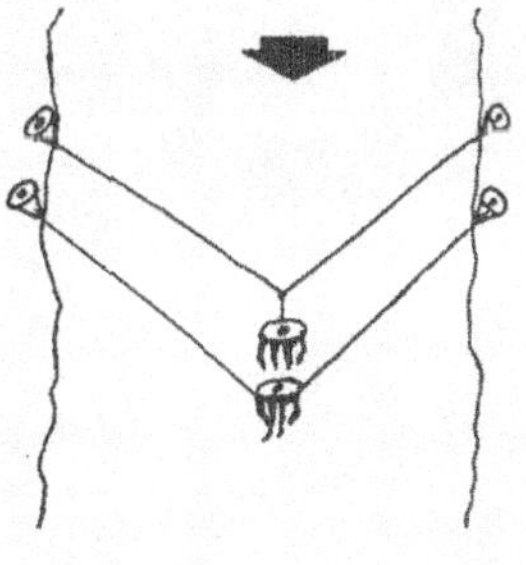

5. Mechanical Advantage Systems. As a last resort you can use a mechanical advantage system. This method is usually used only for body recoveries, as the amount of force exerted on the victim is tremendous and can cause severe injuries.

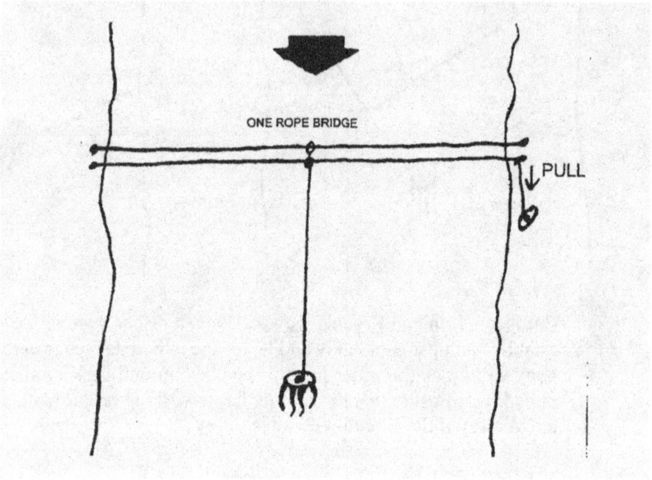
ONE ROPE BRIDGE
PULL

UNITED STATES MARINE CORPS
Mountain Warfare Training Center
Bridgeport, California 93517-5001

FMST.07.28
04/14/02

STUDENT OUTLINE

DEFENSIVE SWIMMING AND FERRY ANGLE

TERMINAL LERNING OBJECTIVES Given a swift water river and throw bags, perform aggressive swift water swimming to meet mission requirements per the reference. (FMST.07.28)

ENABLING LEARNING OBJECTIVES.

(1) Without the aid of reference, demonstrate the defensive swimming position, in accordance with the reference. (FMST.07.28a)

(2) Without the aid of reference, demonstrate use of the ferry angle to cross the river, in accordance with the reference. (FMST.07.28b)

OUTLINE

1. **THE DEFENSIVE SWIMMING POSITION (DSP)**. (FMST.07.28a) The DSP is assumed by lying flat on your back, chin on chest (to read the river & select a route), with your feet up, heels just below the level of your buttocks. Holding your feet in this position will enable you to fend off of shallow rocks and will reduce the likelihood of a foot entrapment or pinning. Your arms and hands should be kept close to the body but may be used like oars on a boat to maneuver. Your head will remain upstream except to cross a strainer or to over take a victim in a swimming rescue.

2. **FERRY ANGLE**. (FMST.07.28b) A ferry angle is used to maneuver diagonally across a river. By placing your body at a 45° angle to the current, the river will push you across the river. This is because the force of the current will travel down the exposed, upstream, side of the body moving it laterally more rapidly then the current pushing straight. This is like traversing a slope with edged skis vice sideslipping down on flat skis. In order to cross faster or to reach an eddy you may roll to your belly and execute the rescue crawl stroke or breaststroke. Remember to maintain your ferry angle with your head upstream this will enable you to travel diagonally across the river.

3. **STRAINER CROSSING**. The first rule is to avoid the strainer or shallow submerged obstacle. Read the river and maneuver to the main channel by using downstream "V's" or tongues. If the obstacle cannot be avoided, roll on your belly, head downstream and use the rescue crawl stroke. You must get your body moving faster than the current. Upon arrival at the obstacle use both hands to "vault" yourself over. A well-timed dolphin kick will assist in keeping your legs clear of entanglements. Once clear of the obstacle, resume the defensive swimming position. Bear in mind that it is usually safer to remain in the main current or to eddy hop downstream than it is to attempt to get out on a flooded riverbank, the outside of a bend or on slippery rocks. Also, use the dolphin kick rather than placing your feet on the bottom and kicking off. This will reduce the likelihood of foot entrapments.

UNITED STATES MARINE CORPS
Mountain Warfare Training Center
Bridgeport, California 93517-5001

FMST.07.25
03/14/02

STUDENT HANDOUT

FEET-WET RESCUES AND C-SPINE ROLLS

TERMINAL LEARNING OBJECTIVES In a swift water rescue scenario, conduct a swimming rescue, in accordance with the reference. (FMST.07.25)

ENABLING LEARNING OBJECTIVES.

(1) Without the aid of reference, while in a river and provided with a victim, perform a contact rescue, in accordance with the reference. (FMST.07.25a)

(2) Without the aid of reference, while in a river and provided with a simulated unconscious victim, perform a C-spine roll, in accordance with the references. (FMST.07.25b)

OUTLINE.

1. **FEET WET RESCUE GENERAL CONSIDERATIONS**. (FMST.07.25a)

   a. Always wear the proper personal equipment, i.e. PFD, helmet, drysuit, etc.

b. Don't make physical contact if it can be avoided. Make him/her "chase" you ashore. Use a life ring, can, or boogie board, and tow it along with the victim. Toss him a throw bag and swim/tow him/her until you reach the shore, then pendulum or haul the victim in. Remember, to a panicked victim YOU are a floatation device.

2. **CONTACT RESCUE**.

a. Approach the victim from upstream. Use the rescue crawl stroke to catch-up to him/her. Talk to the victim and attempt to get him/her to swim to shore or "chase" you to the shore.

b. Explain to the victim the defensive swimming position and attempt to get them in this position. Assume this position yourself, in contact rescues this is called the "Reverse and Ready" position. Tell the victim what you are going to do prior to grabbing them.

c. Gain contact by grabbing the shoulders of the life jacket or utilize a cross-chest carry. If the victim is unconscious and face down use a C-Spine roll down and maintain stability as much as possible.

d. Assume a good ferry angle with the victim downstream of you and use his/her body to protect you from obstacles.

e. If the victim panics and attempts to climb on top of you, break away or lock the victim up and go for an alligator roll. Reestablish control and calm the victim as soon as possible.

f. Once contact is made remember that you may be held accountable under the "abandonment" laws if you quit.

3. **TOWED VICTIM SWIM**. When towing a victim it is essential that you establish a ferry angle. If the victim is on a boogie board or float, explain to him/her what you are doing and have them help by kicking their feet. If you are towing by the PFD grasp the shoulder towards the far bank and remain in the DSP (defensive swimming position) in case your victim panics.

4. **LIVE BAIT RESCUES**. The final go and tow method is the "live bait" rescue. Using a rescue vest with a "blow out" belt and a mule team on the shore to belay you, swim to the victim as explained in contact rescues. Once you establish contact you are, in effect, a throw bag with hands. The belayer will then either pendulum you ashore or hauls you in. In an emergency you can cut away and conduct a normal contact rescue.

5. **C-SPINE ROLLS AND STABILITY**.

a. Basic Rules: (FMST.07.25b)

(1) Suspect C-Spine involvement in any head injury.

(2) Stabilize and immobilize, use C-Collar, Sam Splints, or a field expedient method.

(3) Keep the victim in the water until a spine board or equivalent is available, or there are enough rescuers to move the victim safely.

(4) Observe for other signs, i.e. cranial bleeding, bleeding from ears, eyes, nose (Do not attempt to stop this bleeding) or unequal/unresponsive pupil response.

b. Methods:

(1) First, approach the victim from behind. Reach under the arms and grasp the victim's head with both hands, then roll the victim. This is the least desirable method as you cannot swim properly or evaluate the ABC's of the victim.

(2) Second, approach the victim from the side, extend his/her arms above and alongside his/her head, reach across and roll the victim onto your arm maintaining the victims arms in place to ensure stabilization. Evaluate the ABC's, provide rescue breathing and use your free arm to swim.

(3) Finally, approach a victim wearing a PFD from the head. If face down, cross your arms thumbs inboard and use your forearms to immobilize the neck. Then roll the victim over. If face up, simply approach as if gaining contact on a contact rescue and maintain control of the C-Spine.

UNITED STATES MARINE CORPS
Mountain Warfare Training Center
Bridgeport, California 93517-5001

WMC

04/14/02

STUDENT HANDOUT

BOATS AND BOAT HANDLING

LESSON PURPOSE. The purpose of this period of instruction is to introduce the student to the boats available at MWTC for river rescue evolutions.

OUTLINE

1. **GENERAL**. There are a wide variety of boats and hybrid river craft available to rescue crews. Whitewater rafts, rigid raiders, zodiacs, airboats, and regular pleasure boats can all be used in the proper situations. All have been used both safely and have caused deaths and injuries when used improperly. Wherever you are, learn the equipment available, both its assets and its limitations.

2. **WHITEWATER RAFTS**. The whitewater raft of today is an inflatable, self-bailing, multi-chambered vessel with several tie rings and foot cups. Additionally, the materials used in construction resist rips, tears, and punctures much better than their predecessors. The size most often found on local rivers is 12 to 14 feet long due to the size of the rivers. In the MWTC SAR locker, we have a 13-foot self-bailer. This will be our primary rescue boat.

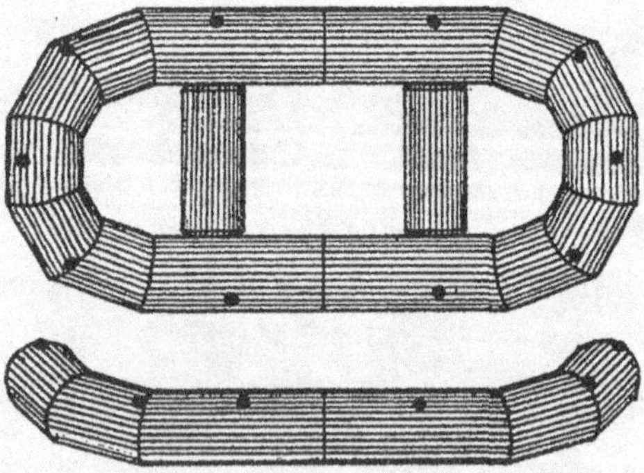

B8

3. **RAFTING COMMANDS**. In order to control the direction of a peddler powered boat the coxswain or guide must use a series of simple, concise, and understandable commands.

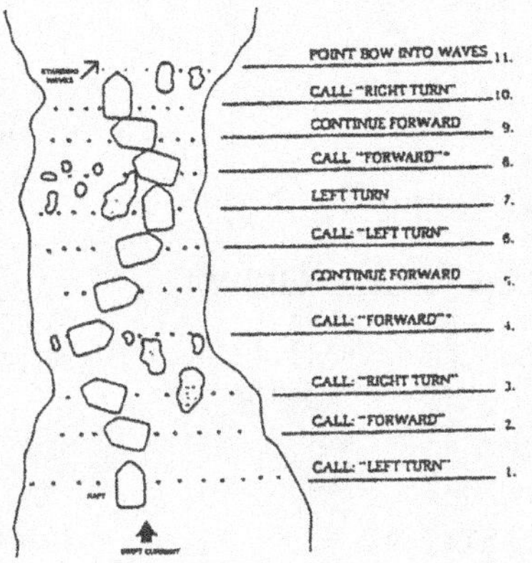

NORMAL TACTICS APPROACH

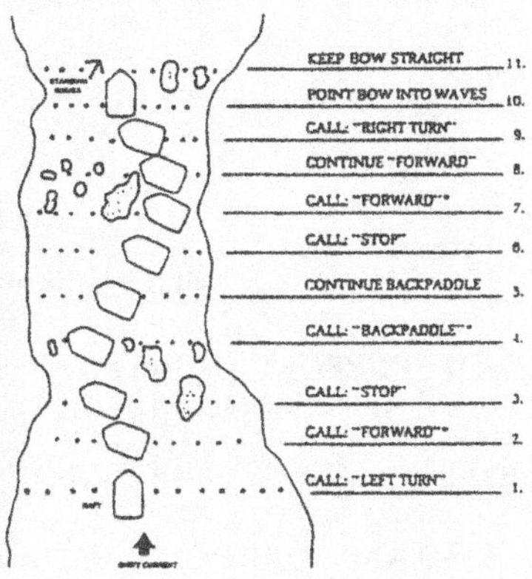

CONSERVATIVE/FAST CURRENT

a. Forward - All hands paddle forward together.

b. Back Paddle - All hands back paddle together.

c. Stop - Stop paddling - Remove blades from water.

d. Left Turn - Left side back paddles, right side forward.

e. Right Turn - Right side back paddles, left side forward.

f. High Side - Move quickly to the down streamside of the boat.

4. **PADDLE STROKES**. In order to control the raft from the coxswain position, either with a crew or alone on a tether system, you will need to apply the following strokes:

a. Sweep - Stroke in a wide arc around the fantail of the boat.

b. Draw - Reach out 90° away from the boat and "draw" it straight back to you.

c. Pry - Put paddle blade beside raft and "pry" the water away from the boat.

d. "J" - Used by lone guide to keep boat straight. Regular straight stroke finished with an outboard stroke or tail.

UNITED STATES MARINE CORPS
Mountain Warfare Training Center
Bridgeport, California 93517-5001

WMC
04/14/02

STUDENT HANDOUT

ROPE SYSTEMS AND BOAT ANCHORS

LESSON PURPOSE: The purpose of this period of instruction is to introduce the student to the rope systems and boat anchors used in swiftwater rescue situations.

OUTLINE

1. **TWO AND FOUR POINT TETHERING SYSTEMS**.

a. With all rope systems the first order of business is to get a rope across the river. Best case is to throw a rope bag across to someone already on the other side. Next best is to use another boat or raft to ferry the rope across. Last we use a swimmer to drag the rope across. Remember to use a smaller line on the swimmer and not to tie it to him/her. When swimming, start high and pendulum across using a good ferry angle.

b. Clip in on the side "D" rings forward for a two point or on the two forward and two aft "D" rings for a four-point tether system. For a two point you now need to simply load in a rescuer (if used) and move the boat into the river by taking slack on the far side and giving slack from the near side. For a four point man the two nearside lines and load a far side aft lineman and a rescuer (if used) into the boat and ferry the boat across. The far side aft belayer will get out and the boat is on a four point system ready for action.

c. This system is good for Class II Rivers with clear banks, few mid stream obstacles, and little floating debris. Use more belayers as required by river speed and boatload.

d. In all rope systems the whistle and hand and arm signals used to communicate are those taught in the communications class earlier in this course.

e. To break down the system simply reverse the procedures. The only difference is that the last man rides in the boat instead of swimming.

B9

2. **THREE POINT AND BOATMEN'S SELF-EQUALIZING ANCHORS.**

a. These anchors have in common the fact that they are self equalizing. This means that, regardless of the direction of pull, all anchor points ("D" rings, thwarts, etc) will remain under equal tension.

b. Using ropes the knots required are the double figure 8 on a bight, a directional figure 8, and a figure 8 on a bight.

c. Using webbing (one inch tubular nylon) a water tape knot is required.

d. If you have an abundance of carabineers, simply clip into the anchor points and back into the small loop on the double figure 8 on a bight, or into the directional figure 8. If not run the rope through the small loop. Remember that this is not preferred as it causes rope on rope stress and rubbing that can melt the rope.

e. When using webbing the same concept applies. Ensure you put a twist in the webbing at the contact point to ensure that a single blow-out will not collapse the entire system.

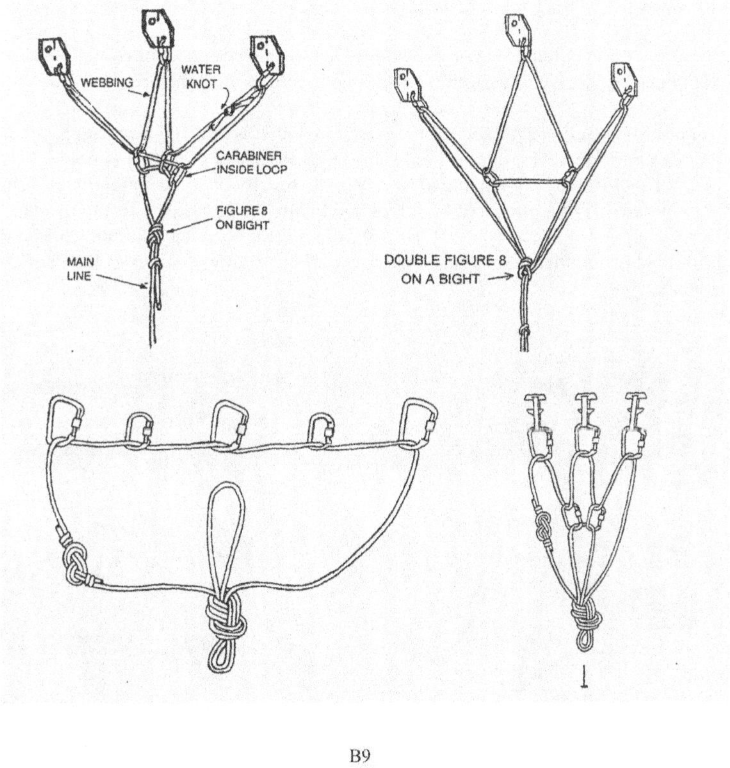

3. **PIG-RIG FOR MECHANICAL ADVANTAGE.** At times you will find yourself with ropes that are long enough to span the river, but not long enough to establish a mechanical advantage system. In these situations, you can use a piggy-back line or "pig-rig" by following the procedures below:

a. Tie off the mainline on either ends, or tie off the far side and set up a belay device on the near side using a munter hitch or rescue 8.

b. Establish an anchor for the pig-rig below the near side anchor.

c. Tie a prussic on the mainline and clip in a carabiner. Tie a figure 8 on a bight into the end of the pig-rig line and clip it into the prussic's carabineer.

d. Run the pig-rig line back to the anchor carabineer and then back through the prussic carabiner.

e. Pull on the running end towards the anchor. This will give you a 4:1 mechanical advantage on the main line.

f. If more tension is required you can either retie the nearside anchor (if tied), or use your belaying device to tension the mainline before moving the prussic.

g. If no prussic cord is available, you can accomplish this with a directional figure 8 on the mainline. However, this will only allow you to tighten the system once. A better method is to use the pig-rig line to tie a prussic directly on to the main line and run your system off of a directional figure 8 tied below the prussic on the pig-rig line. Remember that if the lines are of the same diameter this system will not work as the prussic will slip, therefore, you should use a kragur knot with ropes of the same diameter.

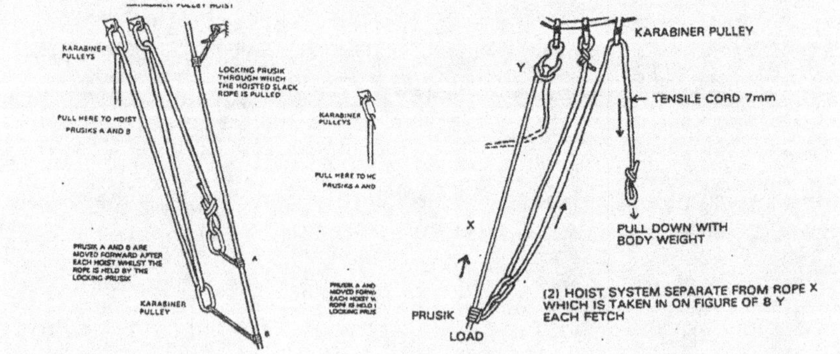

4. **TYROLEAN HIGHLINE AND TELFER LOWERING SYSTEMS** Both of these systems are used to maintain a boat as a stable rescue platform. The lateral movement (side to side/shore to shore) in both systems is controlled by mule teams on the shore. The main difference is that in the tyrolean highline the movement up and down-stream is also controlled from the shore, while in the telfer lowering system, the rescuer in the boat controls that aspect of movement. Both systems have good and bad points and both should be mastered. The situation will dictate which system you will use. Follow the steps below to establish these systems:

a. First you must establish TWO lines across the river. You can use any method taught to accomplish this.

b. Establish one line as the anchor line and tighten it as taught in the "one rope bridge" class or use a pig-rig, if required. If you have uneven mule teams set it at a slight down-stream angle towards the weak team's shore (no more then 10 to 15 degrees) to assist the weak team.

c. Rig the boat with a self-equalizing anchor system.

d. Rig the lowering systems as shown in the diagrams. Either tyrolean or telfer.

e. Load a rescuer or rescue team into the boat and execute. Remember the rescuer can act as coxswain and assist in maneuvering the boat by establishing a proper ferry angle.

f. Communications will be as taught earlier. Ensure you have a shore commander who maintains eye contact with the rescuers and mule teams on both the near and far shores.

g. When using the telfer system, a counter hitch, stitch plate or rescue 8 can be used as a belay device. This enables you to tie-off when you reach the victim so that you can affect the rescue without worrying about controlling the boat.

h. When using either of these systems it is a good idea to set-up a mechanical advantage system between the control (central) pulleys aid the raft/rescue platform.

TELFER LOWERING SYSTEM

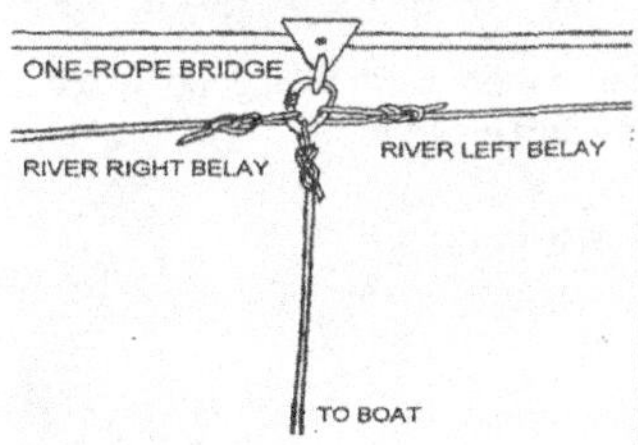

TYROLEAN HIGHLINE

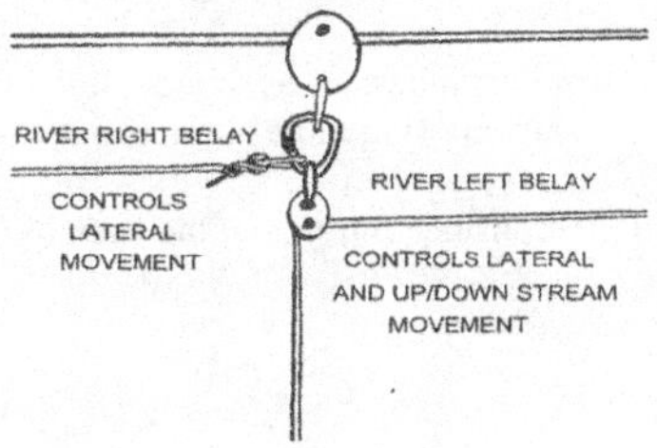

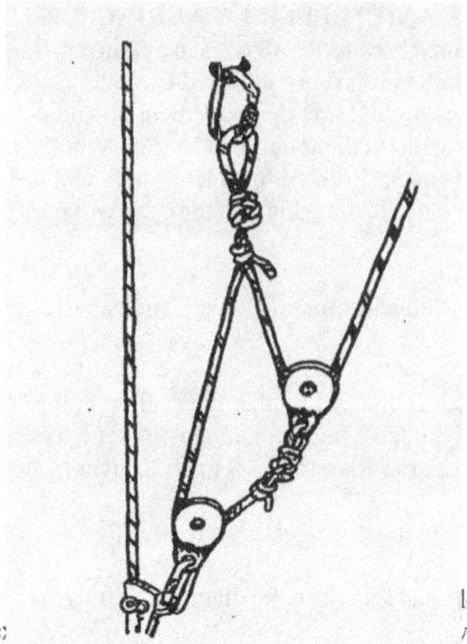

5. **BOAT WRAPS**. As [illegible] logy and Dynamics" class, the river can put an e[illegible] wrapped around an obstacle in the river. You first job is to ensure that NO ONE is trapped between the boat and the obstacle. If you do not have full accountability of personnel, don't hesitate! CUT THE BOAT! The easiest method to do this is to cut the floor out of the older "bucket" boats, or to cut the laces holding in the floor of a "self bailing" raft. Depending on how tightly wrapped the boat is you have several options available:

   a. First ensure the crew is accounted for and treated for any injuries.

   b. Establish a tag-line and attempt to bounce and pull the raft free.

   c. Set a Z-drag inside the boat to dump water.

   d. Use an oar or tree to lever the boat off the obstacle.

e. Set up a mechanical advantage system and attempt to "peal" the boat free.

f. Release air from the tubes. Be aware that this can cause more harm than good by reducing tube strength and exposing more surface area to the water pressure.

g. If all else fails, consider cutting out the floor as discussed earlier.

UNITED STATES MARINE CORPS
Mountain Warfare Training Center
Bridgeport, California 93517-5001

WMC
04/14/02

STUDENT HANDOUT

VEHICLE RESCUES

LESSON PURPOSE. The purpose of this period of instruction is to train the students in the procedures used to rescue personnel from vehicles trapped in swift water.

1. **VEHICLE BEHAVIOR**. There are three types of hazards in a river. Floating objects such as trees and branches, suspended objects or tumbling objects such as sawyers, rocks, and other debris, and stationary objects such as live trees on flooded banks and boulders. A vehicle will become all of these "loads". For the first minute, or so, they will float, as they become partially filled with water, they will be suspended. This is normally when they tumble or roll, and finally they will become stationary as they lodge against another stationary object or fill completely. Once they become stationary, depending on the water depth, they may form an eddy, a pour through or pour over, creating a hydraulic or a standing wave if fully covered. Also, the straighter a vehicle is (head-on) to the water flow the more stable it is. Finally, bear in mind that changes in water flow or displacement (weight) of the vehicle (taking a victim off it) may cause the vehicle to once again become a suspended load and tumble.

2. **RESCUE PROCEDURES**

a. The initial action is to secure the vehicle. Ensure that you have adequate anchors. Trees, guardrails, and bridge abutments are good. Avoid using another vehicle if possible.

b. Approach the vehicle on the downstream side by using the eddy created by the vehicle. Use a boat, either tethered or under power, a strong swimmer or boogie board. Another method is to use the line securing the vehicle as a tension diagonal to reach the vehicle. Then establish a second tension diagonal or endless rope to move the victims ashore. Ensure your upstream spotters are in place to warn of other suspended debris coming your way.

B10

c. If the vehicle is sealed tight, crack one window slightly to allow water to get in the vehicle slowly and equalize the pressure. Warning: If you pop or shatter a window you may cause a sympathetic detonation of the remaining windows.

d. Equip the victims with PFD's and remove to the boat or onto the rope system. Consider the following points:

(1) The victims weight might be the only thing stopping it from rolling. You might need to transfer a rescuer in a one for one exchange.

(2) On hard pack or asphalt the vehicle will shift more easily.

(3) If the vehicle is sideways keep the victims on the upstream side. The high side rule is reversed.

(4) Vehicles roll slowly.

(5) Helo's are a last resort.

UNITED STATES MARINE CORPS
Mountain Warfare Training Center
Bridgeport, California 93517-5001

FMST.07.3
0
04/14/02

STUDENT HANDOUT

STREAM CROSSING

TERMINAL LEARNING OBJECTIVE: Given a mountain swift water environment, and a staff, conduct a mountain stream crossing to meet mission requirements per the reference. (FMST.07.30)

ENABLING LEARNING OBJECTIVES:

(1) Without the aid of references, choose from a given list the considerations in site selection for a stream crossing, in accordance with the references. (FMST.07.30a)

(2) Without the aid of the reference, choose from a given list the safety precautions taken while conducting a stream crossing, in accordance with the reference. (FMST.07.30b)

(3) In a summer mountainous environment, execute the individual preparations taken prior to crossing a stream, in accordance with the reference. (FMST.07.30c)

(4) In a summer mountainous environment and given a staff, cross a mountain stream using the staff method, in accordance with the reference. (FMST.07.30d)

(5) In a summer mountainous environment, cross a mountain stream using team methods, in accordance with the reference. (FMST.07.30e)

1. **SITE SELECTION** (FMST.07.30a)

   a. Mountain streams and rivers are military obstacles and therefore are danger areas for units crossing them. In order to reduce the time in the vicinity of the danger area, a recon team should precede the main body and select the best crossing site. The site selection for a stream crossing should include these eight considerations:

B11

(1) Look for logjams, rocks or fallen trees that will provide a dry crossing if possible.

(2) If a dry crossing is not possible, select a crossing point at a wide and shallow point where the current is slower.

(3) Avoid sharp bends. They can be deep with a strong current on the outside of the bend.

(4) Look for a firm, smooth bottom. This is because large rocks and boulders provide poor footing and cause a great deal of turbulence in the water.

(5) It may be easier to cross several small channels of water rather than one large one.

(6) Do not cross just above rapids, falls, or logjams, taking a fall or slipping could have serious consequences.

(7) Cross in the early morning. The water level will be lower since there has been less daylight for the snow to melt. Also, on sunny days, you will have more time to dry clothing and equipment.

(8) There should be a suitable spot downstream for safety swimmers.

2. **SAFETY PRECAUTIONS** (FMST.07.30b). The following two safety precautions must be taken while conducting stream crossings.

   a. There must be a safety line at a 45-degree angle downstream across the stream. This is for anyone who slips and is swept downstream to grab so that he is capable to stopping himself.

   b. There must be safety swimmers downstream. These are strong swimmers who are positioned downstream to help anyone who is swept downstream. The safety swimmers will use throw bags.

NOTE: The safety line, as well as any other lines that must be taken across the stream, will be taken across by using the lead swimmer method taught in ONE-ROPE BRIDGE.

3. **INDIVIDUAL CROSSING PREPARATIONS**. (FMST.07.30c) Prior to beginning to cross a stream, there are certain preparations that each individual should take. The five preparations are as follows:

   a. Wear pack with shoulder straps fastened snugly. Waterproof the pack for buoyancy if possible.

   b. Weapons will be slung diagonally over the shoulder with the weapon itself being between the pack and the individual's back.

c. Button all pockets and remove blousing garters. This prevents the water from creating added drag against the individual, which it would do if it could flow into open pockets.

d. Wear boots to protect the feet, but remove socks and insoles to keep them dry:

e. Wear the minimum amount of clothing. This reduces the amount of clothing that must be dried after the crossing.

f. Wear helmet in slow moving, shallow water.

NOTE: If you are in a tactical situation, the actual situation will dictate which of the above precautions will be taken.

4. **INDIVIDUAL CROSSING METHODS**. There are three individual methods that may be used.

a. Staff Method. (FMST.07.30d) A strong staff or pole about 6 feet long is used as a crossing aid. It should be strong enough to support the Marine's weight and trimmed clean of any branches. Placing both hands on the pole, the Marine should place the staff just upstream of his intended path. He should use the staff as the third leg of a tripod and should move only one leg or the staff at a time. He should face upstream using the staff to retain his balance. The staff is also used as a probe to discover bottom irregularities that could trip the Marine. The Marine should drag his feet instead of picking them up.

STAFF CROSSING METHOD

b. Swimming. This is an obvious method, if your Marines are good swimmers. This is not always the case, so usually this method is not a preferred one.

(1) In fast, shallow water, the Marine should angle across on his back with his feet downstream and his head up. He should use his hands to tread water and his feet to fend off obstructions.

(2) In fast, deep water, the Marine should angle across the stream on his stomach with his head upstream, to establish a proper ferry angle.

c. Belayed Method. In chest deep or water with a strong current, a rope can assist greatly. A rope will be secured from bank to bank, with the far anchor slightly downstream from the near anchor. The rope will be anchored off so that it lays at a minimum of 45 degrees. The Marine attaches himself to the rope by using a sling rope as safety line, tying a bowline around his waist and a figure-of-eight loop with a steel locking carabiner inserted. He then attaches the carabiner to the crossing line and crosses using the current to assist him.

5. **TEAM CROSSING METHODS** (FMST.07.30e)

a. Line Abreast. Small units (squad to platoon) can cross in moderate currents up to chest deep, by lining arms in a line abreast or chain method. The largest man of the chain is placed on the upstream side of the group. The group will enter the stream parallel to the flow of the stream. The middleman of the chain will control the group's movement and give the command when to step.

LINE ABREAST

b. Line Astern Method. Three or more men can line up facing the current. The upstream man, who should be the largest man in the group, breaks the current while the downstream men hold him steady. The upstream man may use a staff, similar to the individual staff method, to steady himself. All men side step at the same time with one man calling the cadence.

LINE ASTERN

c. Huddle Method. Between three and eight men can face inboard as in a football huddle. They will wrap their arms around each other's shoulders and cross the stream in this formation. The upstream man should change position as they cross because the entire formation will rotate. This prevents one man from becoming exhausted in the upstream position.

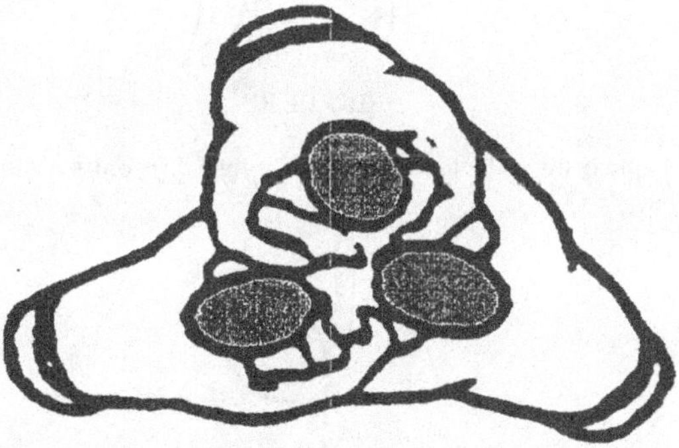

HUDDLE METHOD

UNITED STATES MARINE CORPS
Mountain Warfare Training Center
Bridgeport, California 93517-5001

WMC
04/09/02

STUDENT OUTLINE

ROPE MANAGEMENT

LESSON PURPOSE: To familiarize the student with some of the terms and procedures for working with a rope.

BODY

1. **TERMS USED IN ROPE WORK**

   a. Bight. A simple bend in the rope in which the rope does not cross itself.

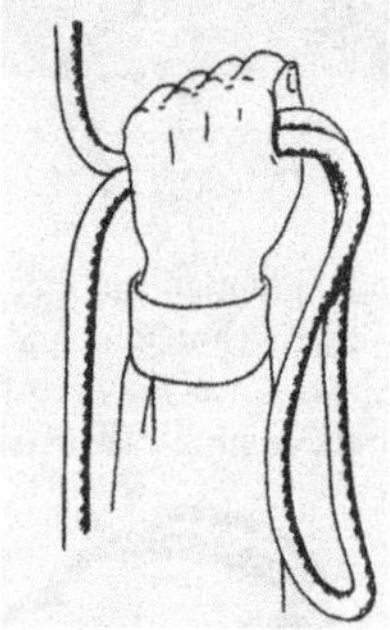

BIGHT OF ROPE

   b. Loop. A simple bend in the rope in which the rope does cross itself.

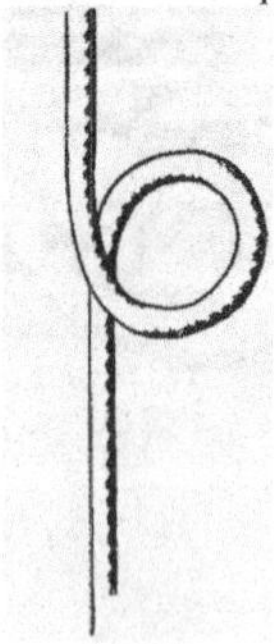

LOOP

c. Half Hitch. A loop which runs around an object in such a manner as to bind on itself.

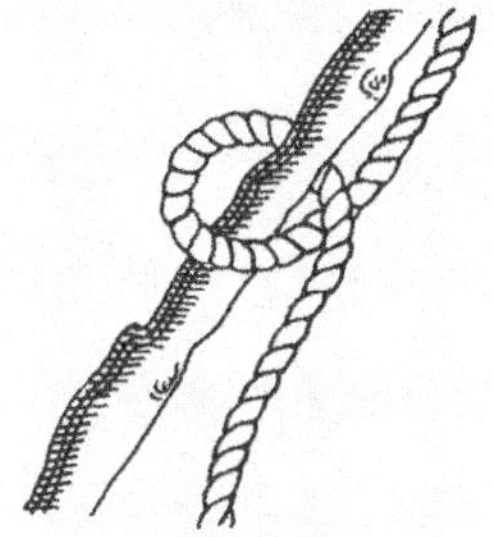

HALF HITCH

d. Standing End. The part of the rope which is anchored and cannot be used, also called the static end.

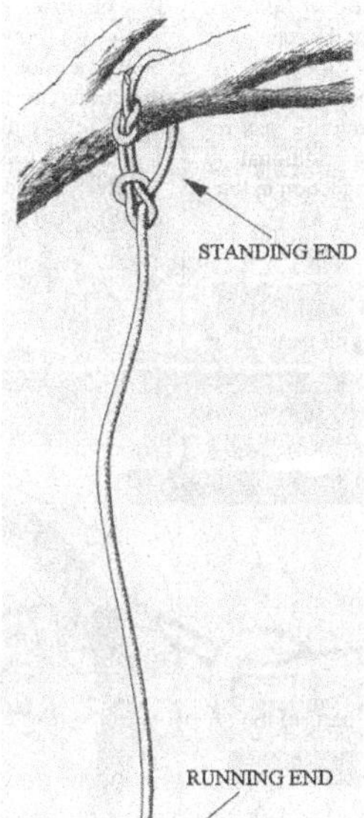

e. Running End. The free end of the rope which can be used.

f. Lay. The same as the twist of the rope. (Applies only to hawser laid ropes, such as manila.)

g. Pigtail. The short length left at the end of a rope after tying a knot or coiling a rope. It may or my not be tied off with a secondary knot, depending on the circumstance.

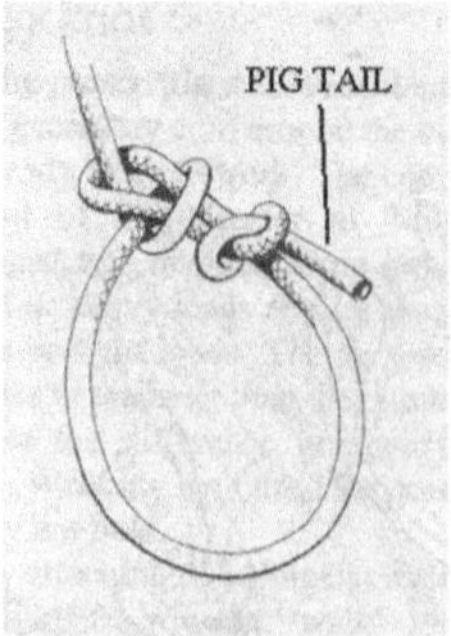

PIG TAIL

h. Stacking (or Flaking). Taking off one wrap at a time from a coil, and letting it fall naturally to the ground.

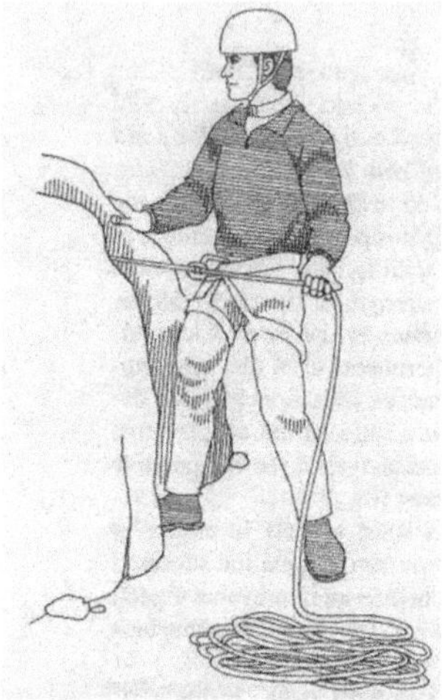

FLAKED OUT ROPEON THE GROUND

i. Dressing the knot. This involves the orientation of all of the knot parts so that they are properly aligned, straightened, or bundled, and so the parts of the knot look like the accompanying pictures. Neglecting this can result in an additional 50% reduction in knot strength.

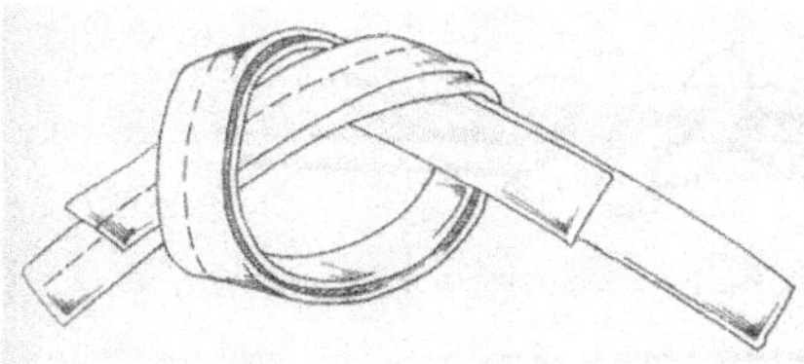

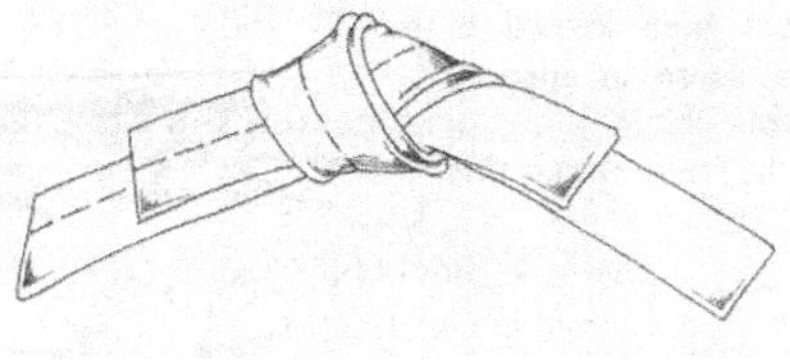

NON-DRESSED DOWN WATER TAPE KNOT DRESSED DOWN WATER TAPE KNOT

j. Setting the knot. This involves tightening all parts of the knot so that all of the rope parts bind upon other parts of the knot so as to render it operational. A loosely tied knot can easily deform under strain and change character.

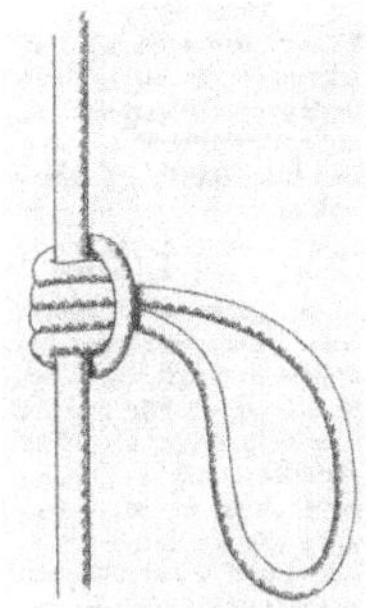

A FULLY SET PRUSIK KNOT

2. **CONSIDERATIONS FOR THE CARE OF ROPE**

a. The rope should not be stepped on or dragged on the ground unnecessarily. Small particles of dirt will get into and through the sheath causing unnecessary wear to the rope within.

b. The rope should never come in contact with sharp edges of any type. Nylon rope is easily cut, particularly when under tension. If a rope must be used around an edge which could cut it, then that edge must be padded or buffed using fire hose if available, or several small sticks.

c. Keep the rope as dry as possible. If it should become wet, hang it in large loops, above the ground, and allow it to dry. A rope should never be dried out by an open flame, or be hung to dry on metal pegs, as this will cause rust to get in the rope thus rendering it unserviceable.

d. Never leave a rope knotted or tightly stretched longer than necessary.

e. When using rope installations, never allow one rope to rub continually against another.

NOTE: With manila ropes this will cause the rope to fray, whereas nylon ropes can melt under the friction that this causes.

f. The rope should be inspected prior to each use for frayed or cut spots, mildew, rot or defects in construction.

g. Mark all climbing ropes at their midpoints to facilitate establishing the midpoint for a procedure requiring you to use the middle of the rope. The rope should be marked with a bright colored adhesive tape.

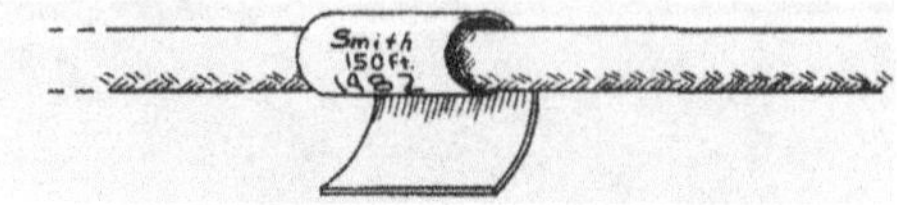

MARKING THE MIDDLE OF THE ROPE

h. The rope should not be marked with paints or allowed to come in contact with oils or petroleum products for these products will weaken it.

i. A climbing rope should NEVER be used for any other purpose except for mountaineering, i.e., towing vehicles.

j. The ends of a new rope or ends caused by a cut should be cut with the rope cutter contained in the MACK and marked with a serial number.

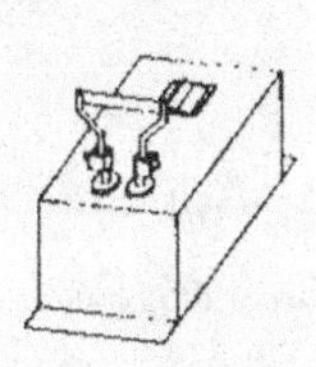

ROPE CUTTER

k. The rope should never be subjected to high heat or flame as this can significantly weaken it.

l. To clean rope use mild soap and rinse thoroughly with water. A rope washer can be used to clean or rinse the rope.

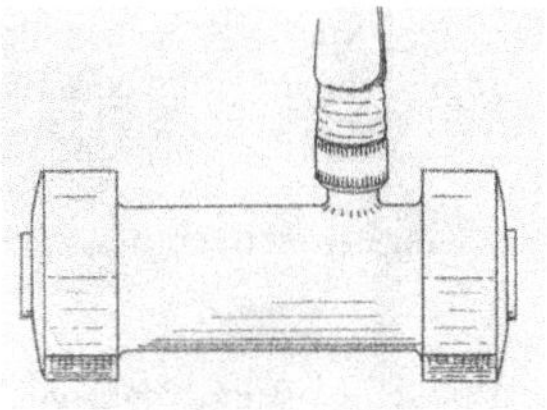

ROPE WASHER

m. When not in use, ropes should be coiled and hung on wooden pegs rather than on nails or any other metal object. They should be stored in a cool place out of the direct rays of the sun.

n. When in areas of loose rock, the rope must be inspected frequently for cuts and abrasions.

o. Always maintain an accurate Rope Log whenever using a rope.

p. Ropes 300-600 foot in length should be Mountain Coiled.

3. **INSPECTION OF ROPE**

a. All ropes have to be inspected before, during, and after all operations. Kernmantle rope is harder to inspect than a laid rope. I.e. green line. The Assault Climber must know what to look and feel for when inspecting a rope. Any of the below listed deficiencies can warrant the retirement of a rope.

(1) Excessive Fraying. Indicates broken sheath bundles or PIC breakage.

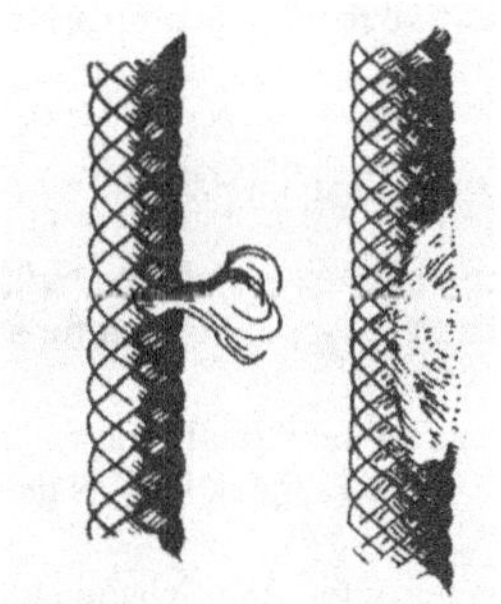

DAMAGED AND EXPOSED ROPE

(2) Exposed Core Fibers. Indicates severe sheath damage. (When you can see the inner core fibers)

(3) Glossy Marks. Signify heat fusion damage, also called a booger.

(4) Uniformity of Diameter / Size. May indicate core damage, noted by an obvious depression (hour glass) or exposure of white core fibers protruding from the sheath (puff).

(5) Discoloration. A drastic change from the ropes original color may indicate chemical change or damage.

(6) Stiffness or Soft Spots. Could signify core damage.

NOTE: Dynamic ropes measuring between 10mm and 12mm are marked at each end of its pigtails with a number "1" indicating that the rope is UIAA approved for single rope lead climbing. Dynamic ropes measuring between 8mm and 9mm are marked at each end of its pigtails with a "1/2" number indicating that two of these ropes are required to conduct a lead climb.

4. **ROPE LOG.** The purpose of the Rope Log is to maintain an accurate record for the use of each rope contained within the Marine Assault Climbers Kit (MACK). Due to the turnover of personnel, and the fact that no one person may have the same rope twice, the rope log is used to ensure the safe use, serviceability, and account of each rope.

a. Serial number. By assigning each rope a serial number, responsible units can determine information about a rope. As soon as the ropes are cut to the desired length for their intended purpose, each rope will be assigned a serial number by the responsible unit. That rope should then be labeled with that serial number in some permanent manner. The best method for this is as soon as the ends of the new rope have been whipped and fused, mark both ends of the rope with the serial number and then dip the ends of the rope, in effect, laminating the serial number to the rope ends.

(1) A rope serial number has five parts. The proper format for a rope serial number is as follows:

1. Type of rope: (S) - Static or (D) - Dynamic.

2. Last two digits of the year the rope was manufactured. Each rope has a shelf life of two years, after that it must not be used for any mountaineering purpose.

3. Four-digit number for that individual rope which is assigned by the responsible unit, and should be assigned sequentially as new ropes are issued out.

4. The length and diameter of the rope. The length may be recorded in feet, and the diameter in millimeters.

5. Responsible unit code. Example: S-96-0001-150/11mm-F2/8

   Example meaning: This rope is a 150 foot, 11mm, static rope, manufactured in 1996. It is the first rope issued by Fox Co 2/8.

b. Recording Information in a Rope Log. Once a new rope has been serialized, the rope log for that rope should be started up. At a minimum, it should contain the following information:

(1) The rope serial number.

(2) Manufacturer. Depending on the manufacturer, ropes of the same type and diameter may vary in tensile strength, stretch factor, and durability.

(3) Date of manufacture. Five years from this shelf life date, ropes are considered to have reached their normal expiration date, and should be destroyed.

(4) Date in service. This is to be recorded for tracking purposes to establish how long a rope is in service. Two years is considered maximum rope life in service.

(5) Each time a rope is used, the using unit is responsible to record how that rope was used and how much use the rope received. Additionally, before checking out a rope and prior to turning it back in, the rope must be inspected by qualified personnel, and initial in the "Inspected By" block of the rope log.

NOTE: ANY TIME A DYNAMIC ROPE IS SUBJECTED TO A FALL FACTOR 2, THAT ROPE SHOULD NOT BE USED AGAIN FOR MOUNTAINEERING.

		DYNAMIC			
ERIAL #					
ANUFACTURER					
ATE OF MANUFACTURE					
ATE IN SERVICE					
	INDICATE THE DATE AND NUMBER OF EACH SPECIFIC USE				
CLIMBS	**FALLS (FACTOR)**	**TOP ROPES**	**RAPPELS**	**REMARKS**	**INSPECTED BY**

ROPE LOG

5. **COILING A ROPE**. There are two types of rope coils frequently used at MWTC. The Mountain Coil and the Butterfly Coil.

a. Mountain Coil. This coil is useful for carrying the rope over a pack or over a climber's shoulder and neck. It can be used for short time storage. The mountain coil can be tied in the following manner:

(1) Sit down with your leg bent at a 90-degree angle, heel on the deck. Starting at one end, the rope is looped around the leg in a clockwise fashion, going over the knee and under the boot sole until the entire rope is coiled.

(2) If coiling a 150-foot rope, use only one leg and offset the other, when coiling a 300-foot rope, use two legs and keep them together.

(3) With the starting end of the rope, form a 12-inch bight on the top of the coils.

(4) Uncoil the last loop and along the top of the coils, wrap 4-6 times towards the closed end of the bight.

(5) The end of the rope being wrapped is then placed through the closed end of the bight.

(6) The running end of the bight is then pulled snuggly to secure the coil.

(7) To prevent the coil from unraveling, the two pigtails are tied together with a square knot.

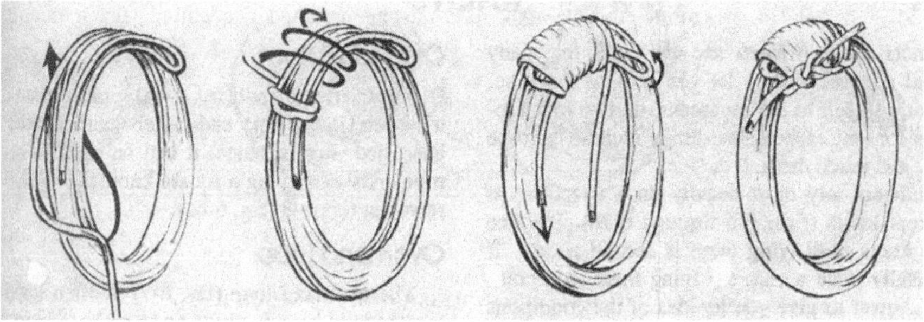

MOUNTAIN COIL

b. Butterfly Coil. This method is used for carrying a rope when the individual needs to have maximum use of his upper body, (i.e. while climbing), without the encumbrance of a large rope coil hanging across his chest.

(1) Coiling the Butterfly Coil

(a) Step 1: Find the middle of the rope, then form a three foot bight laying both ropes in the upraised palm at the two foot point.

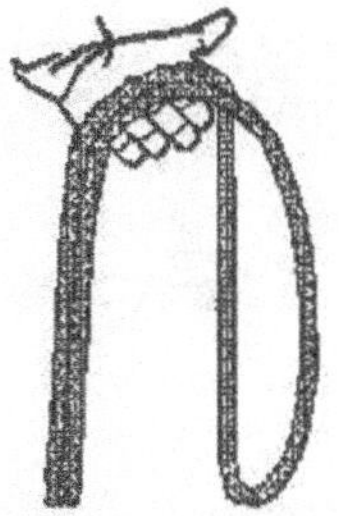

STARTING THE BUTTERLFY COIL

(b) Step 2: Form another two-foot bight with the running end. Place the rope at the two-foot bight along side on top of the original bight ensuring the running end is on the same side as the original bight.

(c) Step 3: Continue making two foot bights, laying them alternately into your palm until there is only six to eight feet remaining. At that point, begin wrapping the two pigtails horizontally four to six times at the mid way point of the ropes in a bight from bottom to top.

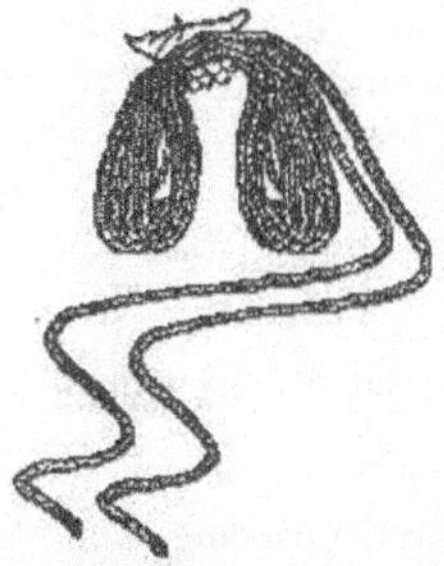

TWO FOOT BIGHTS ON BOTHS SIDES WITH 6 TO 8 FEET REMAINING

(d) Step 4: After completing your wraps, form a bight with the remaining pigtail and then thread it underneath your palm and upwards to one-foot above the coiled rope.

(e) Step 5: With the remaining pigtail, thread it through the one-foot bight in step four.

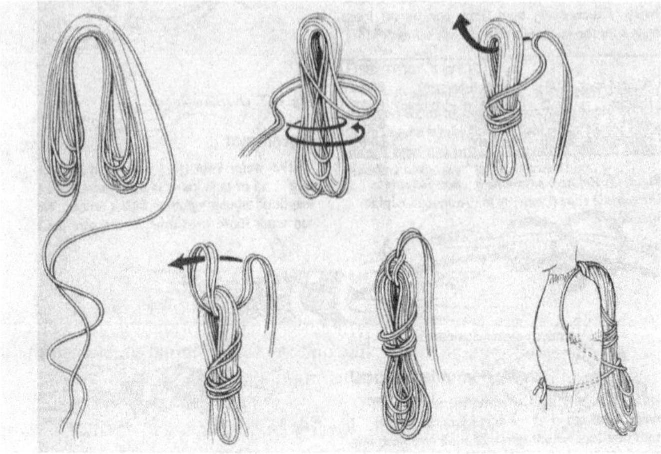

TYING AND CARRYING THE BUTTERFLY COIL

(2) Carrying the Butterfly coil. Separate the running ends, placing the coil in the center of the back of the carrier, then fun the two ends over his shoulders so as to form shoulder straps. The running ends are then brought under the arms, crossed in the back over the coil, brought around the body of the carrier and tied off with a square knot at his stomach.

6. **ROPE THROWING**. To insure that the rope will not get tangled when deployed, certain steps must be taken.

a. With a stacked rope, anchor off the standing end.

b. Take the opposite end of the rope and make 6-8 coils and place them in your strong arm. These wraps will serve as a throwing weight that you can aim.

c. 10-15 feet from the strong-arm coils create a second set of 6-8 wraps and place them in your weak arm.

d. From the edge of the cliff, sound off with the command "STAND-BY FOR ROPE". Just before you release the coils, sound off with the command "ROPE". At that time drop the weak arm coils from the cliff.

e. While taking aim, throw your strong-arm coils overhand or sidearm hard enough to hit your intended target.

f. If the throw was misdirected due to wind, tree, etc., reorganize and attempt to re-deploy the rope.

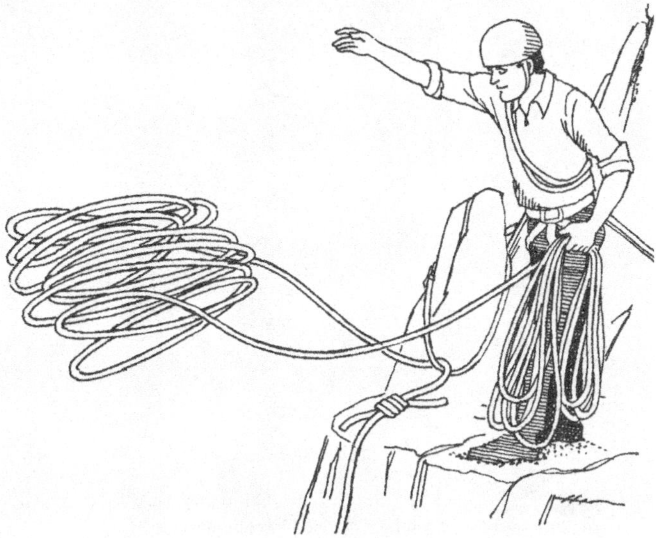

THROWING THE ROPE

6. **MOUNTAINEERING KNOTS.**

a. **Class I** – *End of the Rope Knots*

(1) Square Knot. Used to tie ends of two ropes of equal diameter together. It should be secured by overhand knots on both sides of the square knot.

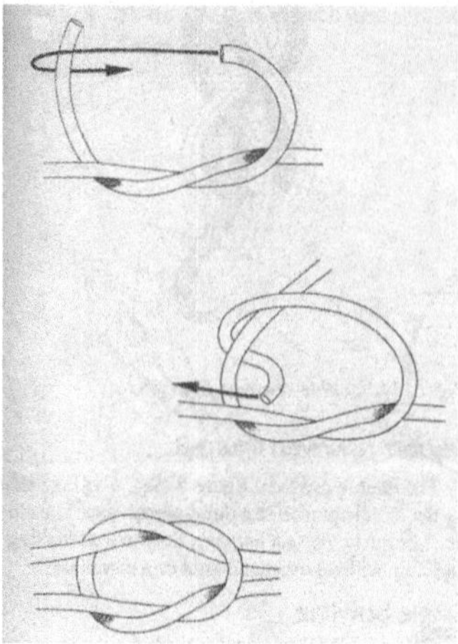

SQUARE KNOT

(2) Double Fisherman's Knot. It is a self-locking knot used for tying two ropes of equal diameter together. It can be tightened beyond untying.

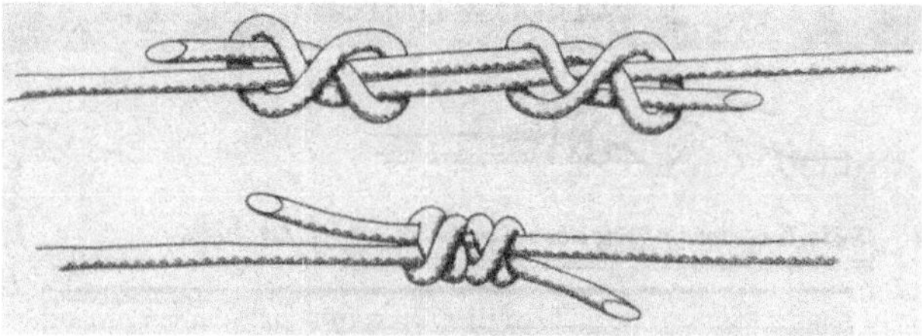

DOUBLE FISHERMAN'S KNOT

(3) Water/Tape Knot. Used to secure webbing or tape runners. It is constructed by tying an overhand knot (without twists) in one end of the tape, and threading the other end of the tape through the knot from the opposite direction. After the knot is dressed down, each pigtail should be a minimum of two inches long.

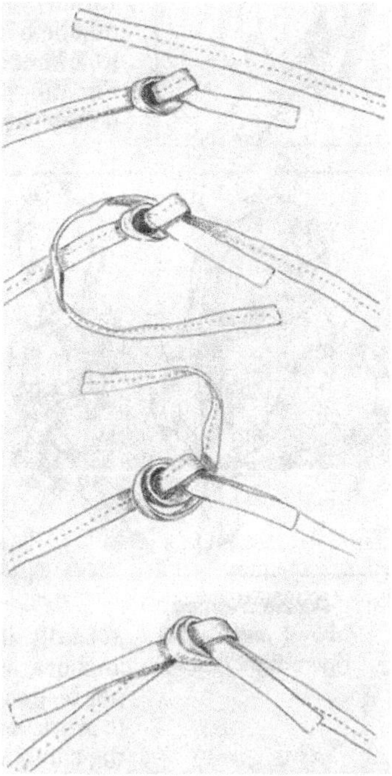

WATER TAPE KNOT

(4) Double Sheet Bend. Used to tie the ends of two or more ropes of equal or unequal diameter together.

SHEET BEND

b. **Class II** - *Anchor Knots*

(1) Bowline. Used to tie a fixed loop in the end of a rope. This knot is always tied with the pigtail on the inside and secured with an overhand knot.

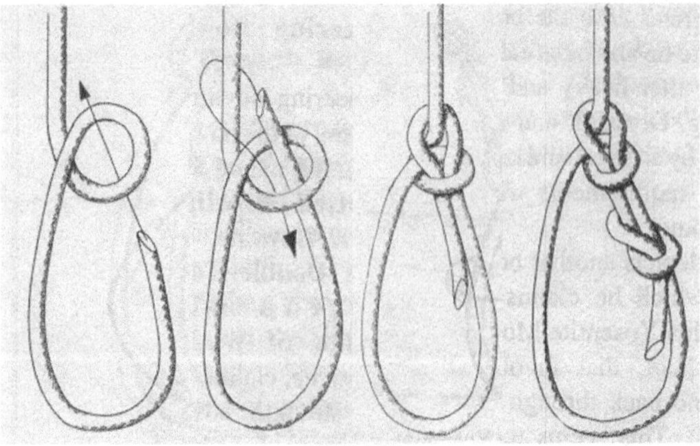

BOWLINE

(2) Round Turn with Two Half Hitches or a Bowline. A loop which runs around an object in such a manner as to provide 360 degree contact and may be used to distribute the load over a small diameter anchor. It will be secured with two half hitches or a bowline.

ROUND TURN AND TWO HALF HITCHES AND

ROUND TURN AND A BOWLINE

(3) Clove Hitch. This knot is an adjustable hitch. It could be considered a middle-of-the-rope anchor knot at the end-of-the-rope when used in conjunction with a bowline or round turn and two half hitches.

(1) Around-the-object Clove Hitch.

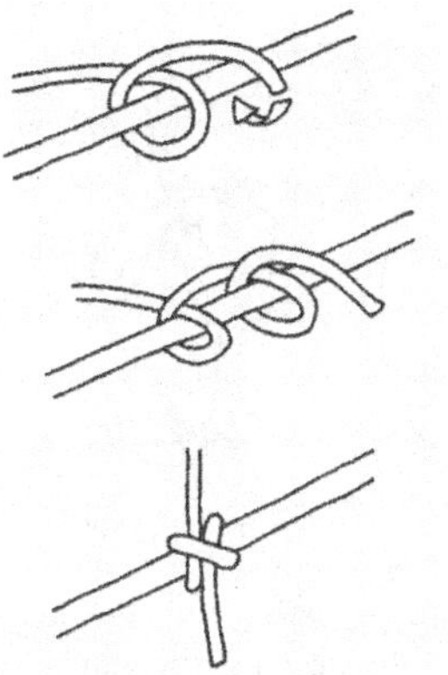

AROUND THE OBJECT CLOVE HITCH

(2) Over-the-object Clove Hitch.

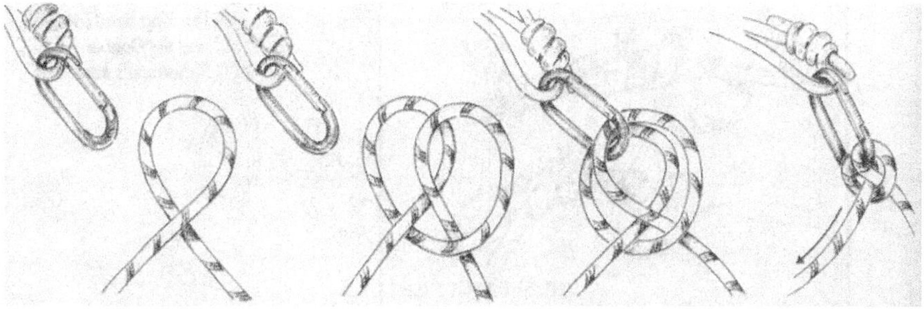

OVER THE OBJECT CLOVE HITCH

c. **Class III** - *Middle-of-the-rope*

(1) Figure-of-Eight-Loop. This is a strong knot that can be readily untied after being under load.

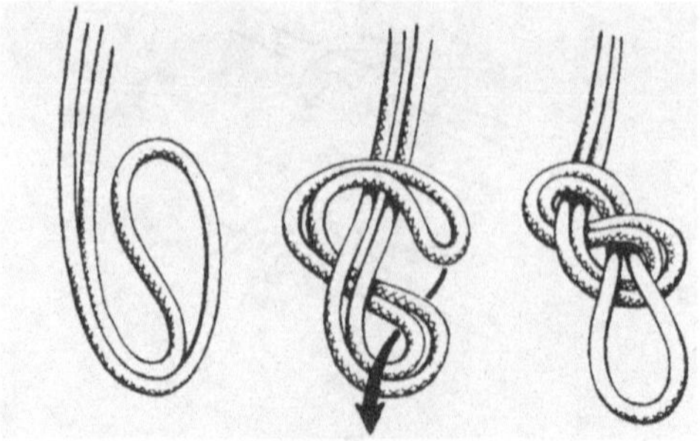

FIGURE 8 LOOP

(2) Double Figure of Eight Loop. The double Figure of Eight Loop is a strong knot and the double loop reduces the wear and strength loss from the rope bending around the carabiner by splitting the load between the two loops.

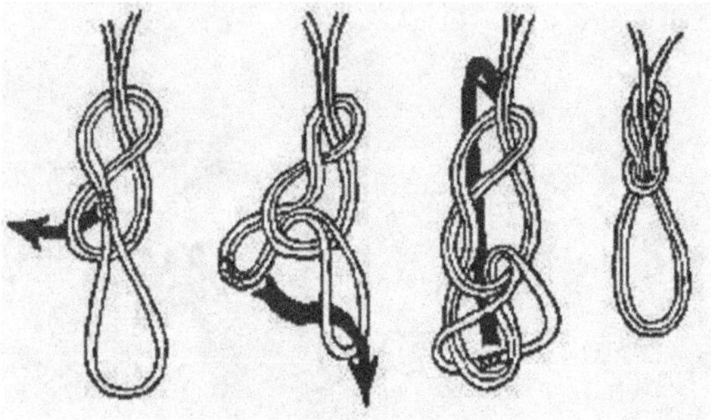

DOUBLE FIGURE 8 LOOP

(3) Two Loop Bowline. This knot can be used to construct a self-equalizing belay or to tie the middle person in on three people on a rope.

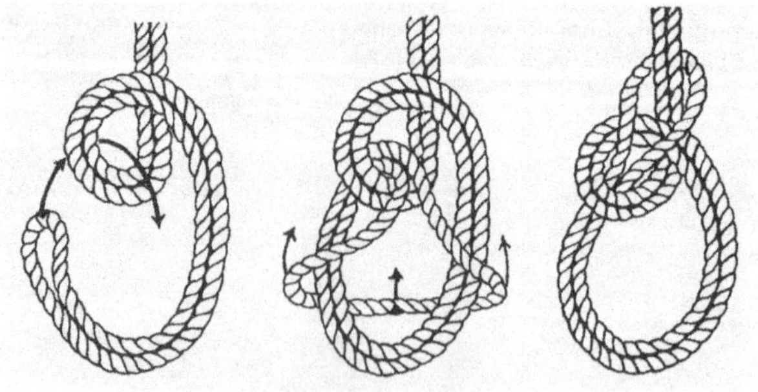

TWO LOOP BOWLINE

d. **Class IV** - *Special Knots*

(1) Prusik Knot. This knot functions by introducing friction that can be alternately set and released. For best results, tie the knot with a smaller diameter cord on a larger diameter cord. If slippage occurs, more wraps maybe used.

(a) Middle-of-the-Rope Prusik. This knot is created with an endless loop also known as a Prusik Cord. Do not tie this knot with tape due to less friction.

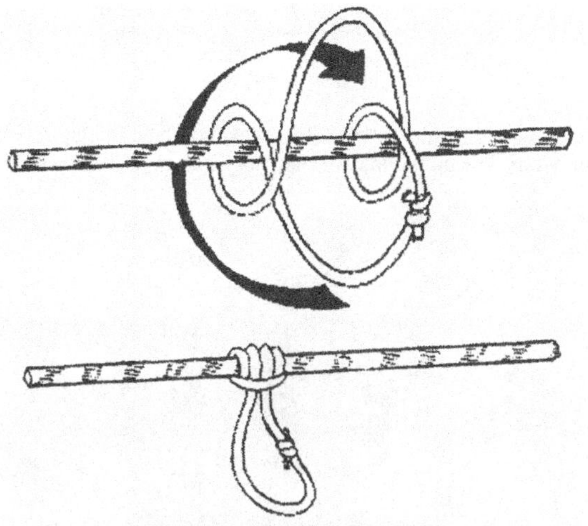

MIDDLE OF THE LINE PRUSIK

(b) End-of-the-Rope Prusik. This knot is always secured with a bowline. Do not tie this knot with tape due to less friction.

END OF THE LINE PRUSIK

(c) French Prusik. This can be constructed with a Prusik Cord or a single strand of cord with figure-of-eight loops on each end.

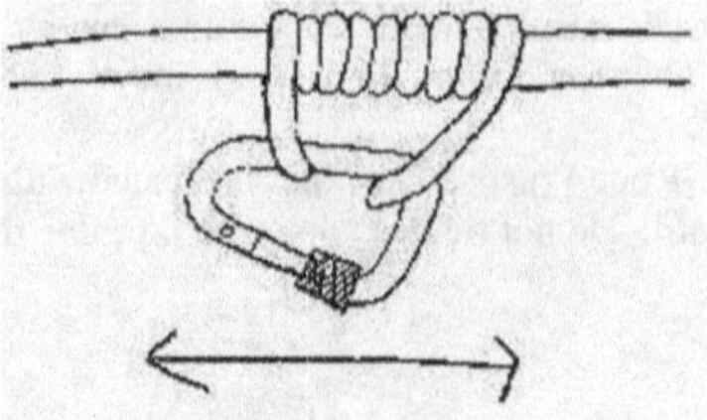

FRENCH PRUSIK

(2) Retraced Figure of Eight. Used to tie the end of the climbing rope into a harness or swammi wrap. The pigtail may or may not be secured with an overhand.

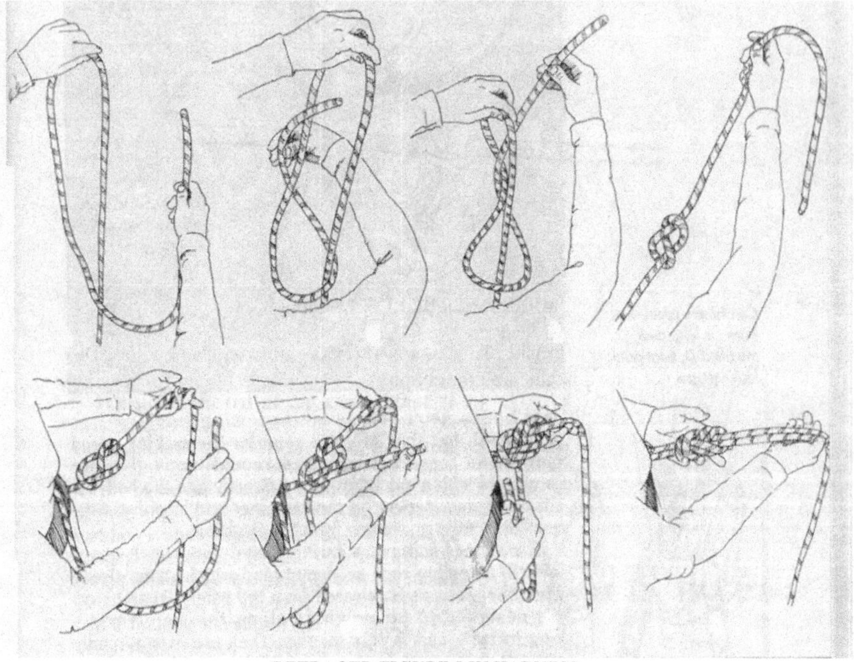

RETRACED FIGURE 8 IN HARNESS

(3) Directional Figure-Of-Eight. When tied and tensioned is applied to the standing and running ends of the rope, the knot will not pull apart. The loop will point toward the direction of pull.

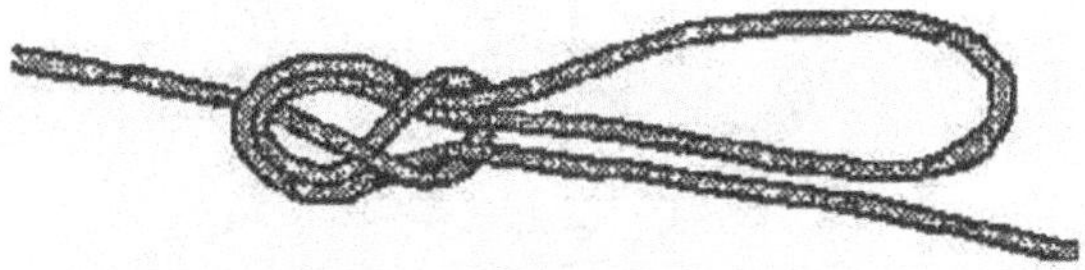

DIRECTIONAL FIGURE 8

(4) Slip Figure 8. This knot is used for retrievable anchors and fixed rope installations for the ease of untying the knot.

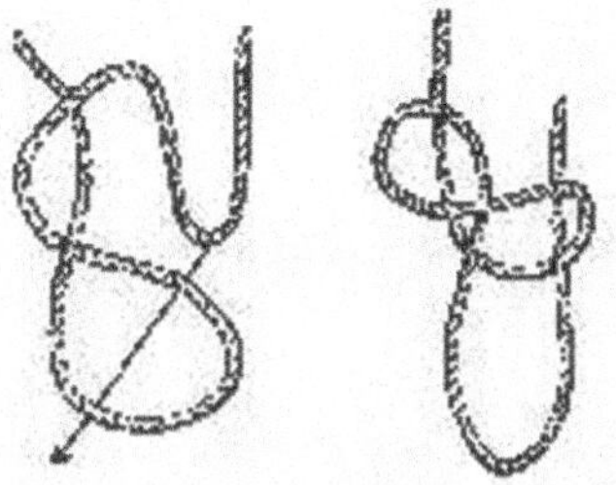

SLIP FIGURE 8

(5) Kliemheist. This is a friction knot.

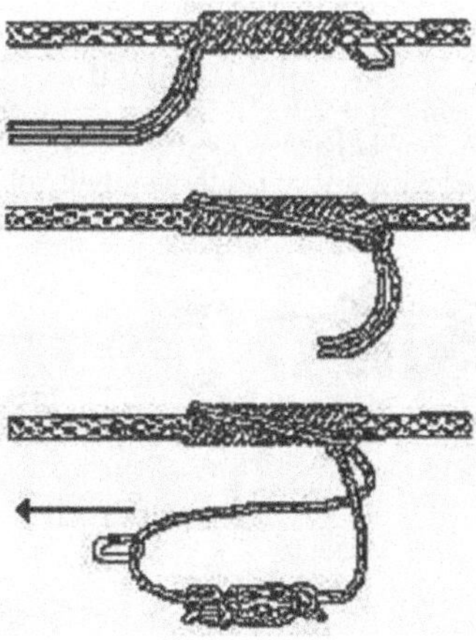

KLEIMHEIST

(6) Overhand Knot. Can be used to secure primary knots on itself.

OVERHAND

(7) M nter Hitch. This is a simple hitch in the rope that is clipped into a carabiner to put friction on the line.

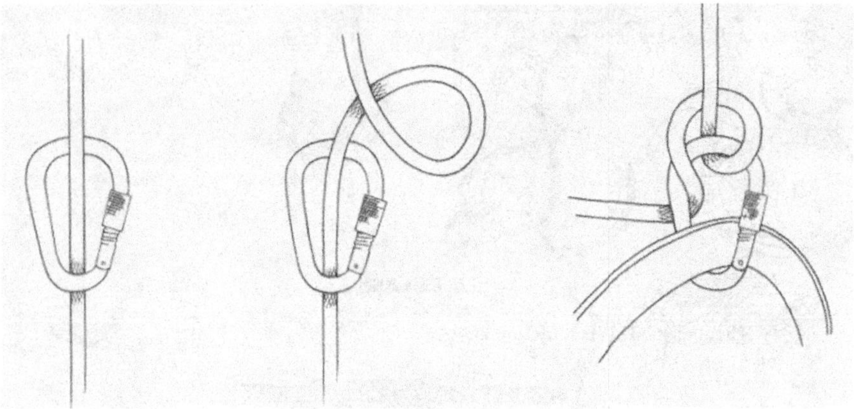

MÜNTER HITCH

(8) Timber Hitch. A Timber Hitch is used to fix a rope to a pole or equivalent for hoisting or towing purposes. It has the capability of casting off easily.

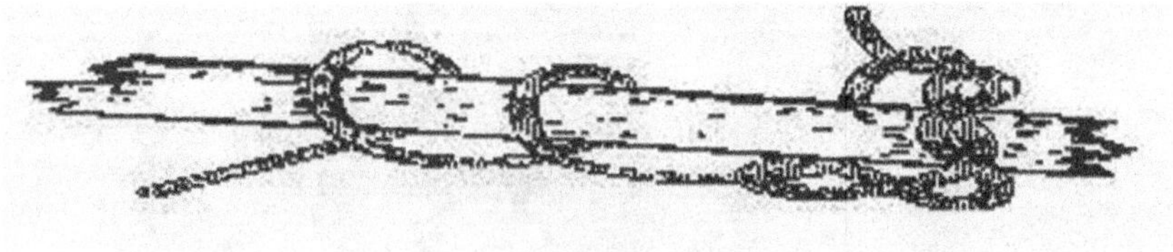

TIMBER HITCH

(9) Mariner's Hitch. This knot's advantage over the prusik is that it can be released under load.

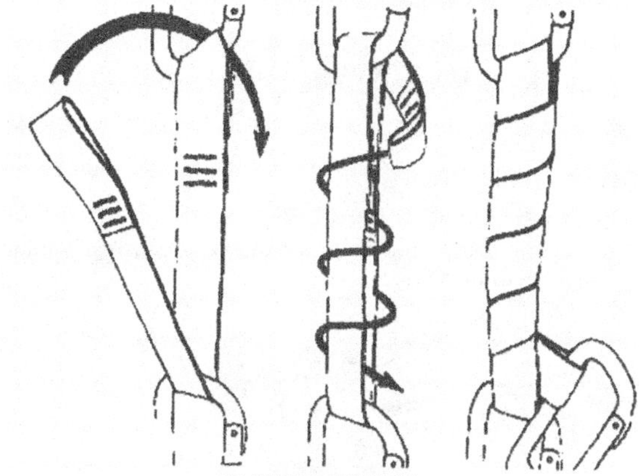

MARINER'S HITCH

(10) Krägur Knot. This friction knot is used to prevent unnecessary slippage when tied onto a rope of the same diameter.

KRÄGUR KNOT

(11) Rappel Seat. This is used as an expedient support harness for rappelling, crossing rope bridges, etc. It is constructed as follows:

(a) Center the sling rope on the left hip.

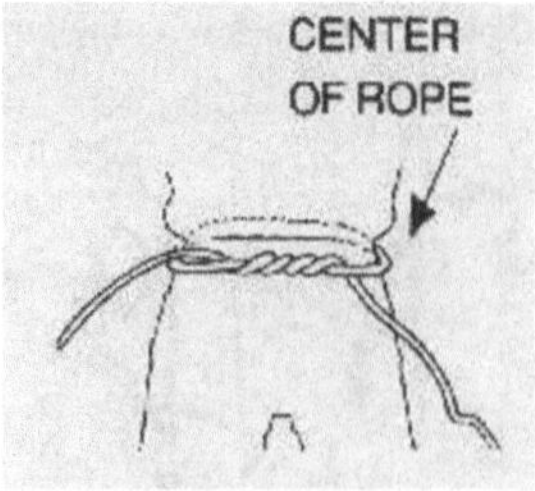

(b) Wrap the sling rope around the waist and tie at least one half hitch around itself (preferably two) in the front.

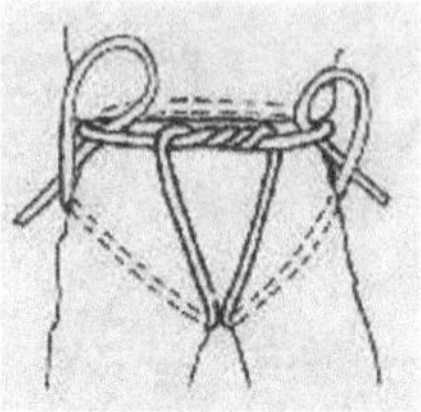

(c) Bring the running ends down through the legs, up over the buttocks, and over the original waist wrap, down between the waist wrap and the waist, and over itself, forming a bight. Cinch this up tightly.

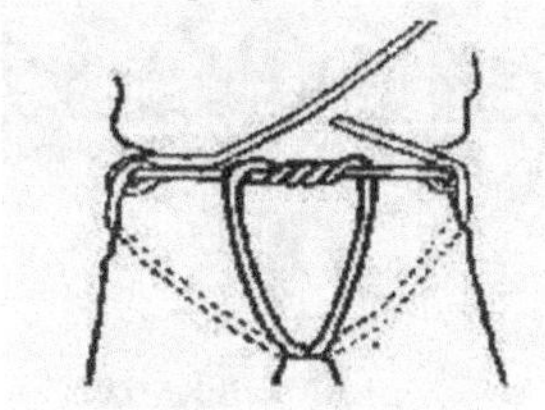

(d) Now tie a square knot with two overhand knots on the left hip ensuring that the overhand knots encompass all the wraps.

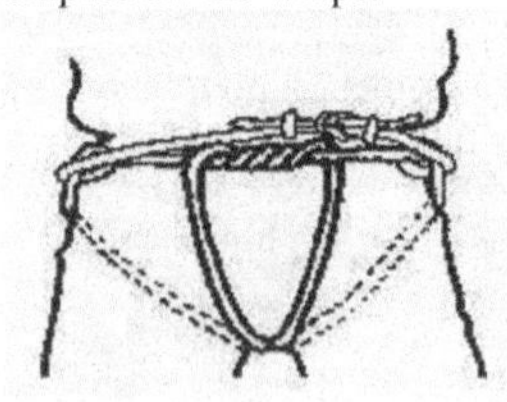

(e) Tuck any excess rope into a pocket.

(12) Bowline on a coil. Used by the first and last men on a climbing rope to tie into the rope. An overhand knot is used behind the knot. It distributes the force of a fall over a larger area of a climber's waist and is preferable to a single bowline around the waist. The bowline on a coil can also be used to take up excess rope. The bowline on a coil should have 4-6 wraps around the waist.

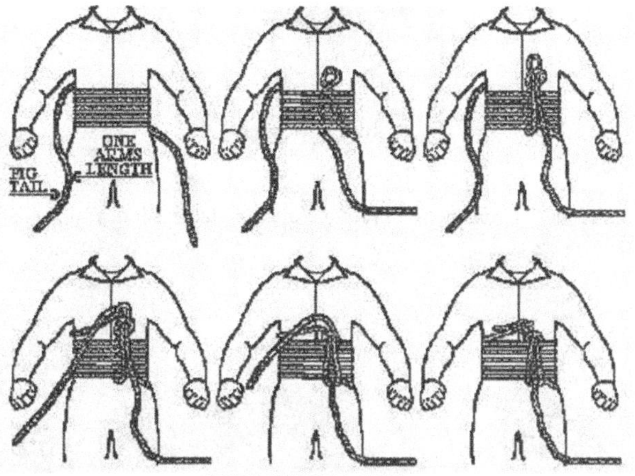

BOWLINE ON A COIL

(13) Swami wrap (Swami belt). The swami wrap (four to six wraps of rope or nylon webbing tied around the waist) popular with rock climbers who wanted to increase the working length of their climbing ropes. It also offers some improvement over the bowline on the coil in distributing the forces sustained in a fall over a larger area of the midsection of the body, but does not eliminate the problem of suffocation when hanging. The Swami wrap is constructed as follows:

(a) Take a sling rope and find the middle.

(b) Place the middle of the sling rope in the small of your back.

(c) Bring both pigtails to the front and continue to wrap around your body until both pigtails are about 15-20 inches long.

(d) Tie the pigtails on the left side with a square knot and two overhand knots.

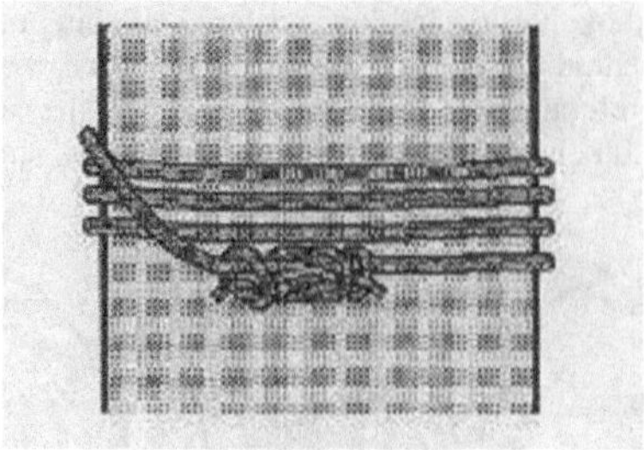

SWAMMI WRAP

(14) Three Loop Bowline. This knot will provide three bights, each of which can be adjusted against the others.

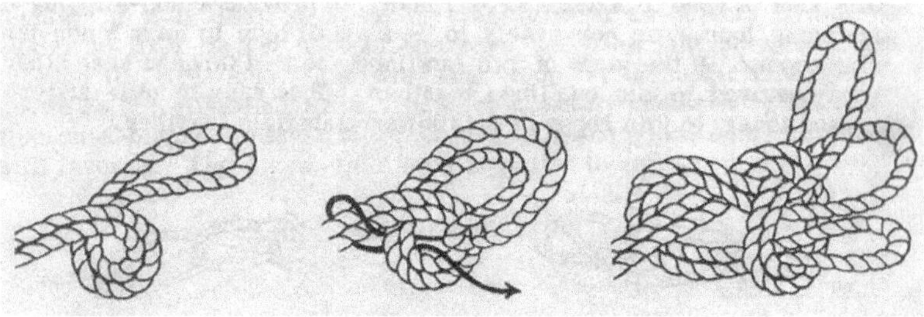

THREE LOOP BOWLINE

7 **KNOT STRENGTH.** Rope, cordage and webbing is strongest when loaded in a straight line. When bending a rope or web to create a knot, the strength of the rope is reduced. All knots should be dressed properly for maximum effective use.

KNOT	RELATIVE STRENGTH OF KNOT
No Knot	100%
Figure 8	75-80%
Bowline	70-75%
Double Bowline	70-75%

Double Fisherman	65-70%
Water Tape Knot	60-70%
Clove Hitch	60-65%
Overhand Knot	60-65%
Square Knot	45%

1. **KNOT TESTING TIME LIMIT.** The following times must be met to pass the knot tying portion of this course.

KNOT	UNIT OPERATIONS TIME LIMIT	SML, ACC TIME LIMIT (BLINDFOLDED)
Square knot	60 Seconds	30 Seconds
Double Fisherman's Knot	60 Seconds	30 Seconds
Water/Tape Knot	60 Seconds	30 Seconds
Round Turn and a Bowline	60 Seconds	30 Seconds
Round Turn and Two Half Hitches	60 Seconds	30 Seconds
Clove Hitch (around the object)	60 Seconds	30 Seconds
Munter Hitch	60 Seconds	30 Seconds
Slip Figure 8	60 Seconds	30 Seconds
Figure 8 Loop	60 Seconds	30 Seconds
Directional Figure 8	60 Seconds	30 Seconds
End of the Line Prusik	90 Seconds	45 Seconds
Retrace Figure 8	90 Seconds	45 Seconds
Rappel (Swiss) Seat	2 Minutes	60 Seconds

UNITED STATES MARINE CORPS
Mountain Warfare Training Center
Bridgeport, California 93517-5001

SMO
05/02/01

STUDENT OUTLINE

NOMENCLATURE AND CARE OF MOUNTAINEERING EQUIPMENT

LESSON PURPOSE. The purpose of this period of instruction is to introduce you to the types of equipment used here at MWTC and how to care for it so as to prevent its untimely failure. This lesson relates to all climbing and installation work that you perform here at MWTC.

OUTLINE

1. **ROPES.** All ropes used in the military must meet UIAA standards or U. S. Federal Test Standard 191A. Most ropes have a 5-year shelf life and maximum 2-year service life.

a. Static. Black in color.

(1) Construction. Kernmantle

(2) Minimum tensile strength. 7500 lbs.

(3) Maximum elongation. 1.5%

(4) Diameter. 11mm

(5) Sizes.

(a) 165 ft ± 5 ft

(b) 300 ft ± 10 ft

(6) Usage. Rescue operations and bridging where a low amount of elongation is desirable under a working load.

b. Dynamic. Olive Drab in color.

(1) Construction. Water-resistant treated Kernmantle to reduce friction.

(2) Minimum tensile strength. 6500 lbs.

(3) Maximum elongation. 6%

(4) Diameter. 10.5mm and 11mm

(5) Size. 165 ft ± 5 ft.

(6) Usage. For lead climbing/party climbing.

c. Maxim Dry Rope. Olive Drab or Multi-Colored.

(1) Construction. Water-repellent treated Kernmantle.

(2) Minimum tensile strength. 3472 lbs.

(3) Maximum elongation. 6%

(4) Diameter. 9mm

(5) Size. 150 ft ± 5 ft

(6) Usage. For glacier travel/ice climbing

d. Gold Line II.

(1) Construction. Eight strand braided nylon plymor.

(2) Minimum tensile strength. 4500 lbs.

(3) Maximum elongation. 20%

(4) Diameter. 11 mm

(5) Size. 300 ft or 600 ft spools

(6) Usage. Sling Ropes and litters only.

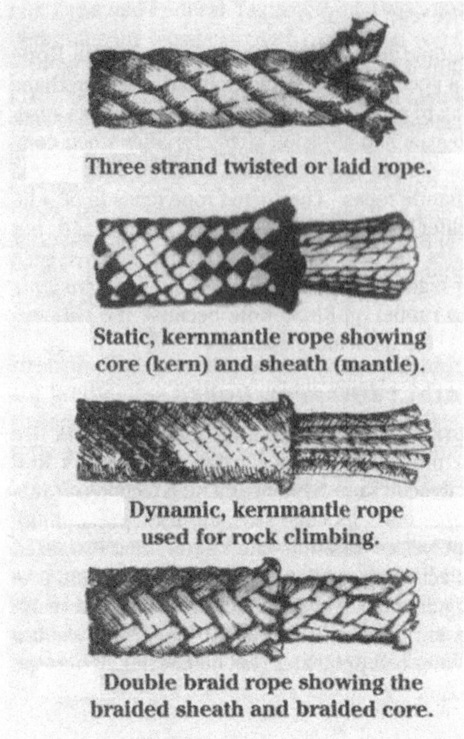

INSIDES AND OUTSIDE SHEATHS OF VARIOUS TYPES OF ROPES

NOTE: Sling ropes are made from 15 foot lengths of plymor or dynamic rope ONLY. Twenty-five foot practice coils should be constructed with static rope, but dynamic rope can be used.

2. ADVANTAGES/DISADVANTAGES

a. Advantages of Nylon Rope

(1) High strength to weight ratio.

(2) Good energy absorption in dynamic ropes.

(3) Flexible.

(4) Rot resistant, not affected by frost.

b. Disadvantages of Nylon Rope

(1) Low melting point. Nylon fuses at 400°F and melts at 480°F.

(2) Susceptible to abrasions and cuts.

(3) Affected by chemicals and light.

c. Advantages of Manila Rope

(1) Easily gripped.

(2) Hard wearing.

(3) Does not deteriorate in heat.

d. Disadvantages of Manila Rode

(1) Heavy, kinks, especially when wet. Absorbs water and swells.

(2) Burns at +300°F.

(3) Edible by rodents.

3. GENERAL INFORMATION

a. Nylon rope stretches under tension and will rupture at between 30% and 70% elongation, depending on construction.

b. Nylon rope loses as much as 30% strength when wet.

c. Temperatures as low as 250°F will damage a nylon rope.

4. NYLON WEBBING

a. The type of nylon webbing available is tubular. Tubular nylon webbing is very strong and flexible. All rules that apply to nylon rope apply to tubular nylon webbing. The size of nylon webbing used is:

(1) 1 inch tubular nylon. Tensile strength approximately 4,000 - 4,500 lbs., depending on the manufacturer.

b. Pre-sewn Spectra Runners. Tensile strength approximately 5,500 lbs.

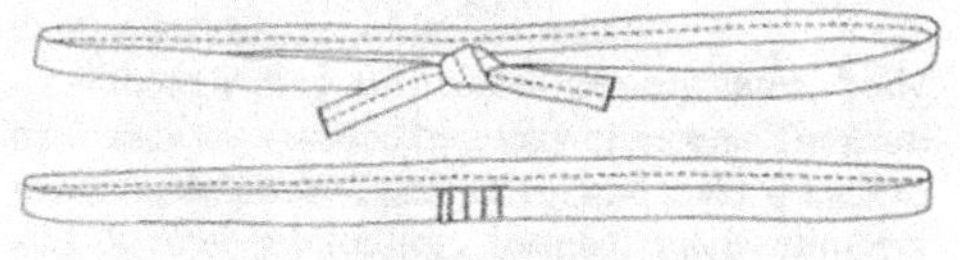

NYLON WEBBING

NOTE: These are minimum strengths. Some manufactures make even stronger webbing.

1. **CARABINERS**. Also commonly known as snap links. There are two different types of carabiners frequently used at MWTC, the steel locking carbine and the non-locking aluminum carabiners.

a. Steel Locking Carabiner. There are two different sizes of carabiners:

(1) Steel Locking Stubai 82.

(2) Large Locking "D" Stubai 85.

b. Aluminum Non-Locking Carabiner.

(1) The standard "D" non-locking carabiner with a tensile strength of 2000 lbs. or more.

2. **NOMENCLATURE OF A CARABINER**.

a. There are four parts to a non-locking carabiner. They are as follows:

(1) Gate.

(2) Gate pivot pin.

(3) Locking pin

(4) Body.

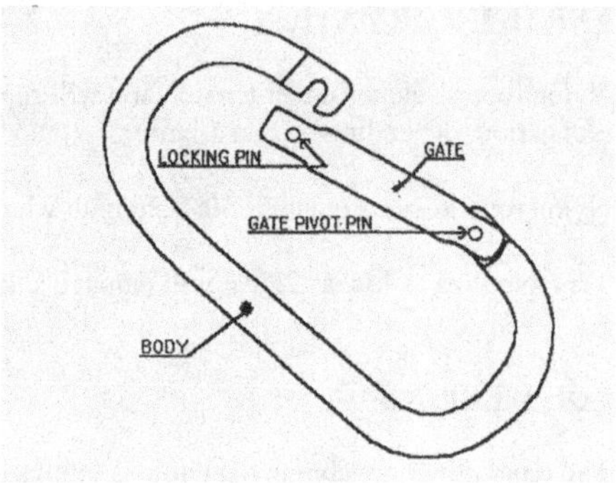

STEEL NON-LOCKING CARABINER

b. There are five parts to a locking carabiner. They are as follows:

(1) Gate

(2) Gate pivot pin

(3) Locking notch

(4) Locking nut

(5) Body

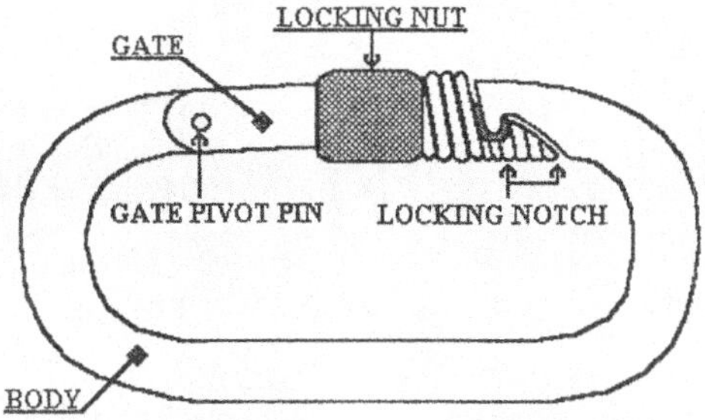

STEEL LOCKING CARABINER

3. **CARE, MAINTENANCE AND SERVICEABILITY OF A CARABINER**. The following inspections and maintenance should be performed on a routine basis:

a. Care.

(1) Avoid dropping of a carabiner onto a hard surface, hairline factures could occur.

(2) Keep the carabiners off of the bare ground because dirt and grit can get into the working parts.

(3) If the carabiner has fallen from a significant height, it should be retired.

b. Maintenance.

(1) Remove all dirt, moisture and grime after use.

(2) Lubricate with tri-flow graphite and wipe off with a clean cloth.

c. Serviceability Checks.

(1) The gate snaps shut with no friction and with no gap between the locking pin and locking notch.

(2) There is no excessive side to side movement of the gate

(3) The pivot pin is tight.

(4) That the spring-loaded gate functions properly.

(5) The locking pin is tight.

(6) The locking nut travels freely and locks securely.

(7) There are no cracks or flaws in the metal.

NOTE: The weakest part of a carabiner is the gate. If an engraver is used to mark a carabiner, it should be applied to the gate and not the load bearing side.

UNITED STATES MARINE CORPS
Mountain Warfare Training Center
Bridgeport, California 93517-5001

WMC
04/11/02

LESSON PLAN

BALANCE CLIMBING

1. LEARNING OBJECTIVES

a. TERMINAL LEARNING OBJECTIVE. In a summer mountainous environment, execute balance climbing, in accordance with the references. (WMC)

b. ENABLING LEARNING OBJECTIVES.

(1) With the aid of references, state orally the safety requirements for balance climbing, in accordance with the references. (WMC.C3a)

(2) With the aid of references, state orally the individual preparations for a balance climb, in accordance with the references. (WMC.C3b)

(3) In a summer mountainous environment, execute the duties of a spotter for a balance climb, in accordance with the references. (WMC.C3c)

(4) In a summer mountainous environment, execute the commands used between the climber and the spotter during a balance climb, in accordance with the references. (WMC.C3d)

(5) In a summer mountainous environment, execute each type of hold used in balance climbing, in accordance with the references. (WMC.C3e)

(6) Without the aid of references and given the acronym "CASHWORTH", describe orally the nine considerations for proper body position while climbing, in accordance with the references. (WMC.C3f)

OUTLINE

1. **SAFETY PRECAUTIONS**. There are two safety precautions that always apply to balance climbing. They are as follows:

C3

a. Never climb more than 10 feet above the ground. By this it is meant that the climber's feet are never more than 10 feet above the ground.

b. A spotter is required for all balance climbs.

SPOTTER WATCHING THE CLIMBER

2. **INDIVIDUAL PREPARATIONS**. Prior to beginning a balance climb there are seven things that the climber must do to prepare himself. They are as follows:

a. Helmet with a serviceable chinstrap must be worn.

b. Sleeves rolled down to give hand and arm freedom of movement. Blouse tucked in to your trousers. In case of fall, it may catch on a rock and cause you to flip over sideways.

c. All watches, rings, and jewelry must be removed before climbing.

d. Gloves will not be worn, as they can slip, and also give a false feel for the rock.

e. Unblouse trousers, if they restrict movement.

f. Soles of boots clean and dry.

g. Select route where vegetation is minimal. Never use vegetation for hand or foot holds.

3. **DUTIES OF THE SPOTTER**. The spotter is the balance climber's partner who, rather than climbing himself, acts as the safety man for the climber during the climb. The five duties of the spotter are as follows:

a. Positions himself directly behind the climber before the climb starts.

b. Maintains his position facing the cliff, directly below the climber and approximately 3-4 feet away from the base of the cliff, for the duration of the climb. He will move diagonally as necessary to remain below the climber.

c. The spotter will stand with his feet shoulder width apart and arms ready to stop the climber if he falls.

d. If the climber falls, the spotter will not "catch" him; he will prevent the climber from falling further down the hill. He will do this by pushing the climber towards the base of the cliff, thereby preventing him from tumbling backwards.

e. At no time will the spotter allow anyone to come between himself and the face of the cliff while a balance climb is taking place. He will require anyone who wants to pass by his position to go behind him.

4. **SPOTTING AND CLIMBING COMMANDS**. The following are the commands used by both the spotter and climber.

COMMAND	GIVEN BY	MEANING
"Last Name, climbing."	Climber	I am ready to climb.
"Climb Climber."	Spotter	I am ready to spot you.
"Last Name, off Climb."	Climber	I am off the climb.
"ROCK!"	Climber or Spotter.	A rock has been knocked off the rock face and is falling.
"FALLING!"	Climber	I am Falling.

a. If the command "ROCK" is given, all personnel in the vicinity will take the following action:

(1) If close to the cliff face, move against the cliff face with your face against the cliff face and your hands between you and the cliff face.

(2) If not close to the cliff face, look up to locate the rock and avoid it.

5. **ACTIONS IF FALLING**. If, while making a balance climb, the climber feels himself slipping and beginning to fall, he will take the following action:

a. Sound the command "falling".

b. Push himself away from the rock face.

c. Maintain proper body position as follows:

(1) Head up.

(2) Hands out toward the rock.

(3) Body relaxed.

(4) Feet kept below the body, slightly apart.

(5) Ensure you face the cliff face as you fall.

6. **TYPES OF HOLDS**. There are five basic holds that are used in balance climbing. They are as follows:

a. Push Holds.

(1) Most effective when hands are kept low.

(2) Often used in combination with a pull hold.

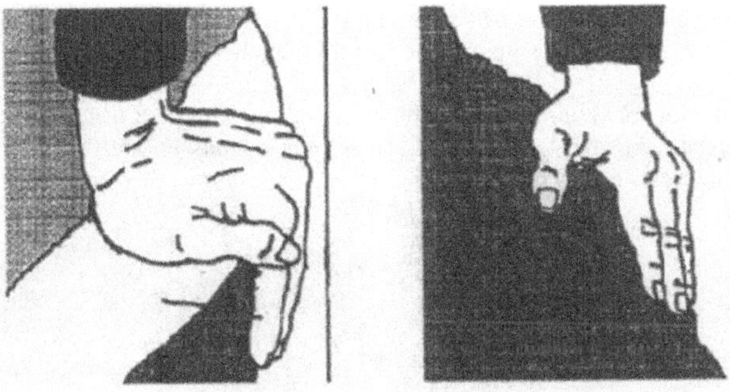

TYPES OF PUSH HOLDS

b. Pull Holds.

(1) The easiest hold to use and, consequently, often overused.

(2) Can be effective on small projections.

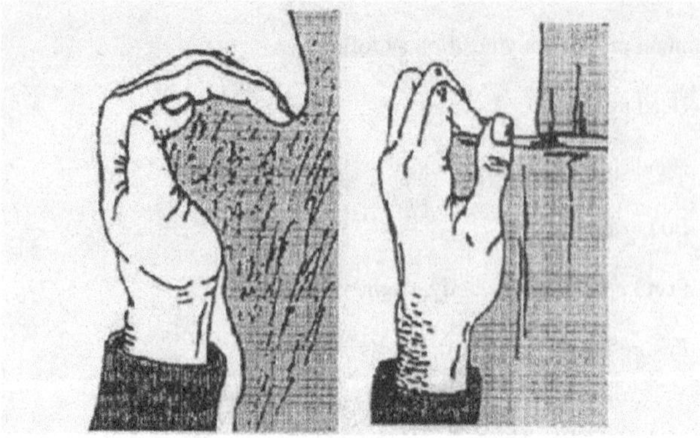

TYPES OF PULL HOLDS

c. Foot Holds

(1) Feet should be positioned with the inside of the foot to the rock.

(2) Use full sole contact as much as possible.

(3) Avoid crossing your feet. If you must cross your feet use a change step. A change step is a method of substituting one foot for the other foot on the same foothold.

(4) Making maximum use of footholds, climbing with your feet, is an effective means of conserving your body strength, since your leg muscles are stronger than your arm muscles.

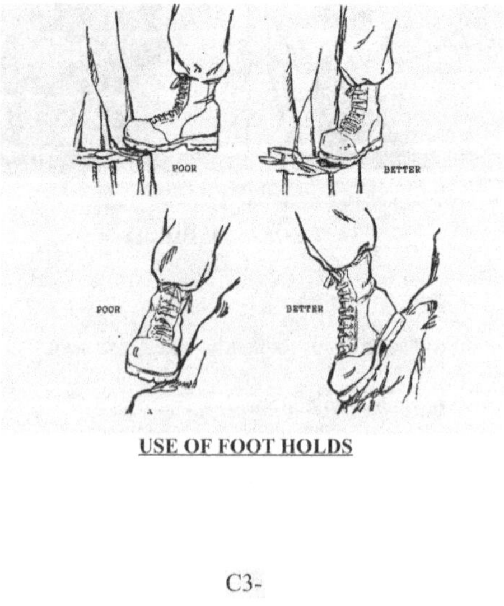

USE OF FOOT HOLDS

d. Friction Holds. A friction hold is anytime you are relying on the friction of your foot or hand against the face of the rock for traction, rather than pushing/pulling against a projection on the face of the rock.

(1) It is a type of hold that feels very insecure to an inexperienced climber.

(2) The effectiveness of this type of hold is dependent upon many things i.e., type, condition and angle of the rock face, type of boot soles, confidence, etc.

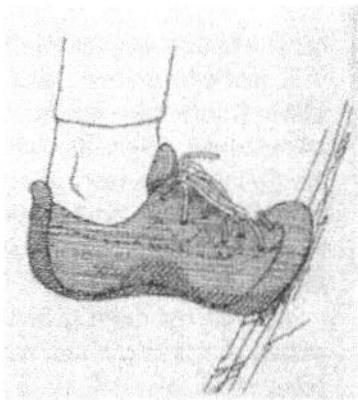

FRICTION USED IN SMEARING

e. Jam Holds. This type of hold involves jamming/wedging any part of your body or your entire body into a crack/opening in the rock.

(1) An important consideration is that you do not jam such that you cannot free that portion of your body after you complete the move. This sounds like a ridiculous statement; however, you must remember that after you complete your move, you may be withdrawing the portion of your body that you used from a different angle than you inserted it.

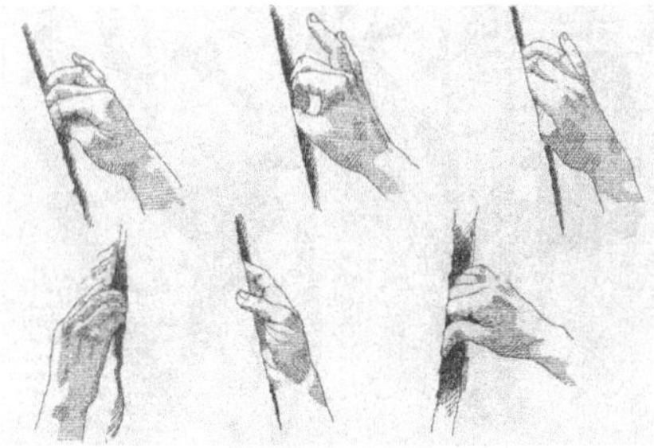

FINGER JAMS

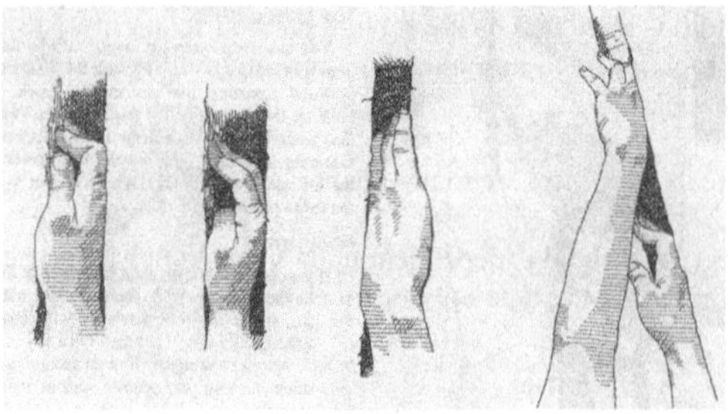

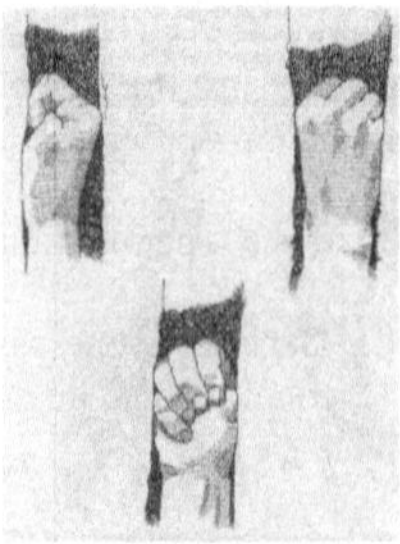

HAND JAMS **FIST JAMS**

7.) **COMBINATION HOLDS.** The five types of holds just mentioned above are not just used individually. They are most often used in combinations with each other. Some examples are:

a. Chimney Climbing. This is when you insert your entire body into a crack in the rock and by using both sides of the opening, and possibly all five types of basic holds, move up' the crack.

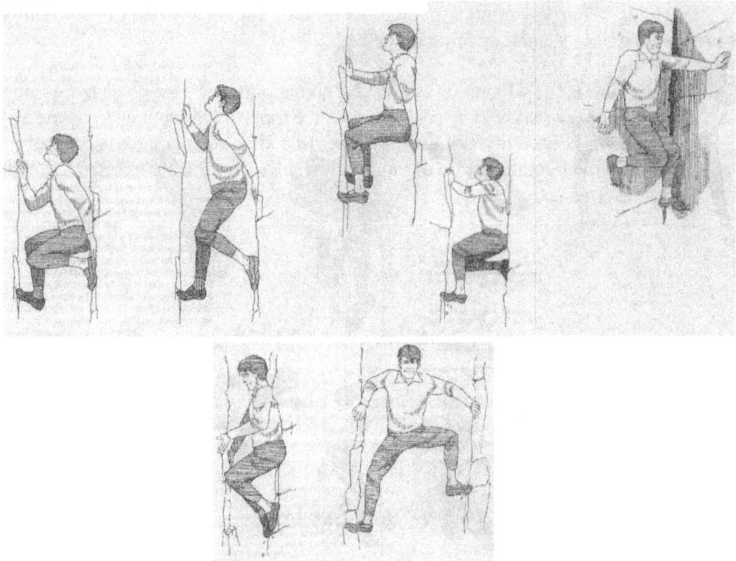

CHIMNEY CLIMBING TECHNIQUES

b. Lie-back. This is a combination of both pull holds with your hands and friction holds with your feet.

LIE BACK TECHNIQUE

c. Push-Pull. As the name implies, this is when you use a push hold and a pull hold together.

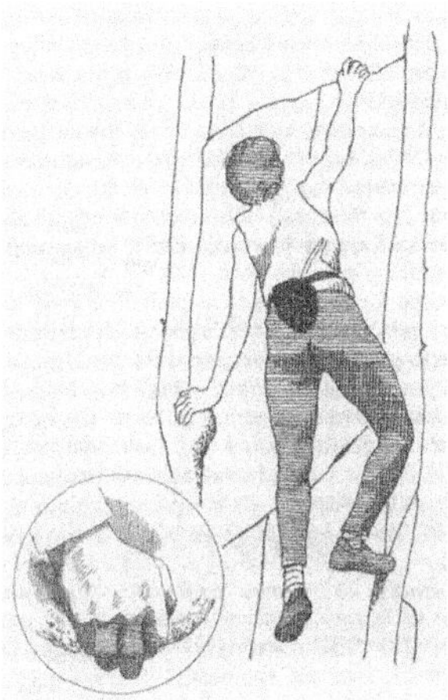

PUSH-PULL

d. Mantling. This is a technique where you continue to climb without moving your hands off a projection by pulling yourself up until your hands are at chest level and then invert your hands and push on the same projection.

MANTELING

MANTELING WITH FACE HOLD

e. Cross-Pressure in Cracks. This is a technique of putting both hands in the same crack and pulling your hands apart to hold/raise yourself.

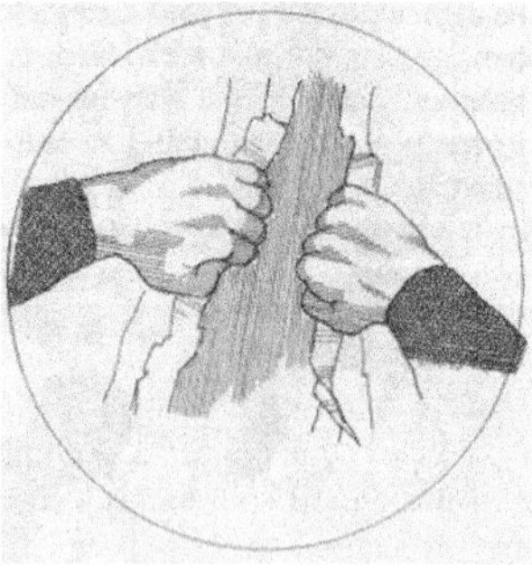

CROSS-PRESSURE METHOD

f. Inverted. Pull or push.

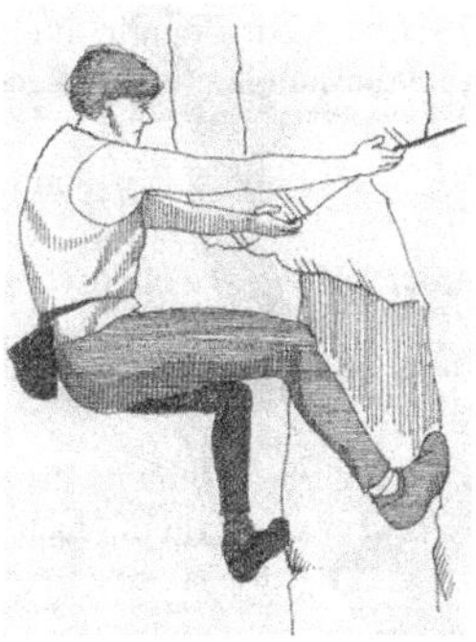

INVERTED TECHNIQUE USING COUNTERFORCE PRESSURE BETWEEN FEET AND HANDS

g. Pinch. As the name implies this is a grip used on tiny little nubbins.

PINCH

h. Stemming. The spreading of arms or legs to maintain a proper body position. (i.e. usually used in a book or chimney.)

STEMMING

8. **GENERAL USE OF HOLDS**. How you use an individual hold is dependent on your experience level, or sometimes, your imagination. Here are some general guidelines.

a. Most handholds can be used as foot holds as you move up the rock.

b. Use all holds possible in order to conserve energy.

c. Even small projections may be used as holds.

d. Do not make use of your knees or elbows due to the reason that it is skin on bone and a slip could occur if pressure is exerted on them. Knees and elbows can be used with the extension of a limb jam.

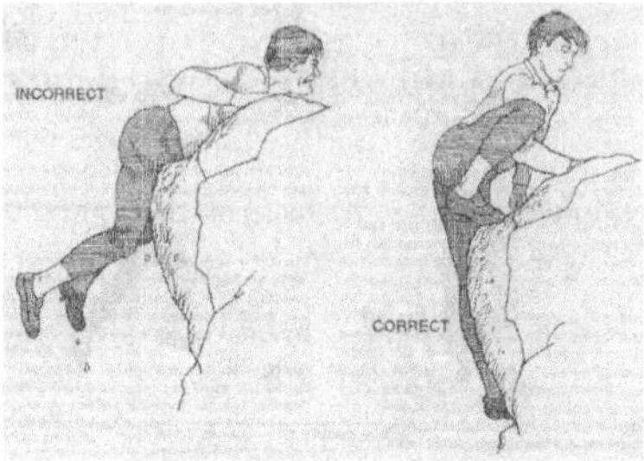

PROPER AND IMPROPER

9. **MOVEMENT ON SLAB**. Movement on slab is based on friction holds.

a. Use any and all irregularities in the slope to gain additional friction.

b. Traversing requires both hands and feet.

TRAVERSING

c. Descending steep slab may require turning inboard to face the slab and backing down.

d. The biggest mistake in slab climbing is leaning into the rock. Maintain maximum friction by keeping weight centered.

SLAB CLIMBING

10. BODY POSITION

a. The climber should climb with his body in balance by keeping his weight centered over and between his feet. Don't hug the rock. Don't over extend and become "spread-eagled". While climbing, keep in mind the acronym "CASHWORTH" for proper body position and movement. (WMC.C3F)

C - Conserve energy.

A - Always test holds.

S - Stand upright, on flexed joints.

H - Hands kept low; handholds should be waist to shoulder high.

W - Watch your feet.

O - On three points of contact.

R - Rhythmic movement.

T - Think ahead.

H - Heels kept low, lower than the toes.

UNITED STATES MARINE CORPS
Mountain Warfare Training Center
Bridgeport, California 93517-5001

WMC.C4
04/11/02

STUDENT OUTLINE

NATURAL AND ARTIFICIAL ANCHORS

TERMINAL LEARNING OBJECTIVE In a summer mountainous environment, establish an anchor system in accordance with the references. (WMC.C4)

ENABLING LEARNING OBJECTIVES.

(1) Without the aid of references and from a given list, choose the two types of anchors, in accordance with the references. (WMC.C4a)

(2) In a summer mountainous environment, establish a natural anchor point, in accordance with the references. (WMC.C4b)

OUTLINE

1. **TYPES OF ANCHORS.** (WMC.C4a)There are two types of anchors that we use. These two types are:

1) NATURAL
2) ARTIFICIAL

C4

2. **GENERAL CONSIDERATIONS FOR ANCHORS**. Anytime you employ natural or artificial anchors; there are special considerations that you must apply. These considerations apply both to the anchor itself and to the material being used to build the anchor. Some examples of these considerations are:

a. Whether using natural or artificial anchors, the installing unit must insure that the anchor is suitable for the load.

b. The anchor position must be relative to the direction of pull on the anchor.

c. The angle between the anchor points should not exceed 90 degrees. This is to ensure that no added stress is put upon the anchors, as well as the equipment being used to construct the anchor.

(1) To decrease the angle between anchor points, materials (i.e. sling ropes, web runners, Prusik cord, etc..) could be used to extend the anchor which will decrease the angle between anchor points.

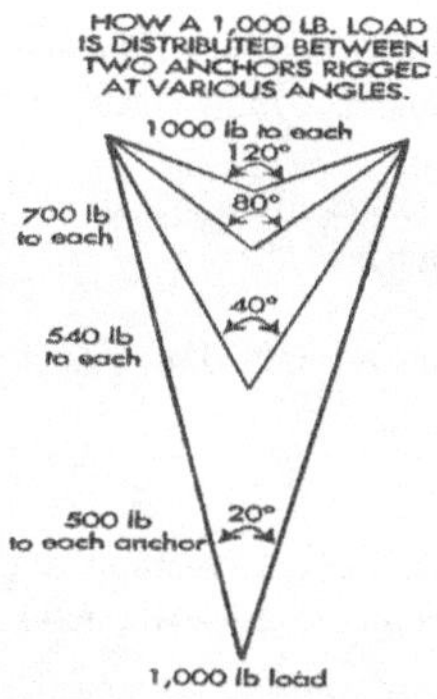

3. **NATURAL ANCHORS.** A natural feature is the preferred type of anchor point. Some examples and considerations are as followed:

a. Types of natural anchors.

(1) Trees.

(a) Select a tree that has not been chopped, burned or is rotten.

(b) The tree should be at least 6" in diameter and strong enough to support the intended load.

(c) Trees growing on rocky terrain should be treated with suspicion, since the roots normally are shallow and spread out along a relatively flat surface.

(2) Shrubs and Bushes.

(a) Select a shrub or bush that is alive and is not brittle, charred or loose.

(b) To avoid leverage, locate the central root and construct the anchor as near to the base as possible.

(3) Rocks and Boulders.

(a) Stability is of prime importance when considering a rock or boulder for an anchor. It must be strong enough and secure enough for the intended load.

(b) All surfaces of the rock or boulder should be inspected for any rough or sharp points. These areas must be padded to protect the rope from being abraded or cut.

(4) Spikes and Flakes.

(a) To check stability of a spike or flake, thump it with the heel of your hand. Anything that sounds hallow is suspicious.

(b) They should be checked for cracks or other signs of weathering that may impair their firmness.

(c) Sharp edges must be padded to protect the rope against cuts and abrasion.

(5) Threads and Chockstones.

(a) A thread is when the rock weathers or cracks to form a hole in the main wall. A chockstone is a rock wedged in a crack.

(b) Check a thread by thumping it with the heel of your hand. Make a visual inspection for cracks and weather. Common sense will prevail for choosing this type of anchor.

(c) When choosing a chockstone, ensure that it has substantial contact with the crack and that the stone's symmetry corresponds with the intended direction of pull.

b. Types of Natural High Tension Anchors. The following are types of Natural High Tension Anchor Systems used to construct installations. These can be tied on any suitable natural anchor point:

(1) Tree Wrap Anchor. The tree wrap is an anchor used to relieve tension on the actual knot itself. This system requires more rope when tied around a large anchor point.

(a) Tie a Figure 8 Loop on the standing end of the rope and wrap the 4-6 times around an anchor point.

(b) Attach a locking carabiner through the knot's loop and clip it onto the running end of the rope.

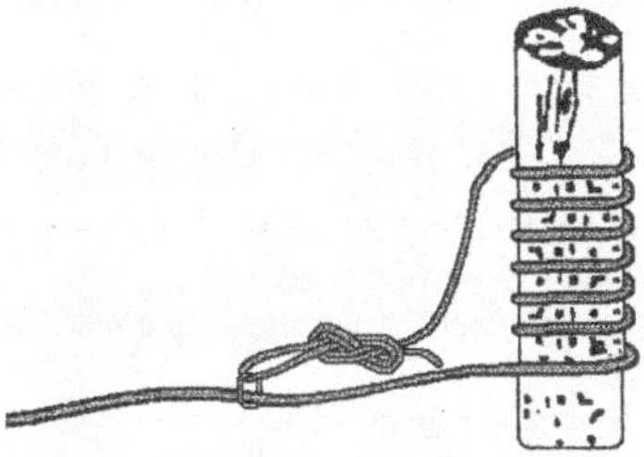

NOTE: If the anchor point to be wrapped is larger then 18' around, then three wraps will suffice.

(2) Swami Wrap Anchor. When tying the Swami Wrap, ensure that the joining knot is kept on the side of the anchor point. The Swami Wrap can be loosened under load if need be. This system can be substituted for a tree wrap when the installation's rope length is an issue.

(a) Select a suitable anchor point and tie a Swami Wrap around it.

(b) Clip a locking carabiner into as many wraps as possible. This will serve as the attachment point for the anchor system.

(3) Figure of 8 Anchor. The Figure of 8 Anchor is a quick and efficient system. This system cannot be loosened under load. It is tied in the following manner:

(a) Tie a Figure of 8 Loop on the standing end of the rope and wrap the knot around a suitable anchor point. Attach a locking carabiner through the knot's loop and clip it onto the running end of the rope.

(b) Before tensioning the anchor system, adjust the rope so the running end of the rope runs smoothly through the carabiner towards the direction of pull. This will prevent any lateral tension.

(b) If two ropes are used, the upper rope's carabiner has its gate upwards, and the lower rope's carabiner has it's gate downward.

c. Primary and Secondary Natural Anchor Point System. The two anchor points should be in line with the direction of pull. The primary anchor is the point nearest the running end while the secondary anchor point is directly behind the primary. It is constructed in the following manner:

(1) The primary anchor knot should be an around the object clove hitch. This is chosen for ease of untying the system after tension has been placed on the rope.

(2) The secondary knot is tied around a suitable anchor point ensuring that the rope is taut between the two anchor points.

4. **ARTIFICIAL ANCHOR SYSTEMS**. Any time we use anything other then a natural feature, we are using an artificial anchor point. Artificial anchors can be constructed in the ground, or on the rock itself. The following are artificial anchor systems:

a. Single Timber Deadman. This system is constructed in the ground and it requires considerable time and effort. The steps of it's construction are as follows:

(1) Dig a trench 6 feet long and 3 feet deep and wide enough to work in at a 90 degree angle to the direction of the pull.

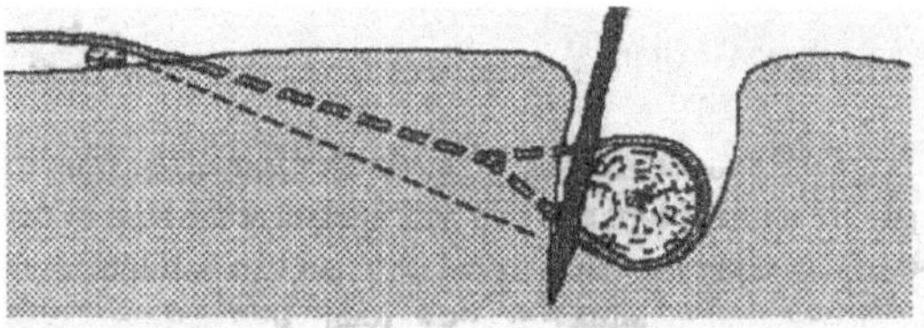

(2) Dig another trench about 12 inches wide. This trench is dug so that it intersects the main trench at a right angle in the middle. The bottom of this trench should be parallel to the direction of pull and should join the bottom of the main trench.

(3) Take an anchoring device (i.e. log, engineer stakes, bundled up branches etc.) that is strong enough to support the intended load. The anchor is placed into the main trench and covered with dirt with the exception of that part of the anchor that joins the second trench. Stakes approximately 3 feet long should be driven approximately 1-1/2 feet into the ground between the dead-man and the slanted side of the trench to assist in holding the dead-man in place if the soil is soft.

b. Picket Hold Fast. The picket hold fast is an easier anchor to construct than the Deadman, and can be used almost anywhere. The strength of the picket system depends on the pickets and the soil or snow conditions. The picket system can be used for both high and non-tension systems. Construction is as follows:

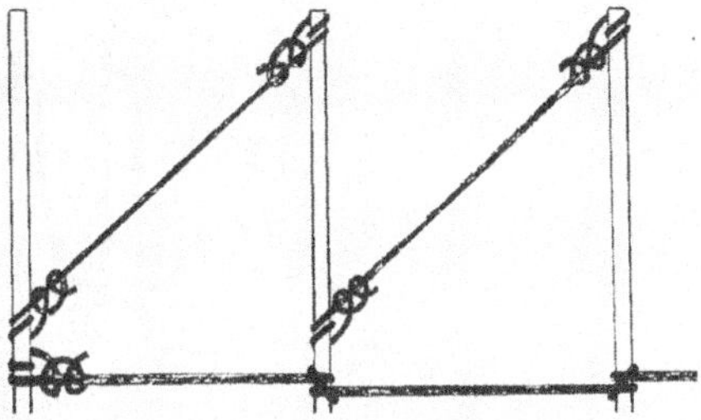

(1) Three stakes (i.e. logs, engineer stakes, etc.) are driven into the ground at a 30 degree angle against the direction of pull.

(2) The line of pickets should be driven in line with the direction of pull. The distance between the pickets can be anywhere from 3-12 feet apart depending on the terrain and soil conditions.

(3) Before tying any rope to the picket anchors, the base of the pickets must be buffed or padded.

(4) To tie a rope to the pickets, go to the furthest picket away from the cliff edge. Tie a round turn and two half hitches at the base of the picket. Take the running end of the rope and tie an over the object clove hitch to the base of the middle picket. Then with the running end of the rope tie an over the object clove hitch to the picket closest to the edge of the cliff face. Ensure that there is tension on the rope in between each picket.

(5) To tie off the pickets to themselves, go to the furthest picket from the cliff edge. And with a sling rope, tie a round turn and two half hitches at the base of that picket. With the running end of that sling rope tie a round turn and two half hitches at the top of the middle picket. Using a second sling rope tie a round turn and two half hitches to the base of the middle picket. Then with the running end of that sling rope tie a round turn and two half hitches to the top of the picket closest to the cliff edge.

c. Equalized Anchor. This system is built with a minimum of three pieces of protection. It can be tied with the standing end of a rope or by utilizing a practice coil.

(1) Construction with the Standing End of the Rope:

(a) Tie a Figure of 8 Loop in the standing end of the rope.

(b) Place a carabiner in each artificial anchor point and attach the knot in either of the outside carabiners.

(c) Clip the rope into the remaining carabiners.

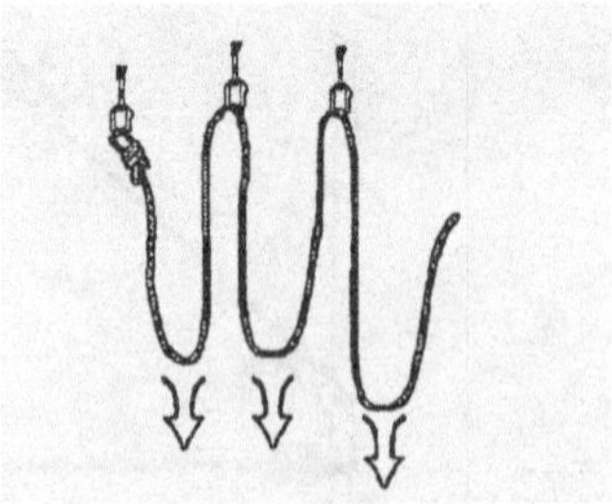

(d) A bight of rope is then pulled down after each carabiner into the anticipated direction of pull. With all three bights, tie an overhand knot around itself to include the running end of the rope.

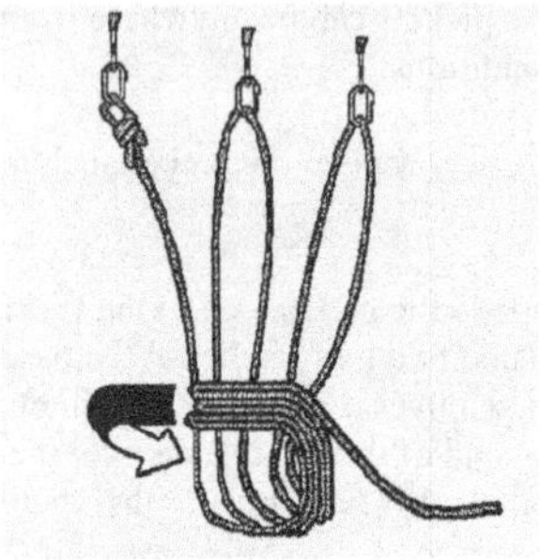

(e) With the running end of the rope coming from the overhand knot, tie a Figure of 8 Loop and attach it to the overhand knot's loop with a locking carabiner.

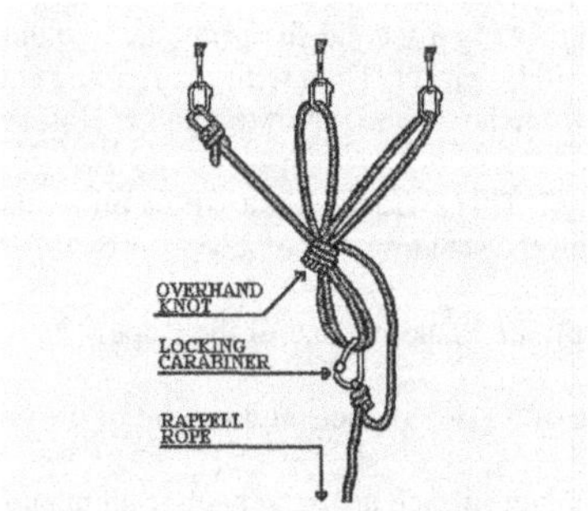

(f) Ensure that there is sufficient slack in the dead rope to prevent disfiguring the knot thus ruining the structural integrity of the knot.

(2) Construction with the Cordalette Method:

(a) Create an endless loop with a practice coil and clip it into each of the artificial anchor points.

(b) A bight of rope is then pulled down between each carabiner into the intended direction of pull.

(c) With all three bights, tie an overhand knot on itself.

(d) Place a locking carabiner into the overhand knot's loop. This will serve as the attachment point for the anchor system.

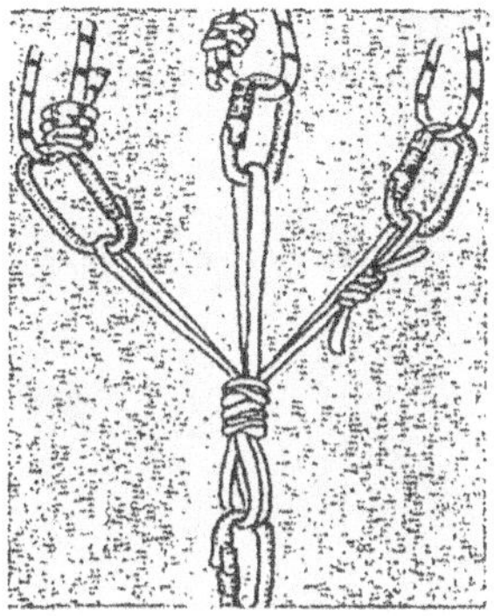

NOTE: If constructing a two-rope high-tension installation, a minimum of three artificial anchors per rope will be used.

UNITED STATES MARINE CORPS
Mountain Warfare Training Center
Bridgeport, California 93517-5001

SML
ACC
02/11/02

LESSON PLAN

SIT HARNESS

LEARNING OBJECTIVES

TERMINAL LEARNING OBJECTIVE: Without the aid of references, properly wear and maintain the MWTC sit harness, in accordance with the references.

ENABLING LEARNING OBJECTIVES

(1) Without the aid of references and given a diagram, name in writing the parts of a sit harness, in accordance with the references.

(2) Without the aid of references, wear a sit harness, in accordance with the references.

BODY

1. (5 Min) **NOMENCLATURE**

   a. Waist belt.

   b. Leg loops (adjustable).

   c. Buttocks straps (adjustable).

   d. Fastex buckle - 2.

   e. Doughnut.

   f. Equipment loops.

   g. D-ring (older models).

   h. Waist belt tie-in point.

   i. Crotch strap.

WEARING OF THE SIT

C5

HARNESS

a. First, disconnect the fastex buckle at the rear of the harness.

b. Hold the harness in front of you, put your feet through the leg loops ensuring that the buckles on the leg loops are outboard, on your thighs.

c. Fasten the waist belt into the buckle, ensuring that it is a tight but comfortable fit. **You must ensure that the waist belt is threaded back through the buckle, this action locks the waist belt to the buckle. Failure to do this will cause the waist belt to slip through the buckle when it is under load.**

d. Adjust the leg loops so that they are high on your thigh, once you have them adjusted, get your buddy to clip the fastex buckle male to the most comfortable female fastex. Then adjust the buttock straps so as the leg loops are held up.

e. If all the above are done correctly, the harness should now be a comfortable but snug fit, after you have fitted it for the first time there is no need to go through the same procedure each time you put it on. Simply hold the harness in front of you, step into it and attach the waist belt to the buckle in the approved manner.

2. **TYING IN THE ROPE (END OF ROPE)**

a. Pass the end of the climbing rope up through the crotch strap, then through the doughnut and through the waist belt tie-in point.

b. Tie the rope using a retraceable figure eight, adjusting the knot to get it as close as possible to the body.

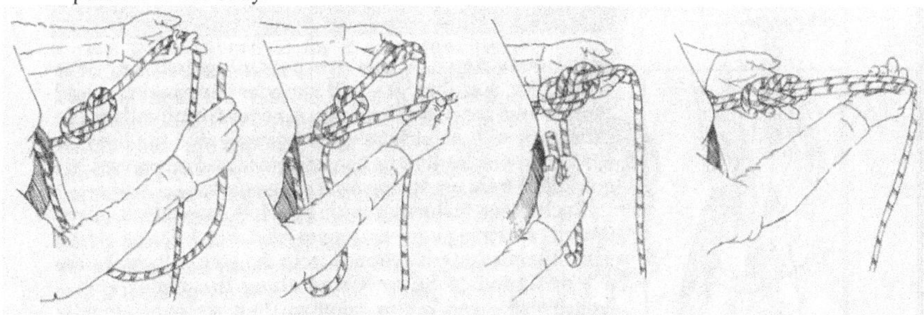

RETRACE FIGURE 8 IN SIT HARNESS

3. **TYING IN THE ROPE (MIDDLE OF THE ROPE)**

a. Take up a bight of rope and tie a figure eight loop, then take a steel locking carabiner and attach the carabiner to the harness by securing the crotch strap, doughnut, and the waist belt tie-in loop. Now attach the figure eight loop to the carabiner.

4. (5 Min) **CARE AND MAINTENANCE**

a. Avoid contact with chemicals, as this will damage the nylon.

b. Regularly inspect for signs of abrasions and normal wear. Pay particular attention to wear points such as the tie-in loops, buckles, and sewn joints.

c. Keep away from heat such as open flames, cigarettes, etc.

d. If soiled by grit and seawater, wash in lukewarm water with pure soap and allow to dry in a warm room away from direct heat.

e. Two to three years of life can be expected during normal climbing use.

f. It is recommended that a harness that has experienced a serious fall should be discarded.

g. Under no circumstances will you ever tie into the equipment rack of any sit harness as a belay, anchor point, etc.

UNITED STATES MARINE CORPS
Mountain Warfare Training Center
Bridgeport, California 93517-5001

FMST.07.33
05/02/01

STUDENT OUTLINE

ESTABLISHMENT OF RAPPEL POINTS AND RAPPELLING

TERMINAL LEARNING OBJECTIVE Given a mountainous environment, mountaineering equipment, helmet, sit harness, rappel gloves, 165 ft static rope, rescue 8, prusic cord, and one or more assistants, conduct rappelling operations to support mission requirements per the reference. (FMST.07.33)

ENABLING LEARNING OBJECTIVES.

(1) Without the aid of references, select from a given list the three criteria involved in site selection for a rappel site, in accordance with the references. (FMST.07.33a)

(2) In a summer mountainous environment, execute the commands used between a rappeller and his belay man when conducting a rappel, in accordance with the references. (FMST.07.33b)

(3) Without the aid of references, select from a given list three of the nine duties of the rappel point NCOIC, in accordance with the references. (FMST.07.33c)

(4) Without the aid of references, select from a given list three of the six duties of the first man down a rappel rope, in accordance with the references. (FMST.07.33d)

(5) Without the aid of references, select from a given list the three types of rappels, in accordance with the references. (FMST.07.33e)

(6) In a summer mountainous environment, rappel down a cliff face using a seat-hip rappel, in accordance with the references. (FMST.07.33g)

C6

OUTLINE

1. **INHERENT DANGER OF RAPPELLING**. Rappelling is inherently dangerous because rappellers rely totally on the equipment.

    a. To ensure a safe training evolution, two ropes will be utilized at MWTC.

    b. All the rappels taught at MWTC can be completed with a single rope if the situation arises.

    c. If utilizing a one-rope system with the carabiner wrap, the rope will be attached to the carabiner with two wraps vice one.

2. **SITE SELECTION**. (FMST.07.33a) when selecting a rappel site consider these three factors.

    a. There must be a good anchor. As previously taught in *NATURAL* and *ARTIFICIAL ANCHORS*, natural anchors are preferred.

    b. The rappel route down should be as free of obstacles (i.e., vegetation, debris) as possible.

    c. There must be suitable loading and unloading platforms.

NOTE: The evaluation of a site for the above factors should be made by the rappel point NCOIC, who should be the most experienced rappeller in the unit.

3. **DUTIES OF THE RAPPEL POINT NCOIC**. (FMST.07.33c)Once a rappelling site has been selected, one person will be appointed to each rappel lane as a rappel point NCOIC. These individuals should have experience as rappellers. The rappel point NCOIC has ten duties and responsibilities, which are:

    a. Ensures that the anchor points are sound and that the knots are properly tied.

    b. Ensures that loose rock and debris is cleared from the loading platform.

    c. Allows only one man on the loading platform at a time and ensures that the rappel point is run in an orderly manner.

    d. Ensures that each man is properly prepared for the particular rappel; i.e. gloves on, sleeves down, helmet secured, rappel seat tied correctly and secured properly.

    e. Attaches the rappeller to the rope and ensures the rappeller knows the proper braking position for that particular rappel.

    f. Ensures that the proper commands or signals are used.

    g. Dispatches each man down the rope.

h. The rappel point NCOIC will be the last man down the rope.

i. The rappel point NCOIC will ensure that the ropes are inspected after every 50 rappels.

j. The rappel point NCOIC will maintain a rope log.

4. **TYPES/USE**. (FMST.07.33e) The three types of rappels and when they are preferred to be used are:

a. Hasty Rappel: It is used when carrying loads down moderate slopes.

b. Seat-Shoulder Rappel: It is used for heavily laden troops over vertical to near vertical cliff faces.

c. Seat-Hip Rappel: It is used when carrying loads over vertical to near vertical faces.

5. **HASTY RAPPEL**. The hasty rappel is the easiest type of rappel to prepare for. It requires no equipment other than a rope and gloves.

a. Conduct. A hasty rappel is conducted in the following manner:

(1) Sleeves will be rolled down and gloves will be put on.

(2) Face slightly sideways.

(3) Place the rappel rope across your back, grasping it with both hands, palms forward, and arms extended.

(4) The hand nearest the anchor is the guide hand. The hand farthest from the anchor is the brake hand.

(5) Lean out at a moderate angle to the slope.

(6) Descend down the hill facing half-sideways, taking small steps and continually looking downhill while leading with the brake hand.

(7) Feet should not cross and the downhill foot should lead at all times.

b. Braking. The steps for braking during a hasty rappel are as follows:

(1) Bring the lower (brake) hand across the front of the chest to brake.

(2) At the same time, turn to face up toward the anchor point.

HASTY RAPPEL

6. **SEAT-SHOULDER RAPPEL**. The seat-shoulder rappel relies on friction as the main effort of controlling the descent. It is very efficient for men with heavy packs because it provides support for heavy loads on the back.

a. Conduct. A seat-shoulder rappel is conducted in the following manner:

(1) Put on your rappel-seat, roll down your sleeves and put on your gloves.

NOTE: To avoid causing a possible injury, it is advised that the rappel seat be constructed at the rear of the rappel site loading platform. If the rappel seat is worn for too long, the rappel seat could loosen up enough to cause you to slip out of your seat while rappelling.

(2) The steel locking carabiner is placed on the rappel-seat so that the gate opens down and away, to prevent the gate from opening once the wraps are placed into the carabiner.

(3) Step-up to the rope with you're left shoulder facing the anchor.

(4) The rappel rope is attached to the rappeller's hard point carabiner as follows:

(a) Snap the rope into the locking carabiner.

(b) Taking slack from the standing (anchor) end of the rope, make one wrap with the rope around the body of the carabiner and back through the gate.

(c) Ensure that the locking nut of the carabiner is fastened to lock the carabiner closed.

NOTE: If you are using only one rope to rappel with due to the tactical situation or equipment availability, the procedures are the same, EXCEPT the individual will make two wraps around the body of the carabiner instead of one.

(5) Take the rope across your chest, over your left shoulder, diagonally across your pack and down to the right (brake) hand.

(6) Descend by walking down the cliff using the braking procedure to control the rate of descent. Look under your brake arm for possible obstacles to avoid.

b. Braking. The steps for braking during a seat-shoulder rappel are as follows:

(1) Lean back.

(2) Face directly uphill while bringing the brake hand across the chest.

7. **SEAT-HIP RAPPEL**. The seat-hip rappel is the most commonly used rappel.

a. Conduct. A seat-hip rappel is conducted in the following manner:

(1) Construct the rappel seat; roll down sleeves and put gloves on.

(2) The steel locking carabiner is placed on the rappel-seat so that the gate opens up and away.

(3) Step up to the rope with your left shoulder facing the anchor.

(4) The rappel rope is snapped into the carabiner as follows:

(a) Snap the rope into the locking carabiner.

(b) Taking slack from the standing (anchor) end of the rope, make one wrap with the rope around the body of the carabiner and through the gate again.

(c) Ensure that the locking nut of the carabiner is fastened to lock the carabiner closed.

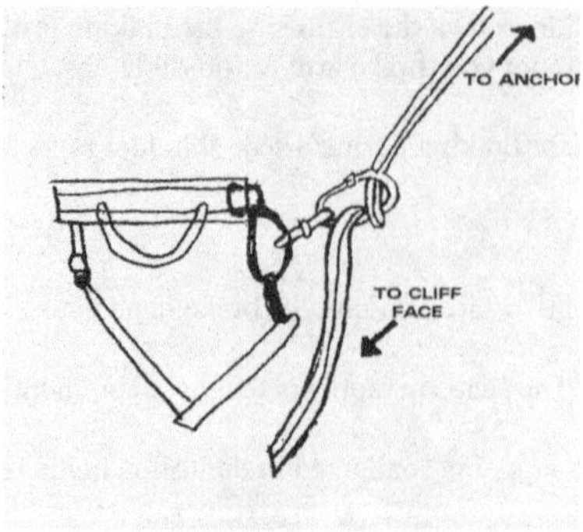

SEAT-HIP RAPPEL

NOTE: If you are using only one rope to rappel with due to the tactical situation or equipment availability, the procedures are the same, EXCEPT the individual will make two wraps around the body of the carabiner instead of one.

(5) The rappeller will grasp the running end of the rope with the brake (right) hand, palm down and turned slightly inboard, near the hip.

b. Braking. The steps in braking for a seat-hip rappel are as follows:

(1) Grasp the rope tightly with the brake hand.

(2) Take the brake hand and place it in the small of the back. This will create enough friction to stop all momentum.

NOTE: At no time will you bound or jump while you are descending. You "walk down" the cliff face using the proper braking procedure to control your rate of descent.

8. **SAFETY OF THE FIRST MAN DOWN**. To ensure the safety of the first man down, the rappeller will:

a. Before deploying the ropes, tie the two ends of the ropes together with an overhand knot.

b. Tie a friction knot on the standing end of the rappel rope with a Prusik cord.

c. Attach the rappel ropes to the harness's hard point with a locking carabiner.

d. Attach the Prusik cord to the harness's hard point with a locking carabiner. This cord is also referred to as a safety Prusik.

NOTE: If the Safety Prusik is too short, it can be extended with a web runner.

9. **DUTIES OF THE FIRST MAN DOWN**. (FMST.07.33d) The first man down the rope has specific duties. They are as follows:

a. Selects a smooth route down for the ropes.

b. Clears the route of loose rocks and debris.

c. The first man down will untie the overhand knot and straighten the ropes out once he reaches the bottom.

d. The first man down belays the next man down the rope.

(1) To belay correctly, the belay man will stand facing the cliff face, arms up with palms facing upward and over lapping. The rope will pass through the opening of the over lapping hand's thumb and index fingers.

(2) To stop a fallen rappeller, the belay man will grab the rope with closed fists and pull straight down bringing the forearms parallel to the deck.

e. Take charge of personnel as they arrive at the bottom to include appointing a belay man.

10. **RAPPELLING COMMANDS.** (FMST.07.33b) In order to conduct rappelling operations safely, it is essential that everyone understands the sequence of events. The following voice commands or rope tugs will be utilized:

VOICE COMMANDS	GIVEN BY	MEANING
"Lane # on Rappel"	Rappeller	I am ready to begin rappelling
"Lane # on Belay"	Belayer	I am on belay and you may begin to rappel
"Lane # off Rappel"	Rappeller	I have completed the rappel and am off the rope
"Lane # off Belay"	Belayer	I have completed the belaying of the rappeller

ROPE TUGS	GIVEN BY	MEANING
Three Tugs	Rappel Point NCOIC	The rappeller is ready to begin rappelling
Three Tugs	Belayer	I am on belay and the rappeller may begin to rappel
Three Tugs	Belayer	The rappeller is off the rappel rope

11. **TYING-OFF**. Occasionally, it may be necessary to stop during a rappel before reaching the bottom of a cliff. The following sequence is used:

a. The commands used in tying-off are as listed below:

COMMAND	GIVEN BY	MEANING
"Lane # Tying-off"	Rappeller	I am ready to tie-off, give me some slack.
"Lane # Tying-off"	Belayer	I have given you enough slack and you may tie-off.
"Lane # On rappel"	Rappeller	I have completed tying-off and I am ready to resume rappelling.
"Lane # On rappel"	Belayer	I am on belay, you may rappel.

b. The procedure used when tying-off is as follows:

(1) The rappeller gives the command, "Lane #, tying-off."

(2) The Belayer gives the rappeller slack and gives the command "Lane # Tying-off". He continues to hold the rope in the belay position, remaining alert and watching the rappeller.

(3) The rappeller quickly brings his brake hand to the twelve o'clock position, so that the running end and standing end are parallel.

(4) The rappeller grasps all ropes with his guide hand as close to the carabiner as possible.

(5) The rappeller steps over the rope so that the running end is going between his legs.

(6) The rappeller releases the rope with his brake hand then reaches under the running end, over the standing end. He then takes up a bight from the running end about two feet long and pulls it over the standing end and under the running end, forming a half hitch.

(7) He pulls the half hitch tightly against the guide hand.

(8) He works the half hitch down snugly against the carabiner while maintaining contact with the guide hand as long as possible.

NOTE*:* A loose half hitch could bind into the carabiner causing difficulties in clearing the knot out. Make sure that the first half hitch is dressed down tightly before moving it against the carabiner.

(9) Place another half hitch above the one already tied.

(10) To untie, reverse the steps. Remove the safety half hitch, then shrink the first half hitch down to a small loop by grabbing the running end of both ropes and pulling them straight out to the left. Place your right hand in the middle of the two bights until they are snug on all four fingers. Then remove one finger and make the bights snug on three fingers. Repeat this process until it's down to one finger and both the bights are equal.

(11) Now, grasp the running end with both hands and smartly jerk the running end of the ropes upward to pop the small loop out from the first half hitch. From this position, keep the guide hand around both of the ropes next to the carabiner; step back over the rope so that the running end is to your right side. Grab just the running end with your brake hand and quickly set the brake behind the small of your back then, readjust you guide hand onto the standing end. You now should be in the seat-hip rappel position.

NOTE*:* The wraps may bind up some after untying, where no further movement down is possible. To alleviate this, keep the brake on and force your body weight down to pop the wraps in the carabiner back to their intended position and then continue on with the rappel.

(12) The rappeller gives the command "Lane #, on rappel."

(13) The Belayer takes up the slack and gives the command "Lane #, on belay."

(14) The rappeller resumes to rappel down.

12. **RESCUE TECHNIQUES**. When conducting rappelling operations, the possibility of a rappeller getting caught on the rope due to either his clothing or equipment can occur. There are two types of rescues that can be performed to free the rappeller from this situation.

a. The Self-Rescue Technique.

(1) After realizing that you are caught up on the rope, check with your belay man to ensure that he has a solid brake set.

(2) Using a safety Prusik, tie a friction knot onto the rappel rope approximately an arms length above the area that is fouled. Anchor the other end of your safety Prusik to your rappel seats hard point. This is the same technique used as discussed earlier with the first man down the rope.

(3) Work the friction knot up until there is no tension on the rappel device. This will give you the necessary slack to free the malfunction.

(4) Once you have corrected the problem, continue to rappel down using the first man down method.

b. The Buddy Rescue Technique. Himself uses this method when the rappeller is unable to correct the problem.

(1) A rescuer will rappel down on another rope to the disabled rappeller and tie-off.

(2) The rescuer will then establish a safety Prusik onto the victim as taught in the self-rescue technique.

(3) After clearing the malfunction, continue with rappelling operations.

c. Rappelling Casualty Rescue. There are three different methods in which to get an injured rappeller to the bottom.

(1) The first method is to allow the belay man to lower the casualty by slowly releasing the tension on the rappel rope.

(2) The second method is used for critical injuries or when the belay man cannot properly control the casualty's descent.

(a) The rescuer will rappel down using another rope. Once the rescuer gets to the casualty, the rescuer will have his belay man brake him off, enabling the rescuer to use both hands.

(b) If necessary, the rescuer will perform the basic first aid needed. Once the casualty is ready to be lowered, the rescuer will call down to both of the belay men to simultaneously lower both men. The rescuer will hold onto the casualty the entire way down so that he doesn't bounce off the rock face.

(3) The third method is called a tandem rappel. Its used to rescue a casualty who has sustained an injury serious enough that he cannot operate the rappel device himself and requires the assistance of a second rappeller. This type of rescue can be conducted from the top of the cliff face or while on rappel, depending on where the injury occurred.

(a) If the injury occurs while on top of the cliff face, the following steps should be taken:

1. Take either a long sling rope or Prusik cord and tie a figure-eight knot offset so that one length of the cord is longer then the other.

2. Tie a figure-eight knot at each end.

3. Take a rappel device and attach it to the offset figure-eight loop with a carabiner and attach it to the rappel rope.

4. Next, attach the casualty's hard point to the short end of the sling rope/Prusik cord and the rescuer to the other end.

5. The rescuer will then utilize a safety Prusik in the same method as the first man down technique, except that the friction knot will be tied below the rappelling device and controlled with the brake hand.

6. The rescuer will then maneuver himself under the casualty to provide assistance and support.

7. At this point you are ready to rappel both rescuer and casualty at the same time.

NOTE: Your rappel device should be far enough in front of you so that the casualty will not be able to reach it.

(b) If the injury occurs during the rappel, the following steps are taken:

1. The rescuer will preset the same system on a separate rappel rope and rappel down to the casualty.

2. The rescuer will then take the short end of the sling rope and attach it to the casualty's hard point.

3. With a knife, the rescuer will cut the casualty's rappel rope away.

4. Rappel down to the bottom of the cliff face.

NOTE: THIS IS THE ONLY TIME THAT A KNIFE WILL BE USED DURING A RAPPEL. EXTREME CAUTION SHOULD BE USED. IT IS NOT TO BE DONE IN TRAINING, BUT ONLY DURING AN ACTUAL RESCUE AND AS A LAST RESORT.

13. **RETRIEVABLE RAPPELS**. Once a unit has rappelled down a vertical obstacle, it may be necessary to retrieve the rope(s). Depending on the height of the obstacle either one or two ropes will used to construct the rappel lane.

a. One-Rope Retrievable Rappel:

(1) Find the middle of the rope and place it directly behind a suitable anchor point.

(2) Join the pigtails of the rope with an overhand knot and deploy the rope down the obstacle.

(3) On one side of the rope in front of the anchor point, tie an over-the-object clove hitch onto a locking carabiner.

(4) On the other side of the rope, tie another over-the-object clove hitch and attach it to the same carabiner.

(5) The first man down will utilize a safety Prusik and untie the overhand knot after he reaches the bottom.

(6) All others conduct proper rappel procedures.

(7) The last man down will disconnect the carabiner from the rope, and ensuring that the middle of the rope is directly behind the anchor point, rappel down the rope.

(8) Pulling on either end of the rope then retrieves the rope.

b. Two-Rope Retrievable Rappel. This type of system is identical to the one rope retrievable rappel except for a few considerations.

(1) The reason for using two ropes vice one rope is that the height of the rappel is greater.

(2) When two ropes are used they should be joined together using a square knot. This knot will be placed directly behind the anchor point.

(3) All else remains the same as the one rope retrievable rappel except:

(a) The last man down will disconnect the carabiner from the rope and will move the joining knot as close to the vertical obstacle's edge as possible. This will prevent the knot from possibly getting caught up while retrieving.

(b) The last man down will then place a carabiner on the rope below the knot. This will enable him to know which line to pull for retrieval.

14. **EQUIPMENT**. Equipment should be worn in accordance with unit SOP. The unit has the responsibility to determine which methods it feels are most beneficial to the mission. Weapons should be worn across shoulder; muzzle down and away from the brake hand with a tight sling securely attached to the weapon. Weapons should also be dummy corded to the individuals.

UNITED STATES MARINE CORPS
Mountain Warfare Training Center
Bridgeport California, 93517-5001

WMC
04/11/02

STUDENT OUTLINE

TOP ROPING

LEARNING OBJECTIVES.

TERMINAL LEARNING OBJECTIVE. In a summer mountainous environment, and given a rock face and a belay man, conduct top roping with cartridge belt and rifle, in accordance with the references. (WMC)

ENABLING LEARNING OBJECTIVE

(1) In a summer mountainous environment, establish a belay point from the top of a cliff for use in a top rope climb, in accordance with the references. (WMC.a)

(2) In a summer mountainous environment, demonstrate the use of climbing commands while conducting a top rope, in accordance with the references. (WMC.b)

OUTLINE

1. **ESTABLISHING A BELAY STANCE FROM THE TOP**.

a. The Belayer will establish a sitting belay stance on the cliff head by:

(1) Constructing a suitable anchor with the standing end of the rope.

(2) With the direction of pull away from the anchor, tie a directional figure 8 loop near enough to the cliff's edge so that the climber can be observed, if possible.

(3) Tying a swami Wrap around oneself, clip a locking carabiner through all the rear wraps and clip a large locking carabiner through all the front wraps.

(4) Secure the rear locking carabiner into the directional figure of 8 loop.

C7

(5) Secure the running end of the rope to the large locking carabiner utilizing a suitable friction belay (i.e., Munter hitch, stitch plate, etc.).

NOTE*:* Gloves **will not** be worn while belaying a climber during a top rope.

2. **SECURING THE CLIMBER TO THE ROPE**.

a. The climber will tie into the end of the top rope by:

(1) Constructing a Swami wrap around oneself.

(2) Tying a retrace figure 8 loop through all of the Swami's wraps or;

(3) A figure of 8 loop clipped into a locking carabiner secured through all the Swami's wraps.

b. Alternative methods.

(1) A bowline on a coil as discussed in ROPE MANAGEMENT or;

(2) Tie into the hard point of the harness as discussed in SIT HARNESS.

3. **ESTABLISHING A BELAY STANCE FROM THE BOTTOM**.

a. Belaying from the bottom is commonly referred to as a "yo-yo" or "Sling Shot" belay. This system can be constructed for either a direct or indirect belay.

(1) Construct a suitable top anchor utilizing either a pulley or one steel locking or two non-locking carabiners as the attachment point for the anchor.

(2) The rope will travel up from the bottom and pass through the attachment point then return to the bottom.

(3) The climber will be secured to one end of the rope and the Belayer from the other end.

4. **CLIMBER'S RESPONSIBILITIES**.

a. The climber ensures that the Belayer is anchored and on belay, by use of commands or prearranged signals prior to beginning the climb.

b. The climber will not out climb the belayer; this will cause slack in the rope between the belayer and the climber.

c. Avoid placing excess pressure/weight on belay man.

d. Weapons will be worn across shoulder, muzzle down and to the left, with a tight sling securely attached to the weapon.

5. **CLIMBING COMMANDS AND SIGNALS.** In order to conduct top rope operations safely, it is essential that everyone understands the sequence of events. The following voice commands or rope tugs will be utilized:

VOICE COMMANDS	GIVEN BY	MEANING
"Lane #, Up Rope"	Climber	Belayer needs to take in the slack.
"That's Me"	Climber	Excess slack has been taken up between us.
"Lane #, On Belay"	Belayer	I am on belay.
"Lane #, On Climb"	Climber	I am ready to climb.
"Climb Climber"	Belayer	You may begin to climb.

ROPE TUGS	GIVEN BY	MEANING
Three Tugs	Climber	Belayer needs to take in the slack.
Three Tugs	Belayer	I have taken up the slack and you may begin to climb.
Three Tugs	Climber	Climbing.

UNITED STATES MARINE CORPS
Mountain Warfare Training Center
Bridgeport, California 93517-5001

FMST.07.40
03/02/02

STUDENT OUTLINE

MOUNTAIN CASUALTY EVACUATIONS

TERMINAL LEARNING OBJECTIVES. In a summer mountainous environment and given a simulated/actual casualty, evacuate casualties, in accordance with the references. (FMST.07.40)

ENABLING LEARNING OBJECTIVES

(1) With the aid of references, state orally the general considerations for medevac procedures, in accordance with the references. (FMST.07.40a)

(2) In a summer mountainous environment, and with the aid of references, construct an expedient litter, in accordance with the references. (FMST.07.40b)

(3) In a summer mountainous environment, and with the aid of references, secure a simulated/actual casualty in a SKEDCO litter, in accordance with the references. (FMST.07.40c)

(4) In a summer mountainous environment, and with the aid of references, attach a belay line to a SKEDCO litter for movement up/down moderate slopes, in accordance with the references. (FMST.07.40d)

(5) In a summer mountainous environment, with the aid of references, and given a simulated/actual casualty, conduct a medevac using the barrow-boy method, in accordance with the references. (FMST.07.40e)

OUTLINE

1. **GENERAL CONSIDERATIONS** (FMST.07.40a) the general considerations are a set of guidelines that can be used no matter how serious the casualty is. They are remembered by a simple acronym, **APASSNGG.**

   a. Apply Essential First Aid. (i.e. splints, pressure bandage, etc.)

   b. Protect the Patient from the Elements. Provide the casualty with proper insulation from the ground. Ensure that he is warm and dry. If there are any natural hazards (i.e. rock fall, lighting, etc.) either move the casualty as quickly as possible or ensure that he is well protected.

   c. Avoid Unnecessary Handling of Patient.

   d. Select Easiest Route. Send scouts ahead if possible, to break trails.

   e. Set Up Relay Points and Warming Stations. If the route is long and arduous, set up relay points and warming stations with minimum amount of medical personnel at warming stations to:

      (1) Permit emergency treatment. Treat for shock, hemorrhage, or other conditions that may arise.

      (2) Reevaluate the patient constantly. If patient develops increased signs of shock or other symptoms during the evacuation, he may be retained at an emergency station until stable.

   f. Normal Litter Teams Must Be Augmented in Arduous Terrain.

   g. Give Litter Teams Specific Goals. This litter teams job is extremely tiring, both physically and mentally. The litter teams must be given realistic goals to work towards.

   h. Gear. Ensure all of the patient's gear is kept with him throughout the evacuation.

2. **ONE -MAN CARRIES**

   a. Sling Rope Carry. The sling rope carry requires two men and a 15-foot sling rope. One as the bearer, and an assistant to help in securing the casualty to the bearer. Conscious or unconscious casualties my be transported in this manner:

(1) Bearer kneels on all fours and the assistant places casualty face down on bearer's back ensuring the casualty's armpits are even with the bearer's shoulders.

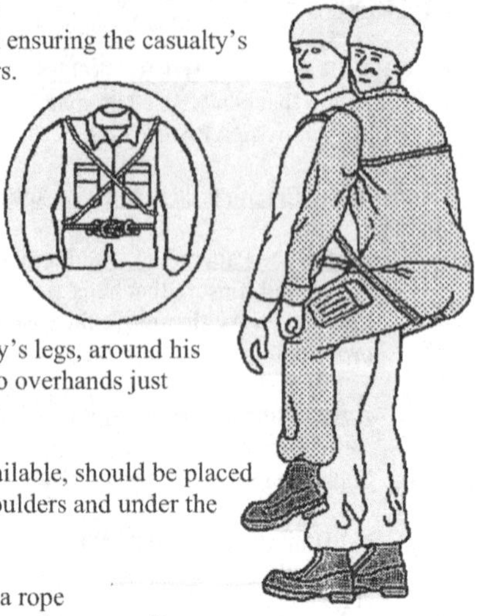

(2) He then finds the middle of the sling rope and places it between the casualty's shoulders and the ends of the sling rope are run under the casualties armpits, crossed and over the bearer's shoulders and under his arms.

(3) Then the ropes are run between the casualty's legs, around his thighs, and tied with a square knot with two overhands just above the bearer's belt buckle.

(4) Ensure the rope is tight. Padding, when available, should be placed where the rope passes over the bearer's shoulders and under the casualties thighs.

b. <u>The rope coil carry.</u> This requires a bearer and a rope coil. It can be used to carry a conscious or unconscious casualty.

(1) Position the casualty on his back.

(2) Separate the loops of the mountain coil into two approximately equal groups.

(3) Slip ½ of the coil over the casualty's left leg and ½ over his right leg so that the wraps holding the coil are in the casualty's crotch, the loops extending upward the armpits.

(4) The bearer lies on his back between the casualty's leg and slips his arms the loops. He then moves forward until the coil is extended. When using the rope coil the bearer ties the coil to himself vice slipping his arms through the loops.

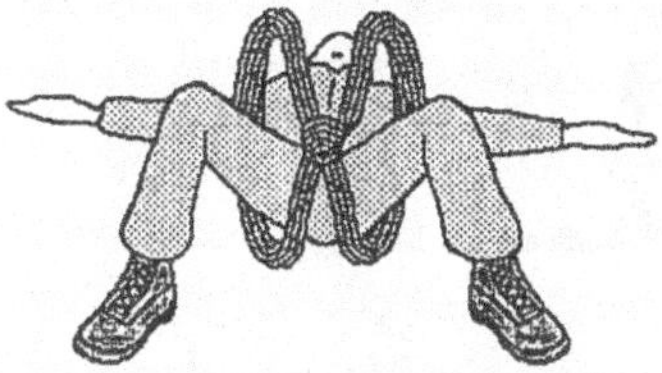

(5) Grasping the casualty's right or left arm, the bearer rolls over, rolling to the casualty's uninjured side, pulling casualty onto the bearer's back.

(6) Holding the casualty's wrists, the bearer carefully stands, using his legs to lift up and keeping his back as straight as possible.

NOTE: The length of the coils on the rope coil and the height of the bearer are to be considered. If the coils are too long and the bearer happens to be a shorter person, it will require the coils to be uncoiled and shortened. If this is not done, then the casualty will hang too low on the bearer's back and make it a very cumbersome evacuation. A sling rope harness can be used around the victim's back and the bearer's chest, which will free the bearer's hands.

3. **TWO-MAN CARRY**

a. Pole Carry. The pole carry method is a field expedient method and should be considered as a last resort only, when narrow ledges must be traversed; vegetation limits the bearers to a narrow trail. This method is difficult for the bearers and uncomfortable for the casualty. Two bearers, four sling ropes and a 12 foot pole - 3 inches in diameter, are required for this carry.

(1) The casualty is placed on his back in a sleeping bag or wrapped in a poncho or blanket, then placed on an insulated pad.

(2) One sling rope is placed under the casualty below the armpits and tied with a square knot across the casualty's chest.

(3) The second sling rope is tied in the same manner at the casualty's waist.

(4) The third sling rope is placed at the casualty's legs below the knee.

(5) The fourth sling rope is tied around the ankles.

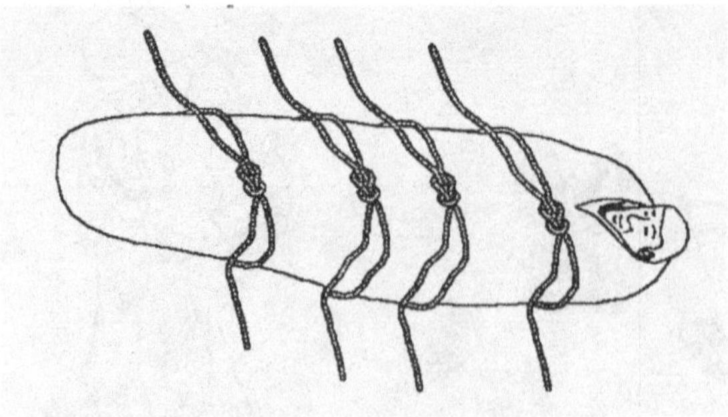

(6) The pole is placed along the casualty's length and secured using square knots with two overhands with the ends of the sling ropes. The square knots should be so tight that the overhands are tied onto themselves

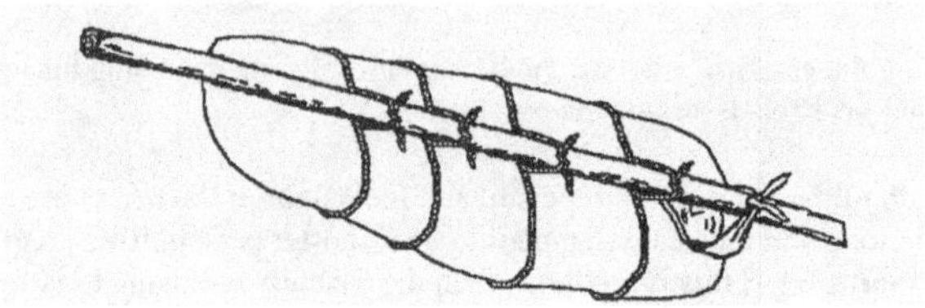

(7) The casualty should hang below the pole, as close to the pole as possible, to prevent swinging during movement.

(8) The casualty's head may be supported using a triangular bandage or a cartridge belt passed around the pole.

(9) For additional support and of movement, two additional bearers may be required, as well as a mountain coil.

(a) Mountain coil is split into two equal coils.

(b) Place knot of mountain coil under casualty's lower back.

(c) Additional bearers slip into each half of the hasty coil one on each side of casualty, aiding in support and movement of the casualty.

4. **EXPEDIENT LITTERS.** (FMST.07.40b) The two types of expedient litters that we will talk about are the alpine basket and the poncho litter.

a. <u>Alpine Basket.</u> To belay the alpine basket, the pre-rigs are attached to the bights formed coming through the loops.

(1) If barrow boy is to be used, the procedure previously discussed will be adhered to.

(2) If barrow boy is not used, then a tag line from the bottom must be implemented to keep the casualty away from the cliff face on the decent.

(3) Construction of the Alpine Basket:

(a) Start by making the same amount of bights as in the rope litter, but start from one end and tie a figure-of-eight loop to run the first bight through.

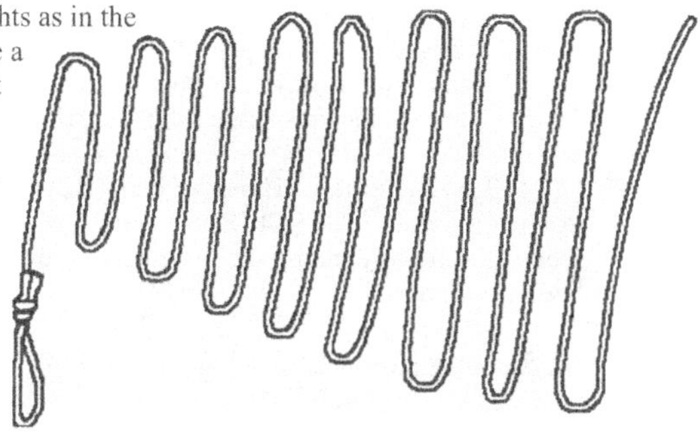

(b) Place padding i.e., isopor mat, on top of the bights and then lay the casualty on the padding and bights.

(c) Start at the casualty's feet and pull the first bight up around the casualty's ankles and through the figure-of-eight loop tied into the starting end of the rope.

(d) Go to the opposite side of the casualty and pull up the second bight and pull it through the loop formed by the bight that was pulled through the figure-of-eight.

(e) Continue until you get to the casualty's armpits, bring the second to the last bight up over the casualty's shoulder and into the bight and then bring the last bight up over the casualty's other shoulder and into the last bight formed.

(f) Secure the end of the bight with an appropriate anchor knot.

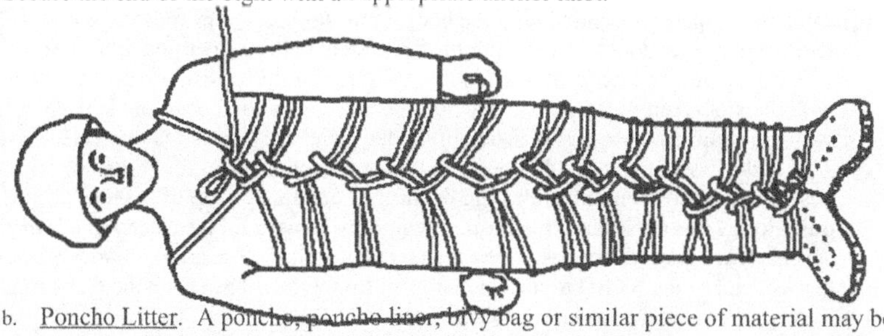

b. <u>Poncho Litter</u>. A poncho, poncho liner, bivy bag or similar piece of material may be used. In addition you will need six individuals with sling ropes.

(1) Lay poncho litter flat on the ground.

(2) Select six rocks about the size of a golf ball. Place one rock in each of the corners and one in the middle on each side in the middle of the litter. The rocks are placed on the under side of the poncho or like material. If a bivy bag is used the casualty should be zipped inside the bivy bag. The rocks should then be arranged in the same manner only on the inside below the zipper.

(3) Tie the sling rope together with an overhand knot. Take the middle of the rope and secure it around the rock with a clove hitch.

(4) An isomat may be laid on the poncho to help make the litter firmer.

(5) The casualty is then placed in the litter. The sling ropes are adjusted by feeding the pigtails of the over hand knot through itself to adjust for length. The loop is then put over the inboard shoulder of the carriers. Insure that the casualty is carried level.

5. **LITTERS.** There are two kinds of litters that are used for casualty evacuations in moderate to vertical terrain. The SKEDCO and the Stokes litter. Each has a variation of procedures for securing a casualty and rigging the litters for either raising or lowering. Let's first look at the SKEDCO litter.

(1) Securing a Casualty to the SKEDCO (FMST.07.40c)

(a) First unroll the litter. The litter must be re-rolled the opposite way to allow the litter to lay flat. Then lay the litter next to the casualty.

(b) If the casualty has any possibility of a spinal injury the Oregon Spine Splint must be used. Secure the splint to the casualty by use of the color-coded buckles. Experienced medical personal are recommended if spinal immobilization is necessary.

(c) Once the casualty is on the SKEDCO use the four body straps to secure the casualty to the litter. Unless injuries prevent, the casualty's arms should be at his sides to prevent further injuries to himself or the rescuers.

(d) Once the casualty is secured with the body straps the feet straps must be secured. The feet straps are secured last to ensure the casualty is in the proper position on the SKED. The feet can be positioned in three ways. The first position is feet together with the straps running on the outside of the feet. The second position is feet apart with the straps running on the inside of the feet. The last position is the feet stacked. This is the most uncomfortable position and not recommended for casualties with possible spinal cord injuries. Placing the heel of one foot on top of the toes of the other forms this position. This position will only be used for a casualty in confined spaces. Once the feet are positioned the feet straps must be secured. Start by bending the feet end of the SKED to form a plat form for the feet. Then loop the feet straps through the second grommets on each side.

(e) The last thing to do is to form the head end to protect the casualty's head. If possible the casualty should wear a helmet. Form the head end tying the pull strap up and secure it to the first body straps.

(2) Medevacing a Casualty. There are many ways to move a casualty once in the SKEDCO. However the medevac team must keep the general considerations in mind. The two methods that we will talk about next are the simplest and require the least amount of additional rigging

(a) The first way is to drag the casualty by the drag strap located at the head end of the SKED. You can also use the SKEDCO's carrying bag as a harness in conjunction with the pull strap and towing harness. If additional people are required, cordage can be added to the pull strap or the front carrying handles.

(b) The second way is to carry the casualty using the carrying handles. By using the set of four removable webbing the litter team can be augmented. To do this each pieces of webbing is tied to make an endless loop. Then pass a bight of the loop through one of the grommets to create additional handles.

(3) Rigging the SKED for Vertical Terrain and Helicopter Lift. In this type of terrain special requirements must be taken to raise or lower a casualty in the SKEDCO.

(a) Rigging the SKED for a Vertical Employment. A vertical raise or lower is used when moving a casualty on steep earth to avoid any further injury to the casualty. On vertical terrain the vertical raise or lower can be used if the terrain is not uniformed or there is a chance of rock fall. Ensure that the casualty's head is always above his feet.

(1) Identify the 30-foot piece of cordage that comes with the SKEDCO. Then tie a figure eight in the middle of the rope. If the rope is worn or missing, the same process can be done with two sling ropes.

(2) Next pass each end of the rope through the grommets at the head of the SKEDCO. Leaving approximately 1-2 feet of rope between the stretcher and the knot.

(3) Continue to feed each end through the grommets and the carrying handles towards the foot end of the SKEDCO. Pass the ends of the rope through the last grommets at the foot end and secure the two ends with a square knot without over hands.

(4) Bring the pigtails up and over the casualty's feet and pass the ends through the
a carrying handles towards the middle of the casualty. Then tie a square knot
b with two over hands.

c

(4) Rigging the SKEDCO for a Horizontal Employment. A horizontal raise or lower is preferred on uniformed vertical terrain. The horizontal employment allows the rescuer to assist the casualty on either a raise or lower. It also allows the rescuer the ability to monitor the casualty's condition and can easily treat the casualty if the need arises.

(a) Identify the two 4 inch nylon straps. They should be two lengths; one four inches shorter than the other. The shorter strap should be marked HEAD STRAPS.

(b) Insert one end of the head strap into a slot near the head of the litter. Then wrap the rest of the straps under the SKEDCO and pass the other end through the opposing slot. Do the same at the foot end of the SKEDCO with the other strap. Ensure that the strap runs smoothly under the SKEDCO.

(c) Connect the strap ends with the large locking caribiner that comes with the SKEDCO. If the caribiner is worn or missing, opposing Stubai 85 locking carabiners will suffice.

6. **STOKES LITTER**. The stokes litter is a litter that is constructed of metal tubing with a plastic covering. The litter is formed in a rectangular basket shape with mesh attached to the frame. Using the stokes for an evacuation (as with any evacuation) it should be padded for the casualty.

   a. Securing the casualty to the stokes litter. In the event that the "seat belts" are missing from the stokes litter, sling ropes can be used to the lash the casualty. The steps involved are:

      (1) Tie two sling ropes together using square knots and two overhands.

      (2) Tie a stirrup hitch around ankles and feet, feed the two pigtails through the right angles of the stokes. Do not cross the ropes at the ankles.

      (3) Lace the sling rope towards the casualty's head by passing the rope through the right angles (not over the top of the rails of the stokes).

      (4) Secure the ends of the sling ropes by tying a clove hitch with two half hitches on the thick vertical bar located by the victim's shoulder.

7. **YOSEMITE PRE-RIG.** The title of this method, pre-rig implies the meaning ready to use, but in this case the rig must be constructed. Its purpose is for attaching a belay line to a litter and to make the litter easily adjusted. This is the one method that is used to secure a litter to a belay line.

   a. Construction: This method normally requires four sling ropes. The steps are: (1) using one sling rope, tie a figure-of-eight loop with one tail.

      (1) Take the remaining tail and run it through the window of the stokes litter or in the stirrups of the collapsible litter.

      (2) Tie a Kragur knot onto the same sling rope.

      (3) Repeat steps (1), (2), and (3) with the three remaining sling ropes.

      (4) Suspend the litter to ensure that the comers are balanced.

8. **ASCENTS OR DESCENT OVER STEEP TO MODERATE SLOPES.** When the litter team is ascending or descending a slope they must consider the potential for further injury to the casualty or to themselves. If the risk of injury is high a belay line may be used to prevent injury to the casualty and the rescuers.

a. Preparing Casualty for Ascents or Decent over Steep to Moderate Terrain. This procedure will be depending on several things. Initially, site selection should contain the following features.

(1) Suitable anchor points.
(2) Clearance for casualty along the route
(3) Loading and unloading points.

b. Additional considerations.

(1) The casualty will always be rigged for vertical employment when on steep to moderate terrain.
(2) The smoothest possible route must be selected.
(3) Ensure that the casualty's head is above his feet.

c. Rescuers positions. There are two methods that can be used for the rescuers for moving a casualty in steep to moderate terrain.

(1) Two to four men will position themselves on each side of the litter. They can then carry the litter by the carrying handles. In steep terrain a second belay line may be used to assist the rescuers. We will discuses the belay line later in the chapter.

(2) The Caterpillar method will require as many personnel as possible. The personnel will split in half and position themselves on each side of the litter forming a tunnel. As the litter is raised or lowered each member will hand the litter to the next member in the tunnel. As the litter passes each person in the tunnel he will peel off and assume the lead either at the top or bottom of the tunnel. This will continue until the litter reaches its desired destination.

d. Belay Line. (FMST.07.40d) For belaying of a casualty, one rope will be used and from the top using one of two methods depending on the application.

(1) Body Belay. This method should only be used over moderate terrain. The belay man will establish a sitting position behind a suitable anchor (i.e., rock, tree, etc.) and pass the standing end of the rope behind his back. The running end of the rope will feed out from the belay man's right side. A figure of eight loop is tied to the end of the running end of the rope. It is then attached to the litter's figure eight loop with a locking carabiner. The belay man will then remove all of the slack between himself and the litter. The standing end of the rope should be stacked on the belay man's left side and run through his left side. As the casualty is lowered, the belay man will feed the rope from behind his back allowing it to run through his right hand. If the belay man needs to stop the casualty, he will clench the rope in his left hand, and

bring the rope to the center of his chest.

(2) Direct Belay. This method is the safest for either raising or lowering a casualty in either moderate to steep terrain.

(a) To Lower a Casualty. First a swami wrap will be tied around a suitable anchor point. Two locking carabiners will be clipped into all of the wraps of the swami wraps, gates up. A figure of eight loop is tied into the end of the static rope and attached to the litter with a locking carabiner. After all the slack has been taken up between the litter and the anchor, the rope must be tied through an appropriate belay device. The belay device is attached to the anchor through one of the two locking carabiners on the anchor. A safety (French) prussic will be tied to the running end of the rope and clipped into the second locking carabiner on the anchor. While the casualty is being lowered, one person will control the rope running through the belay device. The safety prussic will be controlled by a second person. Should the primary belay man lose control, the person operating the safety prussic simply lets go and the prussic will bind onto the rope, stopping the casualty.

(b) To Raise a Casualty. The anchors will be established in the same manner as discussed in lowering the casualty with one minor change. The one change is that instead of running the rope through a belay device, the rope will only run through a locking carabiner. The load will be raised by the use of a mule team. The mule must consist of as many people as possible. The mule team will raise the load in as straight a line from the anchor as possible. If the space does not permit a ninety-degree angle away from the anchor is also an option. The mule team will walk backward until the last man reaches his limit of advance. Once he reaches that limit he will peel off the end and return to the front of the mule team. This process is continued until the casualty reaches the top. If the load becomes unmanageable, the safety prussic will be allowed to bind on the rope while the mule team repositions themselves. If the person operating the safety prussic can not see the casualty a Point NCO will be in charge of communicating with the mule team.

(c) To Belay the Rescuers. If the route is too steep or the footing is poor the rescuers may need some assistance either on the raising or lowering of a casualty. If this is the case a separate belay line will be established for the rescuers. The anchor and the belay line are established in the same manner. The same anchor can be used if it is suitable for the load. The rescuers will then tie either around the chest bowlines or swami wraps. A figure eight loop will be tied into the end of the static rope and connected to the bottom rescuer with a locking carabiner. The other rescuers will connect themselves to the same rope with middle of the line prussic. They will be connected in this manner so that they can adjust their position to the casualty.

9. **BARROW BOY.** (FMST.07.40e) A barrow boy is no more than an assistant to the litter on vertical to near vertical cliff faces. The barrow boy can be used for either the Stokes or the

SKED litters. For this situation the Stokes litter should only be used in the horizontal position. However the SKED can be employed in either the horizontal or the vertical positions.

NOTE: Static load, lowered slowly, little friction, so don't need 2 ropes.

a. Rigging the Barrow Boy.

(1) First the rescuer must ensure that a suitable anchor has been established, a proper belay has been constructed, and that a safety prussic has been constructed.
(2) Then the rescuer must ensure that if an A-frame is used that it has been constructed and anchored properly.

(3) Next the rescuer will tie a rappel seat on. (a sit harness can also be used) Then he will ensure that an around the body bowline is tied onto the casualty. A figure eight will be tied on to the end to act as the casualty's safety.

(4) Then after running the running end of the rope though the carabiners or pulley of the A-frame he will tie a figure eight loop at the end of the static rope. He will then attach the figure eight loop to his hard point with a locking carabiner. Then he will tie a middle of the line prussic above the figure eight and attach it to the same locking carabiner in his hard point. This is called the adjustment prussic. It is used to adjust the position of the Barrow Boy in relation to the litter. Next he will take six to eight feet of slack from the end of the line figure eight and tie a directional figure of eight with direction of pull down. (Note: the prussic should be between the end of the line figure of eight and the directional figure of eight.) The directional figure of eight is the attaching point for the litter and the casualty's safety.

(5) Once the rescuer and the litter are secured, the belay man must take all the slack out of the system. The rescuers will the maneuver the litter through the apex of the A-frame with the help of the point NCO.

(6) Once onto the cliff face the rescuer will then position him self with his adjustment prussic so that he can be of most assistance to the litter on the raise or lower. The rescuer will pull the litter out away from the cliff face so that the casualty rides smoothly up or down the cliff face.

(7) The Point NCO will be in charge of the belay men or the mule team. He will also communicate with the rescuer about the rate of speed, if the rescuer needs to be stopped along the route, and when he reaches the top or bottom of the cliff.

10. **TANDEM LOWERING.** The tandem lowering system can be used for the walking wounded, POW's, or more serious casualties when situation would not permit using the barrow boy.

a. The assistant to the casualty should first tie a rappel seat on himself and then assist the casualty with his.

b. The assistant will take only one belay line and tie a end of the line figure 8 loop, and clip this into his rappel seat.

c. A directional figure 8 will be tied approximately 12 inches up the rope from the figure 8 loop (with the loop pointed down) and this will be clipped to the casualty's rappel seat.

d. If needed, safety prussic cords should be tied above the casualty's directional figure 8. These safety cords should be attached in the same manner as the first man down a rappel.

e. The casualty and the assistant will lower as one, with the assistant helping on the way down the cliff.

11. **OTHER CONSIDERATIONS.** All of the techniques we have discussed for the evacuation of a casualty from top to bottom can also be used on a suspension traverse or rope bridge, with a slight variation in the belay line. Two belay lines may be used for rope bridges and the suspension traverse, if they are available. No matter what type of litter is used, the individuals involved in the evacuation must ensure that the head is always uphill or not lower than the feet.

UNITED STATES MARINE CORPS
Mountain Warfare Training Center
Bridgeport, California 93517-5001

FMST.0
7.34
04/12/02

STUDENT HANDOUT

ONE-ROPE BRIDGE

LEARNING OBJECTIVES

TERMINAL LEARNING OBJECTIVES. In a summer mountainous environment, conduct bridging, in accordance with the references. (FMST.07.34)

ENABLING LEARNING OBJECTIVES

(1) Without the aid of references, list in writing the criteria for site selection for a one-rope bridge, in accordance with the reference. (FMST.07.34a)

(2) In a summer mountainous environment with an assault load , cross a one-rope bridge using the rappel seat method, in accordance with the reference. (FMST.07.34b)

(3) Without the aid of references, list in writing the rescue techniques, in accordance with the reference. (FMST.07.34c)

(4) In a summer mountainous environment transport a SKED stretcher across a one rope bridge, in accordance with the reference. (FMST.07.34d)

BODY

1. **SITE SELECTION**

a. The two criteria for site selection for a one-rope bridge are: (FMST.07.34a)

1) There must be suitable anchors on both sides of the stream.

2) The anchors must offer good loading and unloading platforms.

b. Other considerations involved are:

1) The site chosen for the initial crossing does not have to be at the location for the construction of the bridge, just as long as the rope can be taken to the selected site the

C9

crossing.

2) The site chosen for the lead swimmer to cross should be as free as possible from obstacles in the water, such as large boulders, stumps or logs as discussed in *STREAM CROSSING.*

3) The anchors must be close enough for the 150 foot coil to reach both near side and far side anchors. Keep in mind that it will take approximately 1/3 of the 150 foot rope for tightening and anchoring of the bridge.

2. **DISTANCE ESTIMATION**. The follow methods can be used to determine the distance between anchor points:

a. Azimuth Method. Shoot an azimuth to a point on the far side of the intended obstacle to cross. Then move LEFT or RIGHT (perpendicular to the azimuth) until you get a 15 degree offset of your previous azimuth. Next, measure the distance that you paced in feet from your first azimuth to your second azimuth, and multiply that distance by three. This total will give you the approximate distance across the obstacle in feet.

b. Unit Average Method. Get three Marines to judge (best guess) the distance across the intended obstacle. Add up the total accumulated distance, then divide by three. This will give an estimation of the distance across the obstacle.

Example:	Marine 1	240 ft
	Marine 2	230 ft
	Marine 3	250 ft
		720 ft

Divided by 3 = 240 ft

3. **ORGANIZATION**. The organization of construction is broken down into four groups. They are as follow:

a. The Bridging Team - consisting of the bridge NCO and another Marine.

b. The Safety Line Team - consisting of the lead swimmer and his belay man.

c. The Mule Team - consisting 3 Marines for a 9:1, and 6 Marines for a 3:1.

d. The Security Team – consisting of the rest of the party.

4. **CONSTRUCTION**. This bridge is ideal for squad and platoon sized units, for its quick construction and minimal amount of equipment required.

a. Once the site for the bridge has been designated by the bridge NCO, the Safety Line Team will move up stream from this site to enter the water. Their first step is to flake out the rope. Once this is accomplished, the lead swimmer's will take a bight of the rope and tie a figure 8 loop at the end that will be going to the far side, ensuring the knot has a 18"-24" loop. The lead swimmer's upstream arm will go in it, with the belay man tending his rope the lead swimmer will cross the never using the flow to assist him. (Ferry angle as taught in STREAM CROSSING)

b. At the same time as the lead swimmer is crossing, the bridging team will be preparing the near side anchor. The bridging team will tie a swammi wrap around the anchor, using a sling rope or practice coil, and ensuring the square knot is behind or on the side of the anchor. Once the anchor is secure, flake out the bridging line in order to send it across.

c. Once the lead swimmer is across he will move to the site where the far side anchor will be established and secure his rope. The belay man will then move down to the near side anchor, and as close to the waters edge as possible. He will attach the bridging line to the safety by tying a middle of the rope in a figure 8 knot to the safety line and an end of the rope figure 8 knot to the bridging line connecting the two ropes with an 85 carabiner. The lead swimmer will then pull the rope across. The belay man with the help of another Marine will belay the bridging line across keeping tension on the rope to keep the rope out of the water.

d. After the bridging line is across the lead swimmer will detach the bridging line from the safety line and secure it to the far side anchor using a tree wrap, wrapping from right to left, ensuring there are 3-5 wraps.

e. Once the bridging line is secure. The lead swimmer and the belay man will secure the safety line on both banks, ensuring the rope is creating a ferry angle (minimum 45 degrees). This will be the safety line for the Marines crossing the bridge.

NOTE: If a fall from a rope bridge would result in death or serious injury, then utilize a safety line that will be anchored parallel with the bridge and hand tightened. The Marine crossing will clip into both ropes prior crossing.

f. On the near side bank:

1. The bridging team will need the following gear to build the mechanical advantage:

(a) Four steel locking carabiners & 3 -Three foot prusiks

2. Once the bridging line is secured on the far side, the bridging team will take one 85 and clip it into the near side anchor (swammi wrap). Then they will take the bridging line and clip into that 85, then using one three foot long prusik (16" loop) the bridging line will be secured to the 85 by tying a French Prusik, which will act as a braking knot. The bridging team will now pull the rope taut, and begin construction of the mechanical advantage.

3. The first step is to take one three foot prusik (16" loop) and tie a French prusik on the bridging line as far away from the anchor as possible. Then take one 85 and clip it to the loops of the French prusik.

4. Taking a bight from the running end of the rope clip it into the 85 hanging from the tails of the French prusik.

5. Then bring a bight back from the clipped in rope to the anchor, and taking one 85 clip it into the 85 created in step (2), take the bight of rope and clip into this 85.

6. Take one three foot prusik and tie a French prusik onto the bottom end of the bight clipped into the 85 created in step (3), take one 85 and clip it into the tail loops of the French prusik.

7. From the anchor, take a bight of rope from the 85 created in step (6) and clip it into the 85 created in step (6).

8. You have made approximately a 9:1 mechanical advantage.

NOTE: If the manpower is available the squad/platoon has the option of building a 3:1 mechanical advantage vise a 9: 1, the mule team will just have to add another body having no more then 3 bodies on a 9:1 and 6 bodies on a 3:1

g. Tensioning of the Bridge

1. Now that the bridge is built, the bridge NCO can call up the mule team to tension the bridge.

2. This procedure is facilitated by the mechanical advantage system that was just put into place. The braking knot (French prusik) is used to hold tension on the bridging line while the bridge is being tensioned.

3. The mule team will begin to pull on the running end of the rope coming out of the mechanical advantage. They will pull the rope straight back, trying to keep it in line with the bridge as best as possible.

4. The mule team will tension the rope as much as possible, the bridge NCO will monitor the system. Once the mule team cannot pull anymore tension, the bridge NCO will then have them hold the rope in place, and will reset the braking knot. Once the braking knot is set the bridge NCO will ask for slack from the mule team and cycle the system out. This process will continue until the bridge is tight.

5. On the last cycle the bridge NCO will set the brake knot and with the help of another Marine they will make a bight out of the running end and bring it around the tree while keeping tension.

6. Take a bight and make a complete round turn on the body of the 85 created in step (2). Last, the bridge NCO will tie two half hitches encompassing all the ropes just behind the anchor.

7. At this stage the bridge is tight and secured, the bridge NCO will now call up the remainder of the squad/platoon to cross. The bridge NCO will monitor the crossing of all Marines, and will be the last one to cross.

TRANSITION: Now that we have discussed the construction of the bridge, are there any questions over the material? Next we will discuss the crossing of the bridge.

5. **CROSSING.** (FMST.07.34b) The method used to cross is known as a horizontal traverse. This traverse can be accomplished in the following ways.

a. Rappel Seat Method

1) The Marine ties himself into a rappel seat and inserts a large steel locking carabiner with the gate facing down and away.

2) The Marine faces the bridge with his right hand towards the near anchor to snap into the bridge, once the 85 is locked onto the bridge rope, a helper flips the 85 over so that the locking nut screws down. A carrier rope with a girth hitch may be added to assist the Marine in mounting the rope.

3) The Marine hangs below the bridge from his rappel seat with his head pointing in the direction of the far anchor and allows his legs to hang free.

4) Pulling with his arms makes progress.

5) This method is the safest and therefore the preferred method.

6) If the Marine must take a pack across, he may wear it with the waistband secured: However, the preferred method is to have another carabiner (aluminum non-locking is sufficient) attached to the pack frame at the top and attach this to the bridge behind the Marine putting his legs through the shoulder straps and pulling the pack across with him.

7) One man at a time will cross, although one can load and another can unload concurrently.

8) Weapons will be worn across the shoulder; muzzle down with a tight sling securely attached to the weapon.

b. <u>Pulley Method.</u> This method is used when the one rope bridge is long, uphill, or speed is vital and the Marines crossing it have a lot of heavy personal equipment; M24OG's, radios, etc.

1) Equipment required. One pulley, four steel locking carabiners, and a hauling line twice the length of the obstacle.

2) Setting up the system. Construction of the first suspension point is done by attaching a pulley to the one rope bridge, one 85 carabiner is attached to the pulley, gate down, and a second 85 is attached to the first 85 with the gate facing the near side bank. Into this second 85, attach the hauling rope with a figure-8 loop. The figure-8 is placed halfway along the hauling rope and attached to the 85, which is then locked. Four feet down the line another figure-8 on a bight is placed into the second suspension point.

3) Connection. The Marine clips his 85 into the lower 85, his equipment is clipped into the second 85. The mule team starts to pull.

c. Helmets will be worn for all methods of crossing the bridge. Gloves are optional.

NOTE: There are additional methods of crossing a one rope bridge such as the Commando crawl, Monkey crawl and hand over hand techniques, however, they are not taught at MWTC due to safety reasons.

6. **<u>RESCUE TECHNIQUES</u>** If an individual is unable to complete the crossing on the one rope bridge using the rappel seat method of crossing (i.e. from a fall or exhaustion), a rescue will have to be made in the following manner: (FMST.07.34c)

a. Reach. First try to reach the victim by using an object such as a pole, your hand, if the victim is close enough, etc ...

b. Throw. If reaching the victim is impossible, try throwing a rope to the victim and have him attach the rope to himself, preferably his seat, and pull him back to the desired side of the installation.

c. Tow. If the victim is unable to catch or reach a rope being thrown to him or the victim is unconscious, tie a figure-of-eight loop into the middle of the safety line and then connect the rope to the rope bridge with a steel locking carabiner. The mule team will pull the carabiner up against the victim's seat and begin towing the victim to the desired side of the installation. The rope used to tow the victim should be twice the length of the span of the rope bridge (If necessary, ropes can be tied together to accommodate the span). This will allow towing from either side without having to throw or have a Marine carry the safety line back across the bridge each time a towing rescue is performed.

d. Go. If all else fails, the last option will be to go after the victim. The rescuer will move out onto the rope bridge with a safety line attached to himself at approximately eight feet from the end of the safety line. In the end of the safety line tie a figure-of-eight loop and insert a steel locking carabiner in the end of it. Once the rescuer has made contact with the victim. He will attach the steel locking carabiner to either the victim's seat (preferably the victim's carabiner that is attached to the rope bridge), or to the bridge itself. To ensuring that the carabiner is placed so that it pulls against the victim's carabiner. The mule team now starts to pull both the victim and the rescuer to the desired side of the installation.

e. Cut. If while crossing the rope over water the individual goes underwater and no other rescue technique can be employed, the rope will be cut.

7. RETRIEVING THE ONE ROPE BRIDGE

a. Before the bridge NCO sends the last Marine to cross, they must make the bridge retrievable.

b. The first step is to break down the mechanical advantage, ensuring the braking knot is set before doing so. The system will be broken down until the bridging line is attached only to the brake knot.

c. They will now take the rope around to the back of the anchor, and tie a slip figure 8, with the loop of the slip 8, take it and attach it to the bridging line just in back of the braking knot with a 85.

d. Now pull the running end coming out of the slip 8 until it tightens, and bites on the rope. Once the rope bites, tie a thumb knot behind the slip 8 as close to the slip figure 8 as possible.

e. The bridge NCO, with the help of another Marine, will now untie the swammi wrap, until all of the tension is on the bridging line.

f. The safety line will now be brought up, and attached to the end of the bridging line with an overhand knot.

g. The bridge NCO and the other. Marine will now cross the bridge.

NOTE: It is possible that the bridge may lose tension after breaking the bridge down. The packs of the last two Marines across can be sent over with other Marines, prior to the slip figure being tied.

h. After both Marines have crossed, the safely line will be brought up from down stream.

i. The tree wrap will now be taken off. Once the tree wrap is off, all knots and carabiners will be taken out.

j. The safety line will now be pulled, the bridge is retrieved.

UNITED STATES MARINE CORPS
Mountain Warfare Training Center
Bridgeport, California 93517-5001

WMC
04/12/02

STUDENT OUTLINE

SUMMER MOUNTAIN WARFIGHTING LOAD REQUIREMENTS

LESSON PURPOSE. The purpose of this period of instruction is to introduce the student to the different levels of gear requirements in accordance with MWTC's uniform codes. This lesson relates to mountain walking and tactical evolutions

OUTLINE

1. **BASIC UNIFORM REQUIREMENTS.**

a. Marines operating in mountainous terrain will generally wear the standard issued utility uniform. However, due to the wide range of temperatures, and sudden changes in weather in the mountains, Marines will find the need to be continually adding and removing layers. The amount of clothing worn may vary depending upon the severity of the weather, the activity level of the Marine, and the individual metabolism of the Marine. Every person in that unit should still maintain the same outer camouflage layer.

b. As part of the basic uniform, each man will be required to have in his possession, at all times, seven required pocket items. These seven items should be carried in the pockets of your utility uniform:

(1) Pocketknife

(2) Whistle

(3) Pressure Bandage

(4) Chapstick and sunscreen

(5) Sunglasses

C10

(6) Survival Kit and rations (fire starting material, signaling material, food gathering material, water procuring material, 1st aid, shelter material, and some high energy, lightweight snacks.)

(7) Notebook with pen/pencil

c. Some additional items that should be carried in your pockets at all times:

(1) Contact gloves

(2) Paracord (10 meters)

(3) Flashlight with tactical lens and spare batteries

(4) Chemlights

2. **ASSAULT LOAD.**

a. The Assault Load is equipment in addition to the basic uniform requirements. It is carried in the load-bearing vest (LBV), butt pack and the pack system. This is the equipment carried for short duration missions such as security patrols or during the assault. It is also carried at all times when away from the bivouac site.

(1) An extra insulating layer (Field jacket liner, wooly pully, etc.)

(2) Protective layer (ECWCS parka and trousers)

(3) LBV with 2 quarts of water and first aid kit

(4) Helmet

(5) Rations for the time away from your bivouac site

(6) Cold weather hat or balaclava

(7) Individual mountaineering gear

a. Sling rope

b. Carabiners

c. Rappelling Gloves

(8) Specialized mountaineering equipment

(9) Mission essential gear

3. **COMBAT LOAD**. The Combat Load is the equipment carried for longer duration missions.

The following items are in addition to the items already being carried in the Assault Load:

a. Sleeping bag with bivy bag

b. Isopor mat

c. Stove

d. Fuel bottle

e. Poncho (for expedient shelters or medevac purposes)

4. **EXISTENCE LOAD**. The Existence Load is to be packed for longer duration combat missions, where it is necessary to replace worn out items. These items could be uniforms, boots, personal hygiene gear, etc. The Existence Load should be brought to the forward elements once the situation allows.

5. **PACKING CONSIDERATIONS.** Because most Marines are familiar with how to pack a pack, these are general guidelines only.

a. When not wearing your protective layer or insulating layers, keep it handy at the top of the pack. When taking a break during a movement, you will be able to quickly don a layer to prevent getting chilled.

b. Keep your stove and fuel bottle in the outside pockets of your pack. They may leak and soak your equipment with fuel if stored inside your pack.

c. Keep the climbing rope in the rope bag. This will reduce the rope's exposure to petroleum products, ultra-violet light, cuts, and abrasions.

d. Ice ax is placed through the ice ax loops with spike up, adze outboard, and secured in place.

e. Crampons are attached to the outside of the pack with the tips covered and facing outboard.

U.S. Marine Corps

Physical Readiness Training for Combat

Department of Defense

Pentagon
Publishing

UNITED STATES MARINE CORPS
Marine Corps Combat Development Command
Quantico, Virginia 22134-5001

29 January 1988

FLEET MARINE FORCE REFERENCE PUBLICATION............0-1B

MARINE PHYSICAL READINESS TRAINING FOR COMBAT

1. PURPOSE

Fleet Marine Force Reference Publication (FMFRP) 0-1Marine Physical Readiness Training for Combat is intended for use by all Marines. It provides the information and references necessary to establish and conduct physical conditioning programs to prepare Marines for the physical demands of combat.

2. APPLICABILITY

Although not directive, this handbook is intended for use as a reference by all Marines in developing unit physical readiness training for combat. This FMFRP does not establish policy for the conduct of physical fitness training. Refer to Marine Corps Order 6300.3_ for applicable Marine Corps policy on physical fitness. It may also be used as a basis for instruction conducted by unit leaders or as a reference for instruction or training.

3. RECOMMENDATIONS

To improve this manual, the MAGTF Warfighting Center seeks comments and recommendations from the user. Officers and noncommissioned officers with constructive comments are encouraged to respond. Lack of response will be considered as concurrence. The User Suggestion Form which follows can be detached and forwarded to - -

Commanding General
Marine Corps Combat Development Command (WF 06)
Quantico, Virginia 22134-5001

4. DISTRIBUTION

Initial distribution of this FMFRP 0-1B is shown below. Addressees on this list should ensure that appropriate subordinate organizations are provided copies.

JOHN I. HOPKINS
Brigadier General, U.S. Marine Corps
Assistant Chief of Staff, Warfighting
Marine Corps Combat Development Command
Quantico, Virginia

DISTRIBUTION: TDJ

From:

To: Commanding General, Marine Corps Combat Development Command (WF 06), Quantico, Virginia 22134-5001

Subj: FMFRP 0-1B, MARINE PHYSICAL READINESS TRAINING FOR COMBAT, RECOMMENDATIONS CONCERNING

1. In accordance with the Promulgation Page to FMFRP 0-1B, which invites individuals to submit suggestions concerning this FMFRP directly to the above addressee, the following unclassified recommendation is forwarded:

Page________Article/Para No.________Line No.________Figure/Table No.________

Nature of Change:_____Add_____Delete_____Change_____Correct

2. Proposed New Verbatim Text: (Verbatim, double spaced; continue on additional pages as necessary.)

3. Justification/Source: (Need not be double spaced.)

NOTE: Only one recommendation per page.

MARINE PHYSICAL READINESS TRAINING FOR COMBAT

Chapter 1
PHYSICAL READINESS LEADERSHIP
Section I. Role of Physical Fitness in Combat

1101. PURPOSE OF PHYSICAL FITNESS TRAINING

Physical fitness training in the Marine Corps has one purpose: to prepare Marines to physically withstand the rigors of combat. All other goals of physical fitness training are subordinate to and must support attainment of this goal. The idea that only infantry or reconnaissance units and their attachments normally face physically demanding combat is wrong. This error must not influence the priority commanders of combat support, combat service support, aviation, and headquarters units give to physical fitness for combat. Physical fitness for combat has a high priority for all Marines. A sound, effective unit program, requiring limited time and material, offers a greater payoff in combat than many more expensive and time-consuming training programs. This handbook describes the concept and provides the ingredients of a program for commanders to use in carrying out one of their most serious responsibilities: ensuring that their Marines are physically ready for combat.

1102. SCOPE OF THE MANUAL

This handbook provides guidance for all leaders, trainers, and planners of physical training programs. It describes unit physical fitness for combat training in the following chapters:

a. Physical Readiness Leadership. Chapter 1 provides guidance to leaders in the conduct of physical combat readiness training.

b. Physical Readiness Training Programs. Chapter 2 discusses how to structure programs to meet physical readiness goals in a variety of combat training situations.

c. Physical Conditioning Activities. Chapter 3 discusses the primary physical conditioning activities which commanders may use to attain readiness for combat goals. Sections in this chapter address foot marches under load, strength-building activities, and activities which build confidence and the aggressive spirit needed in combat.

d. Combat Water Survival. As an amphibious force, Marines cannot ignore the importance of combat water survival training. Chapter 4 describes a program for commanders to conduct this training.

e. Competitive Conditioning Activities. Chapter 5 describes competitive activities which stress the relationship to combat actions, featuring team-building types of competition.

f. Evaluation of Performance During Training. Chapter 6 describes tests which measure individual and unit physical fitness for combat.

g. The Human Body. Chapter 7 covers the structure and functioning of the body.

1103. COMMANDER'S ROLE

Major General Commandant John A. Lejeune in the 1921 edition of Marine Corps Manual expressed our philosophy of leadership. Among other things, General Lejeune stated that "it will be necessary for officers ... to endeavor to enlist the interest of their men in building up and maintaining their bodies in the finest physical condition." It is significant that the Commandant prefaced these writings in 1921 with the statement that his thoughts were greatly influenced by the experience of World War I. We have never varied as a Corps from this belief that one of a commander's most serious responsibilities is to have Marines physically prepared for combat. Senior commanders and their staffs carry out this role primarily by providing command emphasis, including planning, support, and supervision. Commanders at the battalion/squadron and company/battery level execute a program of physical fitness for combat by close supervision and, most importantly, leadership by example.

1104. HISTORY OF MILITARY PHYSICAL READINESS

Every war has revealed our military physical deficiencies during the initial periods of mobilization. This realization followed the Civil War and has recurred regularly with each national emergency.

a. Training programs in each war were geared to the physical need of the era. Success was dependent upon the amount of time available during training to physically prepare Marines for battle conditions. Frequently, casualties in initial engagements were attributed to the inability of our Marines to physically withstand the rigors of combat over rugged terrain and under unfavorable climatic conditions. With adequate preparation, Marines have always handled the test of battle.

b. During World War II, the first physical conditioning doctrine that could be scientifically justified by testing was introduced. As the war progressed, this program was effective in the physical conditioning of millions of Marines for combat.

c. Postwar periods have traditionally been a time of consolidation. Unfortunately, some leaders considered the conditioning phase of training to be a wartime tool. With such a philosophy prevalent between wars, physical readiness was relegated to secondary importance resulting in a serious lowering of combat effectiveness. The initial commitment of Marines in Korea dramatically displayed this failure to recognize the extreme physical demands of warfare. Call-up of large numbers of reserve Marines and pressing them into the war within weeks after call-up mainly caused the loss in effectiveness.

d. Costly lessons learned from our military experiences over a period of years and the course of several wars led to an increasing interest in the physical conditioning of the individual Marine. No longer can we afford to emphasize physical fitness during wartime and de-emphasize it during peacetime. It is evident that, in spite of increased mechanization and modern weapons, physical readiness retains a vital place in the life of each individual Marine and in every unit within the Marine Corps regardless of mission.

e. Commanders are well aware of the need for rugged and well conditioned Marines. The daily demands of housekeeping, maintenance, support, training, operations, and other time-consuming tasks make it necessary for commanders to set aside time for frequent, regular, and vigorous exercise periods.

1105. PHYSICAL DEMANDS OF COMBAT

There are three primary elements of effective physical fitness for combat: lower body strength and stamina; upper body strength and stamina; and a competitive, combative spirit. Cardiovascular functioning is not treated as a separate element since it is a necessary foundation for the other three. Exercises and activities designed to improve Marines' condition in the three primary elements will foster good cardiovascular and respiratory condition.

a. Lower Body Strength and Stamina. Some common demands that combat places on lower body strength and endurance are marching long distances under load and functioning effectively at the destination; moving quickly and evasively under fire; and carrying wounded Marines to safety.

b. Upper Body Strength and Stamina. Some common demands that combat places on upper body strength and stamina are rapidly emplacing crew-served weapons; handling large-caliber ammunition for extended periods; climbing walls, cliffs, and other high obstacles; and performing field maintenance on air craft or heavy machinery.

c. Competitive, Combative Spirit. While part of this critical aspect of combat readiness is mental and emotional, robust physical condition and a training program which promotes physical aggressiveness greatly enhance a competitive and combative spirit. Activities which force Marines to overcome natural physical fear, which directly leads to fatigue, are particularly valuable.

1106. PHYSICAL EFFECTS OF COMBAT STRESS

It has been thoroughly documented that the added uncertainty and stress of combat have a major physical effect on Marines as well as the acknowledged psychological effect. In other words, in combat, fear equals fatigue. In training, we cannot easily reproduce this combat stress which reduces the effectiveness of individuals and units. However, we can produce fatigue and psychological doubt in training situations by developing a challenging physical training program which stresses the participants. This compels Marines to exercise their ability to continue to function under difficult and trying conditions. Such activities as long foot marches under load and difficult confidence courses train Marines to overcome their own fear and fatigue. Rugged competitive activities such as martial arts and pugil stick fighting prepare Marines psychologically to overcome an opponent as well as their own fear and fatigue. This sort of physically demanding training builds self-confidence and unit morale. The adage that "the unit that sweats more in peacetime will bleed less in battle" certainly applies to its physical training program. Athletic coaches have long followed this approach by such practices as having teams shoot foul shots after wind sprints in order to strengthen their mental conditioning and toughness. Although pushing Marines to their limit is necessary, commanders often set lower standards initially than ultimately desired. The program should then call for steadily increasing the difficulty. A good guide for a starting point is a standard that will be physically and psychologically demanding for no less than three-fourths of the unit. Expressed another way, no more than one-fourth of the unit should be working at less than their limit. The program should progress quickly, augmented, if needed, by remedial training for those unable to meet the unit standards, until the training is challenging for all but the most exceptional Marines in the command. Commanders must never forget that Marines expect to be challenged and thrive on being pushed to their limit.

1107. LIMITS OF PHYSICAL READINESS

A well-conditioned Marine unit has a significant advantage in combat, but this does not mean that physical conditioning can substitute for technical knowledge or good planning. For instance, units must, as discussed in chapter 3, pursue a continuing and challenging program of marches under loads. This does not disprove the axiom that fighting men, if required to carry over one-third of body weight, will generally become ineffective in battle. Leaders and planners must decide what is essential to be carried and must use an effective logistic distribution system, such as unit trains, to make remaining items available when they are needed. In a similar vein, the valid requirement to foster a competitive and combative spirit in Marines does not excuse leaders from the responsibility to practice tactics which will not needlessly risk lives. Physical readiness for combat is a vital goal, but it is only one part of a unit's overall readiness for the lest of combat.

Section II. Fundamentals of Physical Fitness

1201. FITNESS FOR MARINES

Total combat readiness includes technical proficiency and mental, emotional, and physical fitness. If any one of these attributes is lacking, combat effectiveness suffers proportionately. Without technical fitness, Marines lack the knowledge and skill to fight; without mental and emotional fitness, they lack the incentive and will to fight; and without physical fitness, they lack the physical ability and confidence to fight. Physical fitness in a Marine means a whole and healthy body, the capacity for skillful and sustained performance, the ability to recover from exertion rapidly, the desire to complete a designated task, and the confidence to face any eventuality.

1202. COMPONENTS OF PHYSICAL FITNESS

A sound body, free of disease and defect, does not in itself constitute physical fitness. Before an untrained Marine can be considered physically fit for combat, the following traits must be developed:

a. Strength. Every Marine needs enough strength to perform the heaviest task encountered in routine and emergency activities. The basic areas where strength is required are in the arm and shoulder girdle, abdomen, back, and legs. Muscles increase in size, strength, and firmness with regular and strenuous exercises. Without work, they grow flabby and weak.

b. Endurance. Each Marine needs sustaining power to maintain the maximum ability without undue fatigue. There are two types of endurance:

(1) Muscular Endurance. The Marine needs muscular endurance to fight the enemy under the most tiring combat conditions. Muscular endurance is the ability to perform continuous work over long periods of time. Endurance depends on the bloodstream's ability to deliver large amounts of oxygen and nutrition to the muscle masses and then carry away the waste products quickly.

(2) Cardiovascular-Respiratory Endurance. The development of cardiovascular-respiratory endurance ("wind") is necessary to maintain muscular endurance. Cardiovascular-respiratory endurance depends on the efficiency of the blood vessels, lungs, and heart. The maximum effort a Marine can exert over a period of time is limited by the capacity to absorb oxygen and expel carbon dioxide. The average Marine's cardiovascular-respiratory capacity can be greatly increased by exercise.

c. Agility. A Marine must be able to change direction quickly and as faultlessly as possible. The ability to react instantly and to maintain orientation during rapid changes of body position is important to survival. This important ability may be developed by conditioning exercises that require varied and rapid changes of body position on the ground and in the air, such as obstacle courses.

d. Coordination. Coordination is the ability to move all parts of the body in a smooth, efficient, and concerted effort (commonly called timing). A well-coordinated individual does not make useless movements. An individual who moves with precision and accuracy saves energy. Coordination is best developed by practicing diversified muscular activities and skills affecting all body parts.

1203. TYPES OF EXERCISE

Basically, there are two forms of exercise: isotonic and isometric. Both forms are contained in the activities found in this manual.

a. Isotonic. Isotonic exercises are those in which the expenditure of energy is regulated and released during concentrated efforts. The regulated expenditure of energy is controlled by both the mode of exercise and the individual's effort. This type of effort is common to most exercise and sports. To develop endurance, coordination, and flexibility, isotonic exercise must be used. Strength can also be developed or increased through this type of exercise.

b. Isometric. Isometric exercises are those in which maximum effort is applied and held until the engaged muscle energy is depleted during a single contraction effort. The individual exerts full force against an immovable object for a relatively short period of time (6 to 10 seconds) and then repeats the contraction several times with a short period of rest between each contraction. This type of exercise develops only strength; therefore, it has limited application.

1204. PRINCIPLES OF PHYSICAL CONDITIONING

Men and women vary in their physical makeup. Physiological (body) function and reaction also vary in proportion to heavy demands placed upon the body. To attain the maximum benefit without sacrificing Marines' welfare, fitness must grow with a careful program of conditioning. To allow for adjustments in body functioning as the conditioning program progresses and to ensure attainment of objectives, the following principles of physical conditioning must be applied:

a. Overload. As strength and endurance increase, the physical load must be increased until the desired level of fitness is reached.

b. Progression. In beginning stages, the load must be moderate. Gradual progression from this low state of fitness to a higher state is possible through application of a progressive program.

c. Balance. An effective program utilizes various types of activities and provides for the concurrent development of strength, endurance, and coordination as well as basic physical skills.

d. Variety. Some programs fail because the routine becomes boring. The most successful programs always include conditioning activities, competitive events, and military physical skill development.

e. Regularity. There is no easy or occasional way to develop physical fitness. Regularity of exercise is a must, with daily exercise preferred.

1205. THREE STAGES OF PHYSICAL CONDITIONING

Unconditioned or poorly conditioned Marines pass through the following stages in reaching the desired state of physical condition.

a. Toughening Stage. This stage is approximately 2 weeks long and is usually characterized by muscular stiffness and soreness followed by recovery.

b. Slow Improvement Stage. This stage is approximately 6 to 10 weeks long and is characterized by slow and steady improvement until the desired level, or a high level, of fitness is attained.

c. Sustaining Stage. This stage goes on indefinitely in order to maintain the level of conditioning achieved by passage through the previous stages.

1206. EFFECTS OF CLIMATIC CONDITIONS

Temperature, both atmospheric and body, affects the physical performance of Marines. Proper maintenance of body temperature through warm up exercise, proper dress in cold weather, and removal or adjustment of clothing in hot weather is necessary for effective performance and health. Climatic factors to be considered are the following:

a. Exercise In High Temperatures and High Humidity. Marines can endure strenuous physical activity in extremely hot temperatures if they are given an opportunity to become acclimated and if they take enough salt and water. It is essential to continue physical training programs in hot climates. Marines can better withstand high temperatures when they are well-conditioned. High humidity combined with high temperatures presents a serious danger. These conditions prevent the natural cooling of the body by the evaporation of perspiration. Training schedules should conform to the provisions of the base commander in respect to wet bulb conditions. Those who conduct training under these conditions should monitor weight loss and be careful to make fluid replacement adjustments.

b. Exercise at High Altitudes. Certain problems are encountered in conditioning Marines stationed in high altitudes because the heart undergoes greater exertion during exercise. It is particularly important that only light exercise be given initially at such altitudes. A Marine's body gradually adjusts to high altitudes within a few weeks. After this adjustment, progressively greater amounts of exercise are possible.

c. Exercise in Arctic Regions. Military duty in the arctic is so arduous that a high level of physical conditioning is essential. Because of the difficulties of carrying on physical conditioning exercises in extreme cold, Marines should be conditioned to the highest level possible before they arrive. A sustaining program will then maintain that level. When exercising in cold weather, Marines should be required to remove excess clothing to prevent them from becoming damp with perspiration.

1207. WARMING UP AND COOLING OFF

It is a fundamental physiological principle that Marines should warm up gradually before taking strenuous exercise. Such action speeds up the circulation to prepare the body to take an overload and helps to prevent injury to muscles and joints after exercising. Marines should be kept mildly active, walking, stretching, or performing some other mild muscular activity until breathing and temperature have returned to normal. Marines should never be allowed to cool off too rapidly; in cool or cold weather, they should put on additional clothing during the cooling-off period.

1208. PHYSICAL ACTIVITY AS AGE INCREASES

Combat makes severe physical demands on Marines. All Marines, regardless of age, must be physically ready to meet these demands. There is no physiological reason to cease exercise or exertion with age. Increased age usually brings increased responsibility which, in many instances, leads to a routine that can become almost devoid of physical activity. The key to fitness with increased age is to continue exercising at a reasonable level and to include exercise of a vigorous type in the daily routine. Older persons who have not regularly maintained a reasonable state of physical fitness require more time than younger persons to become fit. Such individuals usually require a longer period of time to recover from physical effort than younger Marines. If general health is good, evidence shows that older Marines can develop and maintain a degree of fitness which permits vigorous activity and proper performance of duties. It is both the individual's and the commander's responsibility that all Marines continue a daily sustaining exercise program.

Section III. Goals of Physical Readiness Training

1301. NECESSITY FOR PHYSICAL READINESS TRAINING

A very important objective of training is attainment and maintenance of operational readiness. Marines must be physically ready for operations at any time and under all conditions of climate and environment. A combination of training to develop proficiency in physical skills and conditioning to improve strength and endurance results in physical readiness for combat. The degree of physical fitness required of Marines can be acquired through physical exercises in a normal well-coordinated and closely monitored training routine. The performance of purely military duties, such as drills and marching, is not enough to build all the desired areas of fitness. Few recruits are physically fit for the arduous duties ahead of them. The softening influences of our mechanized civilization and the difficulties of conditioning Marines make physical fitness more important than ever before. If Marines are to be fully developed to and maintained at the desired standard of physical fitness, a well-conceived plan of physical readiness training must be a basic part of every training program. Marines cannot be adequately prepared in any other way for the hard work and arduous demands required on the battlefield.

1302. OBJECTIVE OF PHYSICAL READINESS TRAINING

The overall objective of the physical readiness training program is to develop individuals and units who are physically able and ready to perform their duty assignments or missions during training and in combat. To attain the objective of physical readiness, exercise activities must develop the following areas:

a. Strength and Endurance. Developing adequate strength to perform required duties and adequate endurance to sustain activity over a long period of time.

b. Muscle Tone. Developing muscle tone adequate to maintain proper posture and reasonable weight control.

c. Skills. Developing military physical skills which are essential to personal safety and effective combat performance. As skill is developed, agility and coordination are attained. The essential skills are-

(1) Marching Under Load. Marching with individual and unit weapons and equipment.

(2) Running. Distance and sprint running on roads and cross-country.

(3) Jumping. Broad jumping and vertical jumping downward from a height.

(4) Dodging. Changing body direction rapidly while running.

(5) Climbing and Traversing. Vertical climbing of ropes, poles, walls, and cargo nets. Traversing horizontal objects such as ropes, pipes, and ladders.

(6) Crawling. High crawling and low crawling for speed and stealth.

(7) Throwing. Propelling objects, such as grenades, for a distance with accuracy.

(8) Vaulting. Surmounting low objects, such as fences and barriers, by use of hand assists.

(9) Carrying. Carrying objects and employing man-carries.

(10) Balancing. Maintaining proper body balance on narrow walkways and at heights above normal.

(11) Falling. Contacting the ground from standing, running, and jumping postures.

(12) Surviving in Water (Or Other Specialized Situations) Using water survival techniques.

d. Character Traits. Jnstilling character traits which help accomplish military missions to include-

(1) Confidence. Confidence develops through achieving progressively more difficult tasks as physical ability grows.

(2) Aggressiveness. Participation in combative activities and contests develops desire and willingness to overcome an opponent.

(3) Reaction Under Pressure. Competitive contests and game situations train Marines to think and to act quickly while under pressure.

(4) Teamwork. Teamwork develops through competitive events in which a number of Marines must coordinate their efforts to accomplish a physical task.

1303. BENEFITS OF EXERCISE

The benefits of exercise are not always understood. Some of the more important results of exercise are listed below:

a. Improved Muscle Tone. Muscular tone improves and, at the same time, muscular strength and endurance are built up.

b. Cardiovascular-respiratory Endurance. Cardiovascular-respiratory endurance, or wind, improves through a process of opening up dormant lung capacity to absorb greater amounts of oxygen.

c. Circulation. Circulation of the blood speeds up and extends to a greater portion of the body as exercise forces the blood to service all parts of the body. The efficiency and effectiveness of the heart, lungs, and blood vessels improve.

d. Flexibility. A wider range of muscular movement is possible and rapidity in physical skills grows.

e. Elimination of Body Waste. Bending and twisting the body and the general speedup of body processes caused by exercise regulate and help eliminate body wastes.

f. Tension. Working off excess nervous energy and relief from daily worries and cares relieve tension. Participation in exercises leaves little time for worry.

g. Sleep. Sleep improves because muscles are healthfully tired after about of exercise. A by-product of sound sleep is relief of tension.

h. Obesity Control.. Control of obesity (fat) is made possible by using up excessive amounts of fat-producing food elements.

i. Injury Susceptibility. Susceptibility to injury is reduced through exercise. Muscles, tendons, and joints are strengthened. Injuries such as hernia, back strain, and joint sprains are less likely to occur if muscles are maintained in proper tone.

Section IV. Leadership Roles

1401. PSYCHOLOGICAL LEADERSHIP

The full development of a Marine's resources is not all physical. To be effective in developing physical readiness, leaders must realize that mind and attitude are also important to success. The more important psychological considerations are to—

a. Promote Understanding of the Value of Physical Readiness. A desire to be physically ready should be created in all Marines. Motivation is increased and Marines take greater interest in their individual physical fitness if they understand the value and benefits of vigorous exercise. When Marines realize their efforts are an investment in their own personal welfare, it should not be difficult to obtain their cooperation. Marines should understand the objectives, the benefits, and the value of each type of exercise activity in their program. They should also understand the relation of physical readiness to survival in combat.

b. Maintain a Positive Approach. Physical readiness training for combat is strenuous and demanding. It is a responsibility of leadership to create an atmosphere where all desire to participate fully. This attitude should be fostered. A negative approach must not be identified with physical readiness training, even with those having difficulty. Only in unusual cases should fear of punishment be the motivating factor behind good performance. For those few who cannot keep up or attempt to malinger, an effective remedial program is essential.

c. Seek Cooperation and Develop Morale. In a program placing maximum physical stress upon individuals, it is necessary to gain their cooperation. Favorable reaction is enhanced by proper planning and organization, challenging requirements, use of competition, and application of a progressive program resulting in physical fitness. As physical fitness grows, morale also grows.

1402. COMMAND AND SUPERVISORY FUNCTIONS

a. Command Functions. Commanders should take the following actions to support physical readiness training:

(1) Lead by personal example.

(2) Instill command interest and indicate to subordinate personnel the importance of this training to the welfare of the organization.

(3) Allot sufficient time for the achievement of objectives and monitor the use of such allotted time. The substitution of other training or routine duties for scheduled physical readiness training is unsound and unwise.

(4) Assign and properly utilize qualified personnel to supervise and conduct physical readiness training. If leaders are not competent, take action to ensure they become competent quickly. Reassign those who do not meet this standard to other duties.

(5) Make necessary facilities and funds available to support a program to develop physical readiness within all personnel.

(6) Measure the physical fitness of individuals and units in order to evaluate progress and to determine if the program is successful.

b. Supervisory Functions. Leaders responsible for planning, conducting, and supervising physical readiness training should take the following actions:

(1) Prepare physical readiness training schedules which apply the principles of physical conditioning and which aim for a particular type of program plan.

(2) Provide for wide participation of as many Marines as possible. All Marines, regardless of position or age, will benefit from regular exercise. In some instances, special efforts are necessary to overcome obstacles to regular and frequent training. Special effort is also necessary to ensure remedial conditioning. Such conditioning should occur for those who are physically substandard and after extended absence due to leave, sickness, injury, and travel.

(3) Prevent waste or unwise use of time allotted for physical readiness training. Time-wasters include unprepared instructors; assignment of one instructor to a group larger than a platoon; progression which does not keep pace with the physical development of the Marines; extreme formality; inadequate equipment or facilities which require waiting turns to exercise; and lengthy rest periods between exercises which interfere with the application of overload.

(4) Ensure that the program contains vigorous physical activity. Such activity places progressively greater demands upon the body during each exercise session and also over the duration of the training program. To be of benefit, exercise must tire the muscles and cause the heart to increase its rate of beat.

(5) Set an overall objective for each physical fitness program.Observe the training as necessary to ensure that the established objectives are being achieved.

(6) Observe physical readiness training to insure the use of a positive approach. To implement a positive attitude, small-unit leaders and instructors should personally set the example; have an understanding, fair, and sympathetic attitude; recognize individual differences; and motivate Marines toward their best effort.

(7) Guide and inform small-unit leaders and instructors concerning approved techniques, directives, and literature. As necessary, arrange for local training of instructors to include clinics, conferences, schools, and demonstrations.

(8) Determine the effectiveness of physical readiness training by personal participation in and observation of training, analysis of field inspection reports, and analysis of individual physical fitness test scores. Scores may be combined to reflect the fitness of the unit.

1403. SMALL-UNIT LEADERS AND INSTRUCTORS

a. Responsibility. The instruction and conduct of physical readiness activities are the function of company/battery commanders, platoonleaders and persons assigned as instructors. Experience has proved the effectiveness of physical fitness development when conducted in company-and platoon-size units under direct control of the leader with overall supervision by the parent-unit commander. For example, all the platoons of a company may exercise at the same time under the general supervision of the company commander, with each platoon conducting the assignment separately and under its own leadership.

b. Leader's Assignment. A small-unit leader or an instructor in a school or training activity is assigned to a combat unit or to a support unit. In this assignment, the leader is responsible for all training to include physical readiness training. In a different situation, the noncommissioned officer (NCO) or officer may beassigned as a full-time physical readiness instructor. This contrasts to a unit leader assignment where only part of the time is devoted to such training. In either case, the leader will hold an important and vital position for the physical fitness of Marines.

c. Leader's Training. Leaders may come to the assignment either fully or partially trained. It may be their first responsibility for the development of physical fitness. If they have had previous training through experience, make certain that their information is supplemented with study of this manual. If they have had professional training in physical education during civilian life, but no military experience, they should also use this manual to learn the methods used by the Marine Corps. A new leader should take advantage of various ways to learn including attendance at leader training courses, self-study, practice, and discussion with more experienced leaders.

d.Leader's Objective. As a physical readiness training instructor, the leader has two general objectives. The first is to motivate Marines to want to be physically fit. The second is to conduct a program that will develop a high degree of physical fitnss. Motivated Marines will react enthusiastically to such a program. It aids greatly in achieving local program objectives.

e. Leader's Personal Fitnes. A unit leader who must instruct and demonstrate physical activities must be in physical condition to do the job without undue physical stress.The leader should be able to do those things that must be demonstrated. The leader's strength, endurance, posture, and skill should set the example. This does not mean that the leader must excel, as other Marines do not expect championship performance. However, they do expect, and deserve, a creditable showing of fitness for the job.

f. Leader's Knowledge. The leader must have three types of knowledge to properly administer physical readiness training. Theyare- -

(1) Knowledge of Marines. The leader must understand Marines, know how to lead and motivate them, understand how they learn, and apply this knowledge wisely in the day-to-day training situation.

(2) Understanding of Body Functioning. A more intelligent exercise program results from understanding and applying the principles which govern physical conditioning of the body. The leader with such knowledge can better prescribe, adjust, and regulate exercise types, amounts, and progression to attain fitness.

(3) Understanding Exercise Activities. The leader needs to understand the contribution each type of physical activity makes to physical fitness, and how to use each activity to develop fitness. Skill to demonstrate and lead the various activities is a necessary part of technique and is invaluable to the instructor or small-unit leader.

Chapter 2
PHYSICAL READINESS TRAINING PROGRAMS
Section I. Development of a Program

2101.GUIDANCE FOR PLANNERS

This chapter instructs planners on physical readiness training procedures. It contains program planning guidance including factors to consider when developing programs; steps in assembling a program; definitions of activity packages and systems of exercise; selection of activity packages; and selection of systems used in implementing those packages.

2102. MARINE CORPS MISSION

The mission of the Marine Corps is to seize and defend advanced naval bases and to perform other missions as the President of the United States may direct. Traditionally, to perform this mission, Marines have been projected into the area of operations by Navy vessels and have attacked hostile beaches over the shore in landing craft, assault amphibian vehicles, and helicopters. Recently, performance of this task has been somewhat complicated by the introduction of the concepts of maritime prepositioning and the air-landed Marine expeditionary-brigade. Under these concepts, Marines, landed at airfields in foreign countries, will join their equipment at commercial ports. The distance between these airfields and ports is often 10 or more miles. In time of international crisis, it can be expected that transportation assets will not be adequate. Marines from all elements of a Marine Air-Ground Task Force may have to march from the airfield to the site of their equipment. All Marines must be physically conditioned to rigors of conducting foot marches with individual weapons and equipment.

2103. EVALUATION OF FITNESS

The Marine Corps mission is such that all Marines must constantly and consistently achieve a high level of fitness that prepares them for the demands of combat. While Marines have many different MOS's, all Marines must be prepared for the demands of marching under load and performing basic infantry tasks, such as rear security and patrolling. While the physical fitness test remains our universal measure of individual fitness, it is the commander's responsibility to observe and evaluate the unit's ability to perform effectively in combat. Often the commander's most effective evaluation tool is weekly physical training sessions. The commander should participate in these sessions and see that they are fully integrated into the unit physical conditioning program.

2104. INTEGRATION OF TRAINING

Training time must be used efficiently and wisely. Every opportunity to integrate physical training into other training activities should be seized. For example, foot marches under load can be integrated into the normal daily activities as a way to move from one training or work site to another. This integration reflects realism in training and should be used to maintain overall proficiency. To accomplish this requires imagination and enthusiasm on the part of the commander. The area of operations in which the training is conducted determines what can be integrated and how. If possible, training should also consider the terrain and climate of the area in which the unit will subsequently conduct training or operations. Training should include familiarization with special equipment and the application of specialized techniques to tactical principles. Activities such as gun drills and command post emplacement drills also have physical conditioning value, particularly if supervised and done competitively or measured against time standards.

2105. ACTIVITIES AT THE UNIT LEVEL

Physical readiness training is a command responsibility and is generally conducted in Marine units at the company and platoon levels. Battalion/squadron commanders must constantly supervise, making sure that all Marines are physically prepared for combat.

a. Physical readiness training should usually be conducted weekly by each battalion/squadron-level unit, using events designed for all personnel (e.g., marches under load, battalion/squadron runs, military field days, etc.).These events enable the commanders to demonstrate personal leadership, and observe and evaluate the physical condition and combat physical readiness of their Marines. This exercise period should be designed to elevate the unit's morale and emphasize unit identity through the wearing of uniforms and the carrying of battalion and company guidons. The battalion/squadron commander should make every effort to be a visible participant in the unit's physical readiness program on a weekly basis. These activities could be conditioning marches under load or a weekly calisthenics/unit run.The suggested times for these activities are early on Monday morning or late on Friday afternoon after the training week has been completed. The uniform may be physical training gear orutilities and boots.

b. Commanders should be cautious of long runs in boots and utilities. This is not to suggest that units should not run in utilities, but to remind commanders that boots are designed for marching, not running. If commanders choose boots and utilities, they should select off-road routes which will provide a cushioning effect.

c. Units should be made well aware of the commander's intent before the unit falls out for exercise. If the exercise period is conducted early in the morning, it need not necessitate an extensive shower/clean-up period afterward, so long as all Marines from the commander on down remain dirty. Commanders should inform their Marines not to fall out for these sessions in clean, pressed uniforms.

d. The physical readiness training of headquarters units is often difficult to manage. Often physical training is left for individuals or sections to develop and manage on their own. This approach can be effective, but if not supervised, it can result in a "paper" program. Headquarters unit commanders should endeavor to conduct weekly integrated training which will enable them to evaluate the physical readiness of their personnel. Often this training has to be scheduled before or after normal working hours to maximize participation of the entire unit.

2106. STEPS IN PLANNING

To implement workable and effective programs (as directed by CMC ALMAR 261/87) the planner must-

a. Determine the Type of Program Needed. Marine units are inherently different in organization and mission. The physical readiness program must be tailored to the mission and to the current physical condition of most unit personnel. Programs to meet this need are of the following types:

(1) **Developmental Programs.** Marines in a beginning or poor state of physical readiness need a program which will develop strength, endurance, physical skills, and character traits which are beneficial to successful accomplishment of military missions. Such programs should be applied progressively to rise gradually to a peak of fitness and skill.

(2) Maintenance Programs. Once Marines reach the sustaining stage of conditioning, their goal is then to maintain this level by participation in a maintenance program.

(3) Remedial Programs. The term "remedial" is usually applied to those individuals or groups who possess substandard physical fitness. For example, a remedial physical conditioning program could be applied to persons who are overweight, who fail to reach physical fitness test standards, or who have missed extended periods of conditioning due to illness, injury, extended hospitalization, or other absence. Saturday training sessions are designed as remedial sessions for all individuals who either miss the daily training or have failed to meet adequate standards. Note that personnel in light duty or no-duty status should be expectedto attend all training sessions, observing, supervising, or assisting as needed.

b. Determine the Time Required. The amount of time for training operations varies considerably. However, every unit can find time to conduct physical readiness training. Frequent (i.e.,daily) physical training of short duration-30 to 60 minutes-is preferred over occasional longer periods. Other demands for training time are so urgent that every minute of time allotted for physical readinesstraining should be used. Determine the time required per week anddivide it into daily blocks.

c. Organize for Various Group Size. It is essential to stress exercise rather than formality. Marines must complete the program where they are- -on the training field, in the motor pool, on the range, next to the classroom, in the office area, in the shop, aboard ship, or elsewhere. Although desirable, it is not always possible to assemble company-size units for physical training. Platoon-size groups are appropriate for the proper conduct of physical conditioning activities. Certain situations may require exercise programs for section- or squad-size units. This manual outlines programs for all situations and types of organizations.

d. Allow for Weather and Exercise Area. In programing and scheduling, the climate and terrain often govern the selection of activities.

(1) Weather changes cause differences in temperature, rainfall, wind chill, and snow. These changes should be anticipated as they dictate the type of program. Alternate plans should be part of the schedule. Nevertheless, normal weather changes and rain should not drive training inside.
(2) Local terrain and available exercise areas may also influence the selection of activities and the type of program which it is possible to support. Some activities can be completed in nearly any area.
e. Plan for Seasonal Change. As most physical readiness training is conducted outdoors, it is necessary to recognize seasonal change. A program should be divided into fall, winter, spring, and summer parts. Seasonal change also causes change in light. For example, an early morning program started in the summer will have ideal lightconditions, yet in the fall or winter, darkness will occur at that same hour and interfere with the conduct of the program. Develop programs in seasonal blocks, and make provisions for anticipated changes.

f. Consider Needed Facilities. An excellent program can be conducted with practically no facilities since there are exercises which require no equipment. However, a better program can be developed when supported by certain facilities and items of equipment. Proper command support, plus ingenuity, will solve this problem. (Items of equipment, when necessary to support the recommended exercises, are included in the chapters on exercise activities.)

g. Specify Appropriate Uniform. The uniform worn for exercising depends upon the season of the year, the state of the weather, and local regulations. Whenever possible, Marines should be dressed alike. Undershirts are preferred as the upper garment when the weather permits. A uniform that restricts the free movement of the body should not be worn when exercising.

h. Consider Availability of Instructors. Leaders who can lead and direct the scheduled activity must be available. Organizational units should train junior officers and noncommissioned officers down to squad or section leaders to instruct and lead the various activities.

i. Select Activity and System. With the type of program needed and the objectives in mind, select an activity package or an integrated training package for each day's scheduled physical training. At this same time, the system to be used in employment of the selected activity or activities must be determined. This selection affects equipment, areas, instructors, transportation, and other support requirements.

j. Secure Command Participation and Support. Prepare and brief the commander to assure full understanding of the objectives and administration of the program. The full participation and support of the commander will greatly improve the level of success.

k. Supervise Execution. Determine needs, publish the program, and supervise its execution as a necessary part of developing the unit's training schedule.

2107. SELECTION OF ACTIVITIES AND SYSTEMS

a. Exercise Activities. The planner must determine exercise activities which will be appropriate to include in the program. An exercise activity is a single means of exercise usually identified by the name applied; for example, running, log exercises, and obstacle course. Many exercise activi ties can become part of activity packages.

b. Activity Packages. Many of the physical activities described in this manual are arranged in prescribed sequences and are known as activity packages. An activity package is a number of exercises of the same type, assembled as a group or a set, and arranged in a specific sequence. Exercise packages are organized in such manner that not more than 15 minutes will be required to complete the execution of any package. Each type of activity is explained in laterchapters. The number of available drills, tables, or circuits; the manner of organization; and the contribution each makes to the total program are covered. Full understanding of this information will greatly assist in developing effective programs. Various designations are used to identify exercise packages; for example, conditioning exercises when arranged in a set order are known as drills, and other packaged activities are designated as tables or circuits. The following are activity packages:

(1) Activity packages for groups:

o Conditioning Drill Two
o Conditioning Drill Three
o Rifle Drill
o Log Drill
o Grass Drill
o Running Tables
o Guerrilla Tables
o Circuit-Interval Table
o Combatives Tables
o Relay Tables

(2) Strength Circuits:

o Fixed Circuit
o Movable Circuit
o Simplified Circuit (Circuit Interval Table)

(3) Activity packages for individuals:

o The 6-12 Plan
o Weight Training
o Isometric Exercise

c. Advantages of Using Packages. These of exercise packages simplifies scheduling and conducting of exercise and results in the following benefits:

(1) Schedule development is simplified as the planner assembles packages which will satisfy the training objective. There is no need to deal with selection of individual activities or to be concerned about the amount of time to be expended on each.

(2) Any 15-minute period, and in some cases less time, can be scheduled or used to perform an activity.

(3) If longer periods of time are available or if the objective demands, several packages can be assembled to provide a more complete period of activity.

(4) Marines are assured a balanced set of exercises or activities as each package is carefully arranged to reach all muscle groups.

(5) The instructor can concentrate on the conduct of a vigorous workout as the type and duration of the activity have already been determined.

d. Nonpackaged Activities. Several types of activity are not packaged. Activities in this category are conditioning marches, unit runs, obstacle courses, combat water survival swimming, teamcontests, and team sports. These activities can be scheduled in combination with packaged activities, or they may be scheduled separately. Often they require a longer period of time. Most nonpackaged activities require a 50-minute period to satisfactorily complete their objective. For conditioning marches, a few hours are required. The benefits of these activities should not be overlooked as some desirable objectives cannot be attained without them.

e. Systems of Exercise. Several methods or systems of organizing exercise and activity packages can be used. Each system is based upon a specific organization as follows:

(1) Single Activity System. The unit leader immediately assumes command of the unit at the beginning of the exercise period. The leader moves the unit to a predesignated exercise site at double time, forms the unit in a circle around the leader, groundsclothing and equipment as appropriate, and quickly moves into the exercise routine. There is usually no time to teach; therefore,the Marines must know the activity to be used. At the conclusion of a 5- to 15-minute period, the unit leader returns the unit to the instructional area at double time and releases the unit for the next scheduled activity.

(2) Progressive Activity System. All Marines (company or platoon) complete activities in the same order during the period. For example, Drill One is followed by dual combatives, and finally a 1-mile run. This system is usually progressive from a warm up activity, such as Drill One, to an activity which contributes in a major way to one of the objectives, such as aggressiveness development through combatives or cardiovascular development, such as running.

(3) Rotating Activity System. Set up the same number of activities or stations as there are platoons in the company. Each platoon rotates through each station in turn. With four platoons in a 50-minute period, about 10 minutes can be devoted to each station. With three platoons, approximately 15 minutes may be spent at each station. Activities must be of a type that can be covered in the time allotted. For example, with three platoons, Station 1 could be Conditioning Drill One; Station 2, running; and Station 3, a team contest.

(4) Circuit System. Set up a number of stations to provide various types of exercise equipment and/or items of apparatus. The idea is to keep all Marines busy and exercising vigorously for a short period of time at each station. The fixed strength circuits and the movable strength circuits are examples of this system. Station changes must be rapid, and the exercise must be started quickly after each change. Since the objective is to exercise at top speed, the motivation comes from frequent change of activity by moving to another station. Rotation by station groups continues until all Marines have covered all stations.

(5) Interval System. This system stresses the development of strength and endurance. It involves heavy work for a given distance within a specified time, alternated with lighter work and recovery, but never stopping during the workout. This procedure is repeated, and the intensity is increased gradually as physical condition improves, but always with adequate recovery. The important concept is stress, recover, stress, recover, and so on. This system is often applied through running, but other activities of a continuous nature may also be used. An example is the Circuit-Interval Table.

f. Steps in Selecting Activities and Systems. The planner should follow these steps.

(1) Evaluate the needs of one unit with emphasis upon unit mission, objective, and time available.

(2) For each day, select an activity package or a combination of activities which will contribute to the objective.

(3) Then determine the system to be used in implementing the selected activities. There are several possible choices and much opportunity for flexibility in program development. For example, the weekly program may contain various exercise packages, systems, and time periods. Additional guidance is contained in following chapters.

2108. SAMPLE WEEKLY PROGRAMS

The variety of weekly programs given shows available alternatives. Programs and schedules other than those illustrated may be assembled to provide schedules to fit any situation. Many other scheduling combinations are possible through use of the packaged activities, varied time periods, and different types of assigned duty.

a. Sample A. Figure 2-1 illustrates a 60-minute time allotment in which the single activity system is used with activity packages that vary from day to day. In this program, 5 hours of training are scheduled for the week. This program needs some type of integrated unit training to supplement it each week. A bimonthly conditioning march of 10 miles in 3 hours is suggested to round out this program. The Saturday session is designed for remedial physical training.

NOTE: A physical training activity must be conducted at a minimum of once every 72 hours. After 72 hours, muscles begin to atrophy.

Figure 2-1 is available in the illustration section at the end of the book

b. Sample B. A more comprehensive weekly program (fig. 2-2) illustrates the scheduling of various-length periods, a variety of activities, and the use of three systems during the week. In this schedule, 3.5 hours are included for the week. The following should be noted concerning this schedule:

Figure 2-2 is available in the illustration section at the end of the book

(1) On Monday, a 60-minute period is available for physical readiness training. Three 20-minute packages are scheduled. Each platoon, under its own leadership, will progress through each scheduled activity in turn.

(2) On Tuesday, Wednesday, and Thursday, only 30-minute periods are available. Here each platoon leader supervises each platoon in a single activity system.

(3) On Friday, the company commander directs training to exhibit personal leadership example and supervision. In this example, only one strength circuit and one cross-country course are available, and each will accommodate only one platoon. For this reason, Conditioning Drill One station is included. A platoon rotates to each station within the 60-minute period.

(4) The Saturday session once again is designed for remedial physical training.
(5) Bimonthly, a 10-mile/3-hour march under load will be necessary to round out this program.
NOTE: Cardiovascular/respiratory exercise to improve endurance must be of long enough duration (20 minutes or more) to induce and maintain oxygen debt.

c. Sample C. A third sample schedule (fig. 2-3) illustrates the use of the single activity system for 5.25 hours of training. This schedule includes sustaining-type activities for a unit that has passed through both the toughening and slow improvement stages of conditioning. The use of competitive activities is featured to holdinterest and provide self-motivation. A bimonthly 10-mile/3-hour march under load will be necessary to round out this program. The Saturday session once again is designed forremedial physical training.

Figure 2-3 is available in the illustration section at the end of the book

NOTE: Thc body will respond to hard work or physical training without soreness and stiffness provided the work or physical training is conducted regularly.
NOTE: Excrciscs of short duration increase strength and bulk but do little to effect endurance.
NOTE: Thc times shown in the charts arc participation times.

Section II. Remedial Physical Training

2201. NEED FOR ATTENTION

Remedial physical training is a process by which physically substandard individuals are conditioned to meet standard levels of performance. To achieve a full degree of operational readiness, it is necessary to bring all Marines up to the prescribed standard. Experience has demonstrated that some Marines have difficulty due to a poor state of fitness, obesity, or lack of motivation. Attention to these individual deficiencies will help improve unit combat readiness.

2202. IDENTIFICATION OF INDIVIDUALS

The company or battery commander identifies persons who cannot achieve the prescribed level of proficiency through the physical fitness test or as a result of observation during training. The commander notes particular weakness of body parts as indicated by failure of certain test events. These Marines are then placed in a special remedial program either at platoon, company/battery, or battalion/squadron level. Extra time is devoted to overcoming the weakness. These extra periods of conditioning may be during or after duty hours, as determined by the commanders. Other persons need reconditioning as result of hospitalization. Restoring physical fitness to damaged areas of the body is accomplished through progressively graded physical activities under professional supervision, not by the physical fitness trainers.

a. Types of Deficiencies. Physical deficiencies which can be corrected by exercise fall into several categories.

(1) Lack of Strength in One or More Body Parts. The major muscle areas concerned are the arms and shoulder girdle, back, abdomen, and legs.

(2) Lack of Overall Endurance. Usually, there is deficient muscular and cardiovascular endurance.

(3) Deficiency in Coordination and Agility. In these cases, physical skill is not developed to a satisfactory degree in activities such as crawling, running, jumping, climbing, traversing, vaulting, pushing, pulling, lifting, and carrying.

(4) Overweight or Underweight. Either condition may interfere with physical fitness and mission accomplishment. Lack of exercise is not always the cause. The cause may be malfunction of normal physiological functions or it may be poor health habits such as over- or under-eating, lack of adequate rest, or overconsumption of alcohol.

(5) Lack of Motivation. Not all persons are motivated to attain or maintain **a** desirable state of fitness. Some Marines do not understand the importance of physical fitness, some find proper exercise too difficult, and others find it inconvenient.

b. Causes of Deficiencies. There are several causes for Marines being deficient in physical fitness:

(1) Absence of exercise.

(2) Exercise which fails to develop all muscle groups and components of fitness.

(3) Exercise which is not vigorous enough or which lacks progression.
(4) Injury or illness which depletes fitness.
(5) Inadequate amounts of sleep or rest.

c. Methods of Detection. The commander has several means by which to detect physical deficiencies:
(1)
from those who pass. Make an analysis to determine the cause of failure as revealed by test scores.
(2) Observe Marines as they perform physical tasks-
o Marines who have difficulty during training or in physical types of work,
o Marines who have become obese and therefore experience difficulty.
(3) Be alert to those Marines who are often on sick call, returnees from hospitalization, or newly assigned.

2203. ADMINISTRATION OF REMEDIAL ACTION

a. Group. Attitude Marines who are singled out as being physically deficient are self-conscious and are not always convinced they need extra help. Within a deficient group, motivation may be low. These Marines must be convinced that a remedial program which is tailored to their needs will help them and will eliminate their deficiencies. Encouragement is often needed and desirable.

b. Leadership. The leader of this group must study individuals and know their deficiencies. The unit leader should counsel Marines individually, observe them closely as they progress through the remedial program, maintain records, and adjust the program as required.

c. Measurement.. Whenever improvement in performance is noted, Marines should be measured by the physical fitness test either individually or by group. When testing reveals an individual to be satisfactory, that individual should be released from the remedial group. There may be exceptions to this policy in the case of Marines who are overweight or in the execution of an exercise program prescribed for some special purpose.

d. Organization of a Remedial Group
(1) A remedial group is usually a small group of Marines within a company or battalion. In some situations, it may be a larger group numbering 50 to 75 Marines. In the case of a larger group, divide the Marines initially into subgroups according to ability. Prescribe exercise loads commensurate with their ability. General conditioning is sufficient in the early part of the program to qualify Marines who are on the borderline. Qualification will reduce the size of the group, permitting more individual and specialized attention to those who remain.
(2) As the program progresses, regroup individuals who have like deficiencies in order that they can concentrate on their weakness. For example—

(a) One group may be weak in the arms and shoulders as revealed by failure in the chin up event. This group, with an assistant instruction charge, can work on pull-ups, rope climb, pushups, rifle or logexercises, horizontal ladder, and similar types of exercise.

(b) Another group may be weak in general endurance as revealed by the3-mile run. This group could profit from participation in conditioning drills, running, grass drill, and strength circuit.

(3) It may be necessary to form some groups to overcome weaknesses in skills, such as an inability to throw, to quickly change direction while running, to crawl rapidly, or to carry a load. Lack of coordination or lack of practice may cause these deficiencies. In this instance, instructors must provide an opportunity to practice and correct poor form and other errors as they are noted.

Chapter 3
PHYSICAL CONDITIONING ACTIVITIES
Section I. Marching Under Load

3101. GENERAL

Few physical fitness activities are as directly related to readiness for combat as foot marches under load. In addition to the obvious physical conditioning and unit cohesiveness benefits, marches under load prepare most Marines for numerous foreseeable tasks in combat. The idea that only infantry and reconnaissance units actually require training marches to prepare them for combat is clearly a mistake. In addition to the very real possibility of Marines from all types of units being used to fill combat-depleted infantry units, it is likely in maritime prepositioning force operations that aviation, combat service support, and command element personnel will have to march with weapons and equipment, from the arrival airfield to the equipment marshalling site. In American coastal cities, the harbor is nearly always 10 or more miles from the airport. It is difficult to imagine any different condition in a lesser developed country .A further combat-related benefit is that foot marching under load is the most effective way to develop leg strength in the context of a unit training program. In his classic study of troop performance in World War II, Men Against Fire, S.L.A. Marshall noted that the greatest single weakness of replacements of all occupational fields was leg strength. Truck drivers must push trucks which are stuck in mud, and headquarters communicators must climb hills to emplace antennas. The demands on the lower body in combat apply to all Marines, and foot marching is one activity which can help build these muscles while being conveniently integrated with the rest of the training program.

3102.TRAINING GOALS

The standard for success of a foot march is very simple to measure: did the unit arrive at the destination at the prescribed time with Marines in condition and required equipment present to accomplish the mission?A progressive program can increase a unit's readiness. It can instill pride by increasing distances and rates of march and by selecting routes over increasingly challenging terrain as the Marines become better conditioned. However, the ability to execute the mission at the conclusion of the march must remain the standard of success.

3103.MOVEMENT PLANNING

The basic considerations in planning a foot march are the mission, tactical situation, terrain and weather, and the units to participate. The success of the march will depend largely upon the thoroughness with which it is planned. A successful march is characterized by adherence to prescribed routes and time schedules, the efficient employment of the means available, and the ability of the unit to accomplish its assigned mission upon arrival at the destination.

a. Movement Orders. Movement planning culminates in the preparation and issuance of an operation order prepared in the standard, five-paragraph format. Necessary annexes are attached to furnish detailed information required for the movement. Written movement orders are rarely prepared at company level. A discussion of the annexes is contained in this chapter.

b.March Planning. March planning, as discussed here, is the planning conducted at battalion level. March planning may be organized into the following steps:

(1) **Preparation and Issuance of the Warning Order.** In order to afford subordinate units the maximum possible time to prepare for a pending move, a warning order containing all available information about the march is issued. The amount of planning time available will determine the time of issuance and the content of the warning order.

(2) Estimate of the Situation. In his estimate, the commander considers the mission, terrain, weather, time, and space factors, available routes, available transportation for the movement of equipment and/or shuttling of Marines, enemy capabilities, disposition of own forces, physical condition and training of Marines, and courses of action available to the command.

(3) Organization and Dispatch of a Reconnaissance. Party very march plan is based on as thorough and complete a ground reconnaissance as time and the situation will permit. Map and aerial reconnaissance are valuable in formulating a plan, but are not a substitute for ground reconnaissance. Route reconnaissance is accomplished by a reconnaissance party which usually consists of a reconnaissance element, an engineer element from the attached or supporting engineer unit, and a traffic control element. Unit standing operating procedures generally establish the basic composition of the reconnaissance party. It is modified as necessary to meet the requirements of a particular march. A recommended method of reporting information obtained by the route reconnaissance party is shown in figure 3-1.The minimum information required from the reconnaissance party is—

(a) Available routes and conditions.(Routes may be specified by higher headquarters.)

(b) Recommended rate of march.

(c) Selection of start point and release point, or confirmation of the suitability of start point and release point previously selected by map reconnaissance.

(d) Confirmation of location of the assembly or bivouac area.

(e) Location of critical points on the route.

(f) Distance between critical points on the route and total distance from start point to release point.

Figure 3-1 is available in the illustration section at the end of the book

(g) Location of obstacles and estimation of necessary Marines and equip- ment needed to repair and maintain routes.

(h) Number of guides required and their location on the route.

(4) Development of Detailed Movement Plan. Some of the elements included in a detailed movement plan are-

(a) Organizing the Column. To facilitate control and scheduling, units will be organized into serials and march units and given an order of march. In determining the order of march, the march planner must consider the enemy situation and the desirable order of arrival of the units at the destination. Where dispersion is required, a unit may be organized into two or more columns, each assigned a different route.

(b) Using Reconnaissance Information. Results of the route reconnaissance will be used to select the route(s); determine the start point, critical points along the route(s), and the release point; and select the rate of march.

(c) Determining March Computations. March unit pass time is based on the strength, formation, and rate of march. The pass time of the marching columns, plus necessary time distance computations, will be used to determine the completion time of the march.

(d) Drafting of Road Movement Table. Using the completed march computations, a draft road movement tabic is compiled.

(e) Checking the Plan. Using the draft road movement table and a road movement graph, the movement plan is checked to ensure that it conforms to the directive of the higher headquarters and the battalion commander's instructions.

(5) Preparation and Issuance of the Road Movement Order. After the plan has been checked and approved by the commander, an operation order is prepared and issued. The operation order may be in written form or issued orally, and is accompanied by a road movement table, overlay and/or strip map, and appropriate administrative details.

(6) Road Movement Table. A road movement table, prepared as an annex to the operation order, provides serial commanders with arrival and clearance times at critical points along the route of march. It also provides the column commander with information as to the proposed location of elements of the column at various times.

(7) Overlay/Strip Map. An overlay serves the normal purpose and should show, as a minimum, the present location of units, route of march, critical points, and the new location of units at the destination. A strip map is a schematic diagram of the route of march and shows landmarks and critical points with the distances between them. A strip map may be issued as an annex to the road movement order, in addition to or in lieu of an overlay.

c. March Computations. Prior to issuance of the operation order, the S-3 must verify time and space computations as they provide him with the necessary data for the preparation of a road movement table.

(1) Time-Distance. Time distance (TD) is determined by dividing the distance to be traveled (D) by the rate of inarch (R):

TD = D/RWhere TD = hours D = distance in kilometers R = kilometers per hour

(2) Length of ColumnThe length of column (LC) is used to determine the pass time of a column. The sum total of the following two parts determines the length of column. (See pars, (a) and (b)).The two parts are the space occupied by the Marines alone (including the distance between Marines) and the sum of the distances between the units of the column (column gap).

(a) The length of column of Marines alone is determined by multiplying the number of Marines by the appropriate factor selected from the table below. The length of column does not include distances between units. LC Marines = No. of Marines x factor.

(b) The total distance in meters between units is determined as follows:

Step One. Determine the number of gaps between serials (total serials minus one).

Step Two. Multiply the number of serial gaps from Step One by the length (in meters) between respective units.

Step Three. Determine the number of gaps between march units (total march units minus one, minus the number of serial distances).

Step Four. Multiply the number of march unit gaps from Step Three by the length (in meters) between respective units.

Step Five. Add the totals from Step Two and Step Four to get the total meters for the column gap.

EXAMPLE:A battalion foot column is organized into 12 platoon-sized march units and 3 company-sized serials. REQUIRED: total column gap distances when there are 100 meters between serials and 50 meters between march units.

(3) Pass-Time.Pass-time (PT) is the time a unit takes to pass a specified point. For foot columns, the pass-time is determined by applying the following formula:PT (minutes) = LC x FACTOR (for appropriate rate of march).

PASS-TIME FACTORS - FOOT TROOPS

.0150 for 4.0 km/h

.0187 for 3.2 km/h

.0250 for 2.4 km/h

.0375 for 1.6 km/h

EXAMPLE: Determine the pass-time of a unit whose length of column is 1,500meters and is marching at a rate of 4 km/h.PT (min) = 1,500 x .0150 (the factor for 4.0 km/h) = 22.5 mm.

(4) Completion Time. Completion time is the time of day that a march will be completed. Completion time is determined by using the following formula: Completion Time = SP time + TD + PT + Scheduled Halts (other than normal hourly halts).

EXAMPLE: A column's starting point (SP) time is 0700.The time-distance is6 hours and 40 minutes. Pass-time of the column is 30 minutes. A 35-minutelunch halt has been scheduled. What is the completion time of the march?Employing the 24-hour clock system, the formula can be applied as follows for simplified addition of the times:

HrMin SP time...............0700

TD.....................0640

PT.....................0030

Lunch halt..............0035

Completion time.........13 105or 1445

The march will be completed at1445.

(5) Experience Tables. Based on previous movements made by a unit, data is accumulated to facilitate march planning. Such data includes approximate pass-times lor various elements of the battalion. The S-3 can utilize these data rather than computing them each time a march is scheduled. Such experience tables serve to reduce the time required to complete the computation phase of march planning. Matter appropriate to the unit standing operating procedure should be integrated therein.

d. Road Movement Graph

(1) A road movement graph is a time-distance diagram used in planning, preparing, or checking road movement tables, and for controlling marches. The graph provides a visual representation of a march plan so that conflicts and discrepancies may be prevented in the planning stage, before congestion occurs on the route. It is not usually issued as a part of the order. Road movement graphs may be applied to small units, to movements of a single column, or to a large organization scheduling separate elements, moving by various means, with different rates of march, over one or more routes.

(2) To construct a road movement graph, use the following steps: (See fig. 3-2.)

Figure 3-2 is available in the illustration section at the end of the book

Step One. Determine the number of hours available for the march. Designate the lower left corner of a sheet of graph paper as the SP time or the earliest even hour before the march is to begin. Select a scale and plot the hours available in sequence from the left on the horizontal axis.

Step Two. Determine the distance to be moved in kilometers. Indicate the SP at the lower left corner of the graph sheet and, using an appropriate scale, plot the number of kilometers on the vertical scale from the SP to the release point. Indicate the location of critical points on the vertical scale.

Step Three. At the proper distance from the SP, draw a horizontal line indicating the location of the RP. Indicate by a vertical line the hour when the movement must be completed. Plot lines representing route restriction, if any, at the proper distances and times on the graph.

Step Four. Determine the pass-times of foot and motor elements of the column.

Step Five. Starting at the SP at the specified hour, plot the movement of the head of the leading element. If the rate of march is 3 kilometers per hour, the lead element will have moved 1.5 kilometers in 30 minutes, 3 kilometers in 1 hour, etc. Plot the trace of the lead man to the RP. The last Marine will cross the SP in pass-times after the lead man. Measure this time on the graph and plot the trace of the last Marine of the column. The lines describing the head and tail of a march serial are parallel. Indicate the time subsequent serials reach the SP and plot the trace of the head and tail. Check to see that the plan complies with all restrictions and orders. If there are any violations or conflicts, the plan may be made to conform by changing the order of march, the starting time of the serials, the planned rate of march, or the organization of the column.

e. Conversion to the Metric System. The adoption of the metric system for expressing linear distances requires a simple means of conversion.

(1) To accomplish this, use the following tables:

o Multiply inches x 2.54 to obtain centimeters. Multiply centimeters x .39 to obtain inches,

o Multiply yards x .91 to obtain meters. Multiply meters x 1.1 to obtain yards,

o Multiply miles x 1.6 (or 8/5) to obtain kilometers. Multiply kilometers x .62 (or 5/8) to obtain miles.

(2) Linear distances can be accurately measured on maps which have been converted to the metric system by using a straightedge graduated in centimeters (cm).The centimeter scale can be used in lieu of the graphic scale when the scale of the map is shown, as follows:

Map Scale:	One cm Equals:
1:500,000	5,000 meters
1:250,000	2,500 meters
1:100,000	1,000 meters
1:50,000	500 meters
1:25,000	250 meters

3104. INDIVIDUAL LOAD

Since the backbone of Marine ground combat forces is the infantry, tactical mobility is largely dependent on the marching ability of the individual Marine. The load that each Marine must carry is the greatest impediment to mobility. In general, the commander must make every effort to reduce the individual load of his Marines to the absolute minimum. While the load has the greatest effect on Marines marching on foot, the value of carrying the minimum essential supplies and equipment applies for all Marine forces.

a. Fighting Load. The fighting load consists of items of clothing, equipment, weapons, and ammunition that are carried by, and essential to, the effectiveness of the combat Marine and the accomplishment of the immediate mission. The fighting load should not weigh more than 40 pounds. The commander must ensure that his Marines' loads- -and his own- -be stripped to the bare minimum. The addition of small, light-weight items in excess of the limit should not be tolerated, since cumulatively, these items will take a toll in energy. Every extra pound a Marine must carry decreases his combat effectiveness.

b. Existence Load. The existence load consists of items other than those in the fighting load that are required to sustain or protect the combat Marine, that may be necessary for increased personal and environmental protection, and that are not normally carried by the individual. The existence load is normally carried by the Marine's unit. Marching, but not engaged in combat, a Marine may be expected to carry a maximum of 50 pounds of supplies and equipment.

c. Considerations

(1) The primary consideration is not how much a Marine can carry, but how much he can carry without impaired combat effectiveness-moral or physical. The combat strength of a unit is not counted simply in numbers of Marines, but in the numbers of willing and physically able Marines.

(2) It is generally better to risk temporary inconvenience from lack of health and comfort items than to exhaust Marines due to overloading.

(3) It is a fundamental truth that men become physically exhausted more quickly when under the stress of combat. Marines must be required to carry less into battle than they are conditioned to carry in training. Marines should be conditioned for carrying weight, but should be equipped in combat for fleetness of foot.

(4) A common mistake is to base the fighting load on the gear and supplies necessary to meet every contingency. The commander cannot reasonably expect the Marines to carry enough gear for every possible eventuality. The items to be carried must be based on reasonable expectations.
(5) It is the commander's responsibility to procure the transport to carry any additional gear. As a rule of thumb, a rifle company, or unit of similar size, requires one 5-ton truck and trailer in support to carry additional gear. In cold weather, or other conditions in which the necessary personal gear increases, this requirement will increase.
(6) The commander must ensure that the supply system provides, on a dependable and timely basis, the balance of essential supplies and equipment not carried by the unit. Marines must feel confident that they will be supported with the necessary supplies and equipment.
(7) In training, commanders must implant in their Marines the pride in operating under austere conditions. To effectively demonstrate the need for this spartan mentality, the commander must personally set the standard. Marines should be trained in field craft, foraging techniques, and the use of caches and field expedients. Maximum use should be made of captured stores.

3105. MARCH TECHNIQUES AND PROCEDURES

a. Organization. A command executing a march is organized into march units, march serials, and march columns, as necessary for control.
(1) March Units. A march unit is a unit of command which moves and halts at the command of a single commander. The march unit normally corresponds to one of the smaller Marine units such as a platoon or company.
(2) March Serials. A march serial consists of one or more march units organized under the senior officer and given a numerical or alphabetical designation to facilitate control. A serial is usually a battalion, but may be a company if the battalion is marching alone.
(3) March Columns. A march column is composed of elements of a command moving over the same route. It may be composed of one or more serials. To facilitate control, a column commander is designated. A column is normally a regiment or larger unit, but may be a battalion if marching alone.

b. Order of March. The order of march is determined by the mission, terrain, and the commander's desired order of commitment of units into action.

c. Control and Coordination Measures. The commander establishes initial control of the march by designating control measures in his march order. The most frequently used measures include:
o Start point and release point.
o Time at which head or tail of column passes the start point.
o Rate of march.
o Order of march.
o Route of march.
o Location of command post.
o March communications.

d. The Pacesetter. The pace-setter is an experienced individual carrying the same load as the majority of the Marines and marching from 4 to 10 meters at the head of the column. (See fig. 3-3.)The pace-setter's primary duty is to maintain the rate of march ordered by the column commander.

(1) Setting the Pace. The pacesetter does this by establishing the pace (length of step) and cadence (steps per minute) to obtain the prescribed rate of march. He should be of medium height so normal strides will be taken. Overstriding or understriding tends to tire leg muscles quickly and affects the combat efficiency of marching Marines. The officer marching at the head of the column supervises the pacesetter to ensure that he takes normal strides and maintains a uniform cadence.

Figure 3-3 is available in the illustration section at the end of the book

(2) Reducing Accordion Effect. Accordion effect occurs when the column alternately lengthens and shortens, causing the rear elements of the column to double time to maintain distance. It is caused by changes in the rate of march by the lead elements, usually after passing a slowing obstacle. To reduce this effect as much as possible, lead elements should slow their rate of march for a sufficient time after crossing obstacles to allow the rear elements to maintain the prescribed distance without running. The rear elements can signal the head when the last man has cleared the obstacle and the rate can be resumed. Since some accordion effect is nearly inevitable, the order of march should be periodically rotated to prevent units in the rear from becoming physically exhausted at the conclusion of the march.

e. Length, Rate and Pace

(1) **Length of March.** The length of march varies depending on terrain and weather, enemy situation, and physical and mental condition of Marines. The normal length of march for a 24-hour period is from 20 to 32 kilometers (12 to 20 miles) marching from 5 to 8 hours at a rate of 4 kilometers (2.5 miles) per hour (km/hmi/h).A march in excess of 32 kilometers in a 24-hour period is considered a forced march. Well-trained units that have been progressively trained will be able to achieve a unit goal of 60 miles in a 3-day period.

(2) Rate of March. The same factors of terrain, weather, load to be carried, and condition of Marines affect the rate of march. The commander considers these and selects a rate which will place his unit at the destination in the shortest possible time in condition to accomplish the mission. Unit standing operating procedure should state the standard rate of march for that unit over normal terrain for both road and cross-country movement and in both day and night. The commander then modifies this rate if the situation requires. Normally prescribed standards are as follows:

Figure is available in the illustration section at the end of the book

(3) Pace and Cadence. The normal pace is 30 inches. A pace of 30 inches and a cadence of 106 steps per minute result in a speed of 4.8 kilometers per hour or 3 miles per hour and a rate of 4 kilometers per hour or 2.5 miles per hour if a 10-minute rest halt per hour is taken. Since the pace of each man may vary, the cadence may have to be adjusted to maintain the prescribed rate. Slope of ground and surface footing affect pace. A normal cadence is relatively simple to maintain on level or gently rolling terrain on firm footing. Mud, sand, loose gravel, and hills all greatly affect pace.

f. Halts. Halts during a day's march are taken at regular intervals to rest personnel and adjust loads .Halts are specified by standing operating procedures or march order. Such factors as midday heat or enemy action may force the adoption of long midday halts or night marches. At long halts, each unit disperses to a previously selected location near the route of march.

(1) Normal Time and Frequency. Under normal conditions, a 15-minute halt is made after the first 45 minutes of marching. Following the first halt, a 10-minute halt is made after each 50 minutes of marching. Variations of this schedule are made when a scheduled halt time occurs when passing through a built-up area or when cover and concealment are required by the tactical situation and none is available. Observation posts may be established, if required for security of the unit during halts.

(2) Actions at Halts. All units in the column should be halted at the same time. At the halt signal, Marines should move to the side of the road, staying within the immediate vicinity of their unit. Marines should remove or loosen gear and sit or lie with feet elevated. Commanders inspect Marines and equipment, and corpsmen administer medical treatment, as required.

g. Communications. The four primary means of communications used in controlling foot marches are radio, visual, sound, and messenger.

(1) Radio. Radios are normally used for communications between platoons and higher headquarters in the march column. Radio transmissions should be held to the absolute minimum necessary for control and should be at minimum power required.

(2) Visual.. Visual communications may include flashlights, luminous markers, panel sets, pyrotechnics, and hand and arm signals. When contact with the enemy is not expected, the loss of light discipline from some of these signals may be a lesser security risk than the risk of disclosure from radio transmissions.

(3) Sound. Sound communications include voice and such signaling devices as horns and whistles. Some of their uses include assembling Marines at the conclusions of halts and providing warnings of air or chemical attacks.

(4) Messengers. Messengers are particularly effective during periods of radio silence and reduced visibility. Messages should be simple and brief to preclude mistakes by the messenger.

h. Security. Both active and passive measures are used against attack by enemy aircraft and indirect fire weapons during movement. Active measures include the use of air sentries and organic and attached weapons in accordance with the unit air defense standing operating procedure. Passive measures include the use of concealed routes and assembly areas, night marches, and extended interval between elements of the column. With imagination and planning, all of these measures can be practiced in training without detracting from the physical conditioning benefit of the march. In fact, such drills as rapidly taking cover during simulated attacks and then quickly resuming the march can prove physically challenging.

i. Reconnaissance. Reconnaissance determines in advance critical points along the march route such as bridges, fords, and obstacles in order that congestion or delay may be held to a minimum and local security provided. Paragraph 3103.C describes organization of the reconnaissance party.

3106.TACTICAL MARCHES

These marches are movements of Marines and equipment not in direct ground contact with the enemy but expecting early ground contact either en route or upon arrival at the objective. Some characteristics of this type of movement include security elements to the front, flanks, and, if necessary, to the rear of the march column. Additional considerations include the selection of concealed areas on defendable terrain for start and release points and rest stops. In training, infantry units of regimental size or smaller will normally derive the greatest benefit from tactical marches. The security requirements usually dictate a slower rate of march and, therefore, reduce the physical conditioning value for other types of units which would seldom move in combat with the expectation of early ground contact with the enemy. Two key points should be remembered regarding tactical marches in a training program.

a. Security. Although security requirements often slow the rate of march, it is a valid and important training objective for infantry units to work toward reducing march times while still practicing sound tactical security. March standing operating procedures, good land navigation skills, and simple but complete march orders contribute as much to this goal as physical conditioning. The idea that proceeding slowly enhances security is seldom true. Keeping the enemy off balance by rapid movement frequently is the best guarantee of security.

b. Training. In peacetime training, it is critical that support personnel who normally are located with infantry units in combat, participate completely with infantry units in forced march training. Not only must they not slow the infantry rate of march because of inferior conditioning, but they must be thoroughly familiar with the infantry unit's standing operating procedure for security on the march. Personnel from within the ground combat element such as artillery forward observer teams and combat engineer units, can be easily identified for this training and should be habitually associated with the same infantry units. Personnel from other elements, such as low-altitude air defense teams and radio battalion detachments, are more difficult to identify and earmark for support of the same infantry unit. Their normal mode of employment is in vehicles and their units frequently provide general support vice direct support. This does not, however, lessen their requirement to be prepared to provide support to dismounted infantry over terrain or in a tactical situation not conducive to vehicles. Such training needs to take place, and planning and cooperation among the elements of the MAGTF can overcome the obstacles.

3107.ADMINISTRATIVE MARCHES

These marches are movements of Marines and equipment made when no enemy interference is expected except from aircraft or indirect fire. During these marches, units are administratively grouped for ease of control and speed of movement.The term administrative foot marches should not be taken to imply that tactical considerations are disregarded.There are simply less stringent security measures used than when ground contact with the enemy is a distinct possibility.The normal formation is the route column with one file moving on either side of the road and with negligible security. Air sentries are used and dispersion between individuals and units is practiced. Although roads usually provide the fastest route, there may be reasons to conduct these marches cross-country (particularly as part of a training program). Administrative marches are the type of foot marches most practical for training programs for non-infantry units. These marches may also play a key role in the physical conditioning program of infantry units, because of the greater speeds possible due to lessened security requirements.

3108. FORCED MARCHES

A forced march is a march which covers a greater distance than 32 kilometer (20 miles) in a 24-hour period. Normally the extra required distance should be achieved by increasing the number of hours marched in a day beyond the norm (8 hours) rather than increasing the rate of march beyond the norm (4 kilometers per hour/2.5 miles per hour). However, there will be occasions when the situation or mission demands an increase in the rate of march.

a. Maximum Recommended Distances. The maximum recommended distances for forced marches are-56 kilometers (35 miles) in 24 hours; 96 kilometers (60 miles) in 48 hours; or 128 kilometers (80 miles) in 72 hours.

b. Sample Time Breakdown. A sample time breakdown for a forced march of 52 kilometers, beginning at first light is as follows:

Figure is available in the illustration section at the end of the book

3109. NIGHT MARCHES

Night marches are characterized by closed formations, more difficult control and reconnaissance, and a slower rate of march than day marches.

a. Control. Control is increased by reducing the distance between individuals and units, and by using connecting files to maintain contact between platoons and companies. Connecting files are normally constituted from personnel from the rear most march unit.

b. Safety on Roads. Because of reduced visibility, night marches on roads used by vehicles require attention to safety procedures to prevent accidents. Commanders should not think these safety measures apply only to peacetime training. In combat, nearly all vehicles operate with blackout lights at night, making visibility even more difficult than in garrison driving. If consistent with the tactical situation, the following measures can reduce the possibility of marchers being struck by vehicles:

(1) Use off-road trails and routes as much as possible.

(2) Guards to the front and rear of march units should be marked with strips of luminous tape and carry red-filtered flashlights. These measures are generally consistent with light discipline, while still affording reasonable warning to approaching drivers.

Section II. Conditioning Drills (One, Two, and Three)

3201. GENERAL

a. Description and Objective. Conditioning drills are calisthenic exercises. Each drill contains seven exercises organized and numbered in a set pattern. Each drill takes 15 minutes to complete. The objective of conditioning drills is to exercise all major muscle areas in order to develop strength, endurance, coordination, and flexibility.

b. Area and Equipment. Any level area is satisfactory for conduct of the drills. Drills One and Two contain ground exercises. If ground conditions are unsatisfactory, Drill Three should be used as it contains no ground positions. Usually, no equipment is required;however, if the group exceeds a platoon in size, an instructor's stand is necessary.

c. Formation. The extended rectangular formation is used in this drill. (See app. A.)

d. Starting Level and Progression. The starting level is six repetitions of each exercise. An increase of one repetition for each three periods of exercise in which the drill is performed is an acceptable rate of progression. This rate is continued until 12 repetitions can be completed. To maintain, continue the drill at 12 repetitions. To progress, move to 6 repetitions at a more difficult drill which exercises the same muscle group. Progression can also be gained by moving from Drill One to Drill Two, as Drill Two is more demanding.

e. Starting Positions. Starting positions vary with the exercise and are explained in each exercise.

f. Leadership. A principal instructor demonstrates and leads the drills. The instructor must be familiar with leadership techniques peculiar to conditioning drills to include the exercises, commands, counting cadence, cumulative count, formation, method of teaching the exercises, and utilization of assistant leaders.

g. Use With Other Programs. Conditioning Drills One, Two, and Three reach all major muscles of the body. They are easy to learn and to perform, and they are simple to administer and supervise. These features, coupled with the short time required for completion, the fact that no equipment is necessary, and adaptability to most areas of execution, make these drills possible in any programs.

3202. CONDITIONING DRILL ONE

This conditioning drill is similar to the calisthenics drill commonly known throughout the Marine Corps as the "Daily Seven."The "Daily Seven" may be used as a substitute for Conditioning Drill One.

a. Exercise l:High Jumper

(1) Starting Position. Feet separated shoulder width, knees flexed, body bent forward at the waist, arms aligned with the trunk and hips, elbows locked, palms facing, fingers extended and joined, head and eyes to the front.(See fig. 3-4, A.)(Elbows remain locked throughout the exercise.)

Figure 3-4 is available in the illustration section at the end of the book.

(2) Cadence. Moderate.

(3) Movement. A four-count exercise: at the count of—

(a) ONE -- Take a slight jump into the air, swinging the arms forward and up to shoulder level.

(b) TWO -- Take a slight jump into the air and swing the arms downward and back, returning to the starting position.

(c) THREE -- Take a vigorous leap into the air, swinging the arms forward and up to an overhead position, momentarily looking skyward, on returning to the ground the knees are flexed, head and eyes return to the front.

(d) FOUR--Repeat the action of count two.

b. Exercise 2: Bend and Reach

(1) Starting Position. Feet spread more than shoulder width, arms overhead, elbows locked, palms facing, fingers extended and joined, head and eyes to the front.(See fig. 3-4, B.)

(2) Cadence. Moderate.

(3) Movement. A four-count exercise: at the count of—

(a) ONE-Bend at the knees and waist, swing the arms straight downward and reach between the legs. Touch the ground as far to the rear as possible and look to the rear. (Elbows remain locked throughout the exercise).

(b) TWO-- Recover sharply to the starting position.

(c) THREE--Repeat the action of count ONE.

(d) FOUR-- Repeat the action of count TWO.

c. Exercise 3: Pushup

(1) Starting Position. Front leaning rest position. To assume this position there is a silent one-two count:on the silent count of one, assume the squatting position, heels together, elbows locked inside the knees, hands flat on the ground directly beneath the shoulders; on the silent count of two, thrust the legs to the rear, toes and heels together, body straight from head to heels.(Seefig. 3-4, C.)

(2) Cadence. Moderate.

(3) Movement. A four-count exercise: at the count of—

(a) ONE -- Flex the elbows lowering the body until the thick portion of chest touches the ground.

(b) TWO -- Raise the body until elbows are straight and locked.

(c) THREE-- Repeat the action of count ONE.

(d) FOUR -- Repeat the action of count TWO. (On returning to position of attention, the silent one-two count is used in reverse).

d. Exercise 4: Trunk Twister

(1) Starting Position. Feet are spread more than shoulder width apart, fingers placed behind neck, thumbs pointing downward, elbows back. (See fig. 3-4, D.)(Elbows remain well back throughout the exercise).

(2) Cadence. Slow.

(3) Movement. A four-count exercise: at the count of-

(a) ONE-- Keeping the knees locked and the back straight, bend forward at the waist sharply, with a slight recovery.

(b) TWO-- Twist the trunk to the left vigorously at the waist, keeping the elbow back. The left elbow is higher than the right.

(c) THREE-- Twist vigorously to the right, so the left elbow comes under the right.

(d) FOUR-- Straighten sharply to the starting position. NOTE: Do not attempt to touch the elbows to the knees on counts two and three.

e. Exercise 5: Squat Bender

(1) Starting Position. Feet are spread less than shoulder width apart, hands on hips, thumbs in small of back, elbows back. (See fig. 3-4, E.)

(2) Cadence. Moderate.

(3) Movement. A four-count exercise: at the count of—

(a) ONE-- Assume the squatting position, maintain balance on the balls of the feet, with trunk erect thrust arms forward to shoulder level, elbows locked, palms down.

(b) TWO-- Recover to starting position. Elbows are well back.

(c) THREE-- Keeping the knees locked, bend forward at the waist, touching the ground in front of the toes.

(d) FOUR-- Vigorously recover to the starting position.

f. Exercise 6: Body Twist

(1) Starting Position. On the back, arms extended sideward on the ground, palms down. The legs are raised to a near vertical position, feet together, knees locked.(See fig. 3-4, F.)

(2) Cadence. Slow-fast.

(3) Movement. A four-count exercise: at the count of-

(a) ONE-- Lower legs slowly to the left until they touch the ground near the left hand, keeping the knees straight and shoulders on the ground.

(b) TWO-- Recover the starting position by quickly raising the legs, keep knees straight and feet together.

(c) THREE-- Repeat movement of count ONE, except the movement is to the right side.

(d) FOUR-- Recover sharply to the starting position.

g. Exercise 7: Stationary Run

(1) Starting Position. Position of attention. (See fig, 3-4, G.)

(2) Cadence. Fast.

(3) Movement.

(a) At the command of execution, start running in place at double-time cadence, lifting the left foot first time cadence. Follow the instructor as he counts two repetitions of cadence;

e.g., 1, 2, 3, 4--1, 2, 3, 4. The instructor then gives informal commands such as FOLLOW ME, running on the toes and balls of the feet, keeping the back straight, speeding up the cadence to a sprint, raising the knees high, leaning forward at the waist, and pumping the arms vigorously.

(b) To halt the exercise, the instructor will count two repetitions of cadence as the left foot strikes the ground: 1, 2, 3, 4 - 1, 2, 3, HALT.

NOTE: When counting cadence, the instructor counts only as the left foot strikes the ground. The duration of the exercise is approximately 1 1/2 minutes.

3203.CONDITIONING DRILL TWO

a. Exercise 1: Jumping Jack

(1) Starting Position. Feet separated more than shoulder width, arms overhead.(See fig.3-5, A.)

(2)Cadence. Moderate.

(3) Movement. A four-count exercise: at the count of~

(a) ONE-- Jump to position with the feet together and assume the squatting position, swinging the arms sideward and downward, placing the hands palms down on the ground, elbows locked inside the knees.

(b) TWO-- Recover to the starting position by jumping to the side straddle and swinging the arms sideward overhead.

(c) THREE-- Repeat the action of count ONE.

(d) FOUR-- Recover to the starting position.

b. Exercise 2: Turn and Bend

(1) Starting Position. Side straddle, arms overhead.(See fig. 3-5, B.)

(2) Cadence. Moderate.

(3) Movement. A four-count exercise: at the count of-

(a) ONE-- Turn the trunk to the left and bend forward over the left thigh, attempting to touch the fingertips to the ground outside the left foot. Keep the left knee straight. On successive repetitions attempt to touch farther and farther to the side.

(b) TWO-- Recover to the starting position.

(c) THREE-- Turn the trunk to the right and bend forward over the right thigh, trying to touch the hands to the ground outside the right foot. Keep the right knee straight.

Figure 3-5 is available in the illustration section at the end of the book.

(d) FOUR-- Recover to the starting position.

c. Exercise 3: Eight-Count Pushup

(1) Starting Position. Position of attention. (See fig. 3-5, C.)

(2)Cadence. Moderate.

(3)Movement. An eight-count exercise: at the count of—

(a) ONE-- Assume the squatting position, palms on the ground directly beneath the shoulders, elbows locked inside the knees.

(b) TWO-- Thrust the legs to the rear assuming the front leaning rest position.

(c) **THREE**-- Flex the elbows until the thick portion of the chest touches the ground.

(d) **FOUR**-- Raise the body on a straight plane until the elbows are locked.

(e) **FIVE**-- Repeat the action of count THREE.

(f) **SIX**-- Repeat the action of count FOUR.

(g) **SEVEN**-- Recover to the squatting position as in count ONE (elbows locked inside the knees).

(h) **EIGHT**—Return sharply to the position of attention.

d. Exercise 4: Turn and Bounce

(1) **Starting Position.** Feet spread more than shoulder width apart, arms sideward at shoulder level, palms up. (See fig. 3-5, D.)

(2) **Cadence.** Slow.

(3) **Movement.** An eight-count exercise: at the count of—

(a) **ONE**-- Turn sharply to the left as far as possible, then recover slightly.

(b) **TWO**-- Again turn to the left as far as possible and recover as in ONE.

(c) **THREE**-- Repeat the action of count TWO.

(d) **FOUR**-- Recover sharply to the starting position.

(e) **FIVE**—Turn sharply to the right as far as possible, then recover slightly.

(f) **SIX**—Again turn to the right as far as possible and recover as in FIVE.

(g) **SEVEN**-Repeat the action of count SIX.

(h) **EIGHT**-Return to the starting position.

NOTE: The head and hips remain to the front throughout the exercise and the knees and elbows are locked at all times.

e. Exercise 5: Squat Stretch

(1) **Starting Position**. Attention. (See fig. 3-5, E.)

(2) **Cadence**. Moderate.

(3) **Movement**. A four-count exercise: at the count of—

(a) **ONE**-- Squat, placing the hands on the ground about 12 inches in front of the feet.

(b) **TWO**— Keeping the fingertips on the ground, straighten the knees completely and raise the hips.

(c) **THREE**-- Recover to position ONE.

(d) **FOUR**-- Recover to the starting position.

f. Exercise 6: Leg Circular

(1) **Starting Position.** On the back, arms stretched sideward, palms down, feet raised foot from ground, knees straight. (See fig 3-5,F.)

(2) **Cadence.** Slow.

(3) **Movement** A four-count exercise: at the count of—

(a) **ONE--** Swing the legs as far as possible to the left, keeping the knees straight and the legs together.

(b) **TWO--** Swing the extended legs over head with the thighs as close as possible to the trunk.

(c) **THREE--** Swing the legs as far as possible to the right.

(d) **FOUR--** Recover to the starting position.

g. Exercise 7: Stationary Run

(1) Starting Position. Position of attention. (See fig. 3-5, G.)

(2) Cadence. Fast.

(3) Movement

(a) At the command of execution, start running in place at double-time cadence, lifting the left foot first. Follow the instructor as he counts two repetitions of cadence; for example: 1, 2, 3, 4--1, 2, 3, 4. The instructor then gives informal commands such as FOLLOW ME. Running on the toes and balls of the feet, keeping the back straight, speeding up the cadence to a sprint, raising the knees high, leaning forward at the waist, and pumping the arm vigorously.

(b) To halt the exercise, the instructor will count two repetitions of cadence as the foot strikes the ground: 1, 2, 3, 4-1,2, 3, HALT.

NOTE: When counting cadence, the instructor counts only as the left foot strikes the ground. The duration of the exercise is approximately 1 1/2 minutes.

3204. CONDITIONING DRILL THREE

a. Exercise 1: Side Straddle Hop

(1) Starting Position. Position of attention. (See fig. 3-6, A.)

(2) Cadence. Moderate.

(3) Movement. A four-count exercise: at count of-

(a) ONE-- Take a slight jump into the air, moving the legs sideward (more than shoulder width apart); at the same time, swing the arms overhead (to an overhead position) clapping the palms together.

Figure 3-6 is available in the illustration section at the end of the book.

(b) TWO-- Take a slight jump into the air, swing the arms sideward and downward returning to the starting position.

(c) THREE-- Repeat the action of count ONE.

(d) FOUR-- Repeat the action of count TWO.

b. Exercise 2: Back Bender

(1) Starting Position. Standing, feet 12 inches apart, fingers placed behind the head. (See fig. 3-6, B.)

(2) Cadence. Slow.

(3) Movement. A four-count exercise: at the count of-

(a) ONE-- Bend the upper trunk backward, raising the chest high, pulling the elbows back, and looking upward. Keep the knees straight.

(b) TWO-- Recover to the starting position.

(c) THREE-- Repeat the action of count ONE.

(d) FOUR-- Recover to the starting position.

c. Exercise 3: Squat Thrust

(1) Starting Position. Position of attention. (See fig. 3-6, C.)

(2) Cadence Moderate.

(3) Movement. A four-count exercise: at the count of—

(a) ONE-- Assume the squatting position; heels together, placing the hands flat on the ground, shoulder width apart, elbows locked and inside the knees.

(b) TWO-- Thrust the legs to the rear, assuming the front leaning rest position, body in line from head to toe, heels and toes together.

(c) THREE-- Return to the squatting position as in ONE.

(d) FOUR-- Return to position of attention.

d. Exercise 4: Side Bender

(1) Starting Position. Feet are spread more than shoulder width apart, arms are raised sideward and overhead, thumbs interlocked palms to front, fingers extended and joined, elbows locked. (See fig. 3-6, D.)

(2) Cadence. Slow.

(3) Movement. An eight-count exercise: at the count of—

(a) ONE-- Bend to left as far as possible, then recover slightly.

(b) TWO-- Again bend to the left as far possible, then recover slightly.

(c) THREE-Repeat the action of count TWO.

(d) FOUR-- Recover sharply to the starting position.

(e) FIVE-- Bend to the right as far as possible, then recover slightly.

(f) SIX-- Again bend to the right as far as possible, then recover slightly.

(g) SEVEN-- Repeat the action of count SIX.

(h) EIGHT-- Recover sharply to the starting position.

NOTE:Keep the elbows and knees locked throughout the exercise. The bend should occur to the side and not the front.

e. Exercise 5: Knee Bender

(1) Starting Position. Feet are spread less than shoulder-width apart, hands on hips, thumbs in small of back, elbows back. (See fig 3-6, E.)

(2) Cadence. Moderate.

(3) Movement. A four-count exercise: on the count of—

(a) ONE-Do a knee bend, lean trunk forward at the waist, thrust arms between legs until the extended fingers touch the ground palms to the ground, hands 6 inches apart.

(b) TWO~Recover sharply to the starting position.

(c) THREE-Repeat the action of count ONE.

(d) FOUR-Repeat the action of count TWO.

f. Exercise 6: Bottoms Up

(1) Starting Position. Front leaning rest position. A silent one-two count is used as in the pushups. (See fig. 3-6, F.)

(2) Cadence. Moderate.

(3) Movement A four-count exercise: at the count of-

(a) ONE-- With the weight on the hands, and knees locked, jump forward bringing the feet as close to the hands as possible; look to the rear.

(b) TWO--Keeping the knees locked, thrust the legs backward assuming the front leaning rest position.

(c) THREE-- Repeat the action of count ONE.

(d) FOUR-- Repeat the action of count TWO.

g. Exercise 7: Stationary Run

(1) Starting Position. Position of attention. (Sec fig. 3-6, G.)

(2) Cadence. Fast.

(3) Movement

(a) At the command of execution, start running in place at double time, lifting the left foot first. Follow the instructor as he counts two repetitions of cadence; for example: 1, 2, 3, 4--1, 2,3, 4. The instructor then gives informal commands such as FOLLOW ME, running on the toes and balls of the feet, keeping the back straight, speeding up the cadence to a sprint, raising the knees high, leaning forward at the waist, and pumping the arms vigorously.

(b) To halt the exercise, the instructor will count two repetitions of cadence as the left foot strikes the ground: 1, 2, 3, 4 - 1, 2, 3, HALT.

NOTE: When counting cadence, the instructor counts only as the left foot strikes the ground. The duration of the exercise is approximately 1 1/2 minutes.

Section III. Rifle and Log Drills

3301. RIFLE DRILL

a. Description and Objective. Rifle exercises are conditioning exercises performed with a rifle. Each drill contains six exercises (fig. 3-7) and they are numbered in a set pattern. The drill takes 15 minutes to complete. The objective of rifle drills is to exercise the arms, shoulders, and back muscles in order to develop strength and endurance, particularly in the upper body. In units without rifles, log drills may be substituted.

b. Area and Equipment. Any level area is satisfactory for conducting this drill. Each Marine completes these exercises from a standing position and no ground contact is required. Each Marine will need a rifle and, if the group exceeds a platoon in size, then the instructor will need an instructor's stand.

c. Formation. The extended rectangular formation is used in this drill. (See app. A.)

d. Starting Positions. Starting positions vary with the exercises and are explained in each exercise. As in all set conditioning drills, the command used to start the exercise is STARTING POSITION, MOVE. The following directions apply to rifle drill.

(1) In those exercises which start from the rifle downward position, on the command MOVE, execute port arms as prescribed in NAVMC 2691, Drill and Ceremonies Manual, and then assume the starting position. The command to return to the men to the position of attention at the conclusion of the exercise is POSITION OF ATTENTION, MOVE.

(2) In exercises which terminate in the rifle downward position, on the command MOVE, execute the position followed by order arms as prescribed in NAVMC 2691.

(3) In the exercises which terminate in a position other than the rifle downward position, Marines first assume the rifle downward position before executing port arms and order arms.

(4) These movements are executed without command. This procedure promotes uniformity, but precision is not expected. To be effective, rifle exercises must be strenuous enough to tire the arms, but not to the point where the arms cannot move with precision.

e. Leadership. A principal instructor demonstrates and leads the drill. He must be familiar with leadership techniques for conditioning exercises and the peculiar techniques for rifle drill.

3302. EXERCISES PERFORMED WITH RIFLES

The exercises of rifle drill are outlined in the following paragraphs,

a. Exercise 1: Foreup, Behind Back

(1) Starting Position. Rifle downward, feet together.(See fig. 3-7, A.)

Figure 3-7 is available in the illustration section at the end of the book.

(3) Movement. A four-count exercise: at the count of—

(a) ONE-- Swing the arms forward and upward to the overhead position. Inhale.

(b) **TWO**-- Lower the rifle to the back of the shoulders. Exhale.
(c) **THREE**-- Recover to position ONE and inhale.
(d) **FOUR**-- Recover to the starting position and exhale.

b. Exercise 2: Lunge Side, Turn and Bend
(1) Starting Position. Rifle downward, feet together.(See fig. 3-7, B.)
(2) Cadence. Moderate.
(3) Movement. An eight-count exercise:at the count of—
(a) **ONE**-- Lunge sidewards to the left, swing the rifle forward and upward to the overhead position.
(b) **TWO**-- Turn the trunk to the left and bend forward over the left hip. At the same time, swing the rifle to a **low** horizontal in front of the left ankle.
(c) **THREE**-- Recover to position ONE.
(d) **FOUR**-- Recover to the starting position.
(e) **FIVE, SIX, SEVEN, and EIGHT**-- Repeat on the right side.

c. Exercise 3: Foreup, Back Bend
(1) Starting Position. Rifle downward, feet together.(See fig. 3-7, C.)
(2) Cadence. Moderate.
(3) Movement. A four-count exercise:at the count of—
(a) **ONE--** Swing the arms forward and upward to the overhead position.
(b) **TWO**-- Bend backward, emphasizing the bend in the upper back. The face is up.Keep the knees straight.
(c) **THREE**-- Recover to position ONE.
(d) **FOUR**-- Recover to the starting position.

d. Exercise 4:Up and Forward
(1) Starting PositionRifle downward, feet together.(See fig. 3-7, D.)
(2) Cadence. Fast.
(3) Movement. A four-count exercise:at the count of—
(a) **ONE**-Swing the arms forward and upward to the overhead position.
(b) **TWO-** Swing the arms forward to shoulder level.
(c) **THREE** Recover to position ONE.
(d) **FOUR**-Recover to the starting position.

e. Exercise 5: Foreup, Full Squat
(1) Starting Position. Rifle downward, feet in narrow stance. (See **fig.** 3-7, E.)
(2) Cadence. Moderate.
(3) Movement. A four-count exercise: at the count of—
(a) **ONE**-- Swing the arms forward and upward to the overhead position.
(b) **TWO**-- Swing the arms down to shoulder level and assume the squatting position.
(c) **THREE**-- Recover to position ONE.
(d) **FOUR**-- Recover to the starting position.

f. Exercise 6: Arms Forward, Side Bend

(1) Starting Position. Sidestraddle, regular stance, rifle forward.(See fig. 3-7, F.)
(2) Cadence. Moderate.
(3) Movement. A four-count exercise:at the count of-

(a) **ONE**-- Bend the trunk to the left. Keep the knees straight.

(b) **TWO**-- Recover to the starting position.

(c) **THREE**-- Bend the trunk to the right. Keep the knees straight.

(d) **FOUR**-- Recover to the starting position.

NOTE:Keep the rifle on the same level as the shoulders throughout the exercise.

3303. LOG DRILL

a. Description and Objective. Log exercises are conditioning exercises performed with a log. Each drill contains six exercises and they are numbered in a set pattern. The drill takes 15 minutes to complete. The objective of log drills is to develop strength and muscular endurance and, in this instance, under maximum loads. Log exercises also develop teamwork. Log may be used in lieu of conditioning drills after the Marines have become somewhat conditioned.

b. Area and Equipment. Any level area is satisfactory for conducting this drill. Each Marine completes these exercises from a standing position and no ground contact is required. If the group exceeds a platoon in size, then the instructor will need instructor's stand. Each six-person group or eight-person group will need a log. The logs should be from 6 to 8 inches in diameter. They may vary in length from 14 feet (for 6 people) to 18 feet (for 8 people).The logs should be skinned, smoother, and dried. The 14-foot logs should weigh approximately 300 pounds and the 18-foot logs, approximately 400 pounds. Rings should be painted on the logs to indicate each person's position. When not in use, the logs should be stored on a rack to keep them off the ground.

c. Formation. All the Marines assigned to the same log team should be about the same height at the shoulders. The recommended method of dividing the platoon is to have the Marines form a single file or column with short people to the front and tall people to the rear. Have the Marine assume their positions in the column according to shoulder height, not head height. When they are in position, they are given the command COUNT OFF BY SIXES (OR EIGHTS), COUNT OFF, to divide them into six- or eight-person log teams. Each team in turn, can then proceed to the log rack, shoulder a log, and carry it to the designated exercise area. The log teams form in columns in front of the instructor. With the Marine holding the log in the chest position, have them face the instructor and ground the log at least 10 yards between columns and 10 yards between log teams within the columns.

d. Starting Positions. The Marines fall in, facing the log, with their toes about 4 inches from it. The basic starting positions and commands are as follows:(See fig. 3-8.)
(1) **RIGHT-HAND STARTING POSITION, MOVE**. It the command MOVE, move the left foot 12 inches to the left, and lower the body into a flat foot squat. Keep the back straight, head up, and arms between the legs. Encircle the far side of the log with the left hand. Place the right hand underneath the log.(See fig. 3-8, A.)

(2) LEFT-HANDED STARTING POSITION, MOVE. These commands are executed in the same manner as in paragraph a. except that the left hand is underneath the log and the right hand encircles its far side. (See fig. 3-8, B.)

Figure 3-8 is available in the illustration section at the end of the book.

(3) RIGHT SHOULDER POSITION, MOVE. At the command MOVE, pull the log upward in one continuous motion to the right shoulder. At the same time, move the left foot to the rear and stand up, facing left. Balance the log on the right shoulder with both hands.(See fig. 3-8, C.)This movement cannot be performed from the left-hand starting position because of the position of the hands.

(4) LEFT SHOULDER POSITION, MOVE. These commands should be given from the left hand starting position. At the command MOVE, pull the log upward in one continuous motion, to the left shoulder. At the same time, move the right foot to the rear and stand facing right. Balance the log on the left shoulder with both hands.(See fig. 3-8, D.)This movement cannot be performed from the right-hand starting position.

(5) WAIST POSITION, MOVE. From the right hand position pull the log waist high. Keep the arms straight and fingers laced underneath the log. The body is inclined slightly to the rear and the chest is lifted and arched. (See fig. 3-8, E.)

(6) CHEST POSITION, MOVE. This command should be given after the waist position has been assumed .On the command MOVE, shift the log to a position high on the chest, bring the left arm under the log and hold the log in the bend of the arms.(See fig. 3-8, F.)Keep the upper arms parallel to the ground.

(7) To move the log from the right shoulder to the left shoulder, the command is: LEFT SHOULDER POSITION, MOVE. On the command MOVE, push the log overhead and lower it to the opposite shoulder.

(8) To return the log to the ground from any of the above positions, the command is: STARTING POSITION, MOVE. At the command MOVE, slowly lower the log to the ground. The hands and fingers must be kept from under the log.

e. Leadership. A principal instructor demonstrates and leads the drill. He must be familiar with the leadership techniques for conditioning exercises and the peculiar techniques for log drill.

3304. EXERCISE PERFORMED WITH LOGS

The exercises of log drill are outlined in the following paragraphs. Figure 3-9 graphically explains log drill.

a. Exercise l:Two-Arm Pushup

(1) Starting Position. Right or left shoulder position. Regular stance. (See fig. 3-9, A.)

(2) Cadence. Moderate.

(3) Movement. A four-count exercise:at the count of—

- **(a) ONE--** Push the log overhead until the elbows lock.
- **(b) TWO--** Lower the log to the opposite shoulder.
- **(c) THREE--** Repeat the action of count ONE.
- **(d) FOUR--** Recover to the starting position.

b. Exercise 2: Forward Bender

(1) Starting Position. Chest position. Regular stance.(See fig. 3-9, B.)

(2) Cadence. Moderate.

(3) Movement. A four-count exercise:at the count of-

(a) ONE-- Bend forward at the waist, keeping the back and legs straight.

(b) TWO-- Recover to the starting position.

(c) THREE-- Repeat the action of count ONE.

(d) FOUR-- Recover to the starting position.

c. Exercise 3: Straddle Jump

(1) Starting Position. Right or left shoulder position, feet together, fingers interlaced on top of the log.(See fig. 3-9, **C.)**

(2) Cadence. Moderate.

(3) Movement. A four-count exercise:at the count of—

(a) ONE-- Jump to a side-straddle. Pull down on the log with both hands to keep it from bouncing on the shoulder.

(b) TWO-- Recover to the starting position.

(c) THREE-- Repeat the action of count ONE.

(d) FOUR-- Recover to the starting position.

Figure 3-9 is available in the illustration section at the end of the book.

d. Exercise 4: Side Bender

(1) Starting Position. Right shoulder position, feet regular stance.(See fig. 3-9, D.)

(2) Cadence. Moderate.

(3) Movement. A four-count exercise:at the count of—

(a) ONE-- Bend sideward to the left as far as possible, bending the left knee.

(b) TWO-- Recover to the starting position.

(c) THREE-- Repeat the action of count ONE.

(d) FOUR-- Recover to the starting position.

(4) After completing the required number of repetitions, change shoulders and execute an equal number of repetitions to the other side.

e. Exercise 5: Knee Bend

(1) Starting Position. Right or left shoulder position. Narrow stance. Fingers interlocked on top of the log.(See fig. 3-9, E.)

(2) Cadence. Slow.

(3) Movement. A four-count exercise:at the count of—

(a) ONE-- Flex the knees to a quarter-squat position.

(b) TWO-- Flex the knees to a half-squat position.

(c) THREE-- Lower the body to a three-quarter squat position.(Lean slightly forward.)

(d) FOUR-- Recover to the starting position.

NOTE:Pull forward and downward on the log throughout the exercise.

f. Exercise 6: Overhead Toss

(1) Starting Position. Right or left shoulder position, regular stance. The knees are bent to a quarter-squat.(See fig. 3-9, F.)

(2) Cadence. Moderate.

(3) Movement. A four-count exercisc:at the count of-

(a) ONE-Straighten the knees and toss the log into the air approximately 12 inches overhead. Catch the log with both hands and lower it toward the opposite shoulder. As the log is caught, lower the body into a quarter-squat.

(b) TWO-- Again toss the log into the air and when caught return it to the original shoulder.

(c) THREE-- Repeat the action of count ONE.

Section IV. Grass Drills

3401. GENERAL

a. Description and Objective. Grass drills are extremely strenuous exercises and are performed at top speed for only short periods of time. No cadence is counted but the Marines continue to execute the multiple repetitions of the command until the next command is given. The grass drill consists of two drills:Drill One and Drill Two. Each drill contains six exercises. The objective of the drills is to decrease reaction time, to develop cardiovascular endurance, and to provide a vigorous workout for all major muscles.

b. Area and Equipment. Any level area suitable for ground contact and of a size to accommodate the group is adequate. No equipment is needed.

c. Formation. All movements are executed in place. The extended, rectangular formation is recommended for a platoon- or company-size unit. The circle formation is suitable for groups of squad or section size. At the beginning of an exercise program, 2 to 3 minutes of grass drills will insure a good workout.

d. Progression. Progression is gained by gradually increasing the length of time devoted to the drills. As the physical condition of the Marines improves, the periods should be gradually lengthened to 5 minutes. As the second drill is more difficult than the first, some progression can be attained by initially executing Drill One; then as the program and the Marines progress, introduce Drill Two. To extend the duration of the drill, it may be necessary to repeat the drill.

e. Starting Position

(1) The drills are started from the GO position. Other basic positions are FRONT, BACK, and STOP. (See fig. 3-10, A.)

(a) GO. Running in place (top speed): on the toes and balls of feet, knees raised high, arms pumping, body bent forward at waist.

(b) FRONT. Prone position: elbows bent (along body), palms flat on ground directly under the shoulders, legs together and straight.

(c) BACK. Supine position (flat on back): arms extended near side on ground with palms down, legs together and straight, feet toward the stand or instructor.

(d) STOP. Football lineman stance: feet spread and staggered, left arm across left thigh, right arm straight, knuckles on ground, head up, back parallel with ground.

(2) To assume the FRONT or BACK position from the STANDING, GO, or STOP position, vigorously get into the prescribed position as quickly as possible. (See fig. 3-10, B.)

(3) To change from the FRONT to the BACK position, quickly do a pushup, move the feet several short steps to the right or left, lift the arm on the side toward which the feet move, and thrust the legs vigorously to the front. (See fig. 3-10, C.)

Figure 3-10 is available in the illustration section at the end of the book.

(4) To move from the BACK to the FRONT position, sit up quickly, place both hands on the ground to the right or the left of the legs. Move the feet several short steps to the rear on the side opposite the hands. When the feet are opposite the hands, thrust the legs vigorously to the rear and lower the body to the ground.(See fig. 3-10, D.)

f. Leadership. A warm up activity of lesser intensity should proceed grass drill. During the instructional phase and conduct of these drills, the following points should be applied.

(1) The instructor executes only GO and STOP with the Marines. This allows the instructor to supervise the drill.

(2) The commands for grass drills are given in rapid succession without the usual preparatory command.

(3) To prevent confusion, the instructor should give the commands sharply to distinguish them from comments or encouragement.

(4) As soon as the Marines know the drill, they should respond to the instructor's commands and perform all exercises vigorously and as rapidly as possible. All exercises are executed continuously until the next command is given. Insist on top speed performance; anything less is not effective.

(5) The commands peculiar to each exercise are identical to the name of the exercise.

(6) Marines are not to be required to assume the position of attention once the drills are started. To halt the drill for instructions or for rest, the command UP is used. At this command, the Marines assume a relaxed standing position. Do not demand formality. At the conclusion of a fast and vigorous 5-minute grassdrill, it is physically impossible for people to stand at attention.

(7) The sequence of commands for the execution of grass drills should occur as follows:Drill One. GO, FRONT, Bouncing Ball; GO, BACK, Bicycle; GO, Full Squatter; GO, BACK, Situps; GO, FRONT, Mountain Climber; GO, FRONT, Roll Left; GO, STOP, UP.

g. Use With Other Programs. Since grass drills can be executed in a short period of time, they may be executed where only a few minutes are available for exercise or in conjunction with another type of activity. Grass drills are an excellent substitute for running when time is a factor.

3402. GRASS DRILL ONE AND TWO

a.Grass Drill One

(1) Bouncing Ball. From the FRONT position, push up, supporting the body on the hands (shoulder-width apart) and feet. (See fig. 3-ll.)Keep the back and legs in line and the knees straight. Bounce up and down by a series of short, upward springs from the hands, hips, and feet simultaneously,

(2) Bicycle. From the BACK position, raise the legs and hips. Keep the elbows on the ground and support the hips with the hands. Move the legs vigorously as if pedaling a bicycle.

(3) Full Squatter. From the STOP position, assume a full knee bend, the feet on line, hands on hips. Bounce up and down in place by short, bouncing jumps.

Figure 3-11 is available in the illustration section at the end of the book.

(4) Situps. From the BACK position and with arms stretched overhead, sit up, reach forward, and touch toes. Return to the supine position.

(5) Mountain Climber. From the STOP position, place both hands on the ground directly under the shoulders. Thrust the right leg to the rear, knee straight. The left foot should be close to the left hand, the left knee outside the left arm. Shift the weight to the hands, thrust off with the rear (right) foot and bring that foot up close to the right hand, the right knee outside the right arm. At the same time, thrust the left leg vigorously to the rear, knee straight. Continue at a fast cadence, alternating the legs.

(6) Roll Left. From the BACK or FRONT position, make one complete roll in the direction commanded. On completing the roll, return to the FRONT or BACK position.

b. Grass Drill Two

(1) Legs Over. .From the BACK position and with arms stretched overhead, palms up, raise the legs upward and then swing them backward over the head until the toes touch the ground behind the head. Return legs to the starting position. (See fig. 3-12.)

(2) V-Up and Touch Toes. .From the BACK position, raise the legs with the knees straight, sit up until the trunk and legs form a V, and touch the toes with the hands. Return to the BACK position.

(3) Rocker. .In the FRONT position, clasp the hands behind the back, arch the body, holding the head back. Start rocking, using the front part of the trunk as a rocker.

(4) Bounce and **Clap Hands**. The procedure is the same as for bouncing ball, but while in the air, clap the hands. This requires a more vigorous bounce or spring. (See fig. 3-12.)

(5) Leg Spreader. .From the BACK position, raise the legs so that the heels are 10 to 12 inches from the ground, spread them apart as far as possible, then close them together. Open and close legs as rapidly as possible.

(6) Forward Roll. .For forward roll from the STOP position, place both hands on the ground, tuck the head, and do one complete forward roll, keeping the legs tucked as you roll, and come back to the STOP position.

Figure 3-12 is available in the illustration section at the end of the book.

Section V.Guerrilla Exercise

3501.GENERAL

a. Description and Objective. Guerrilla exercises are individual exercises of various basic skills that are performed rapidly while moving forward in a circle formation. There are two tables of guerrilla exercises. Each table takes 15 minutes to complete. The objective of this exercise is to increase strength and endurance, aid flexibility, and develop coordination.

b. Area and Equipment. Any level area is suitable for conducting guerrilla exercises. No ground contact is required other than the hands. There is no equipment requirement.

c. Formation

(1) The circle formation (app. A) is used for guerrilla exercises. Each platoon forms its own circle and engages in guerrilla exercises under a platoon instructor. If the platoon exceeds 30 Marines, double or concentric circles may be used.

(2) When the circle is formed, the instructor steps into the center of the circle and moves clockwise in a small circle. He commands:QUICK TIME, MARCH, 1-2-3-4. (Rapid cadence of approximately 130 counts per minute. Cadence and step are maintained between exercises.)

(3) To reform the platoon after completing guerrilla exercises, the instructor halts the Marines and places the base man or platoon guide where he wishes and commands:

(a) BASE MAN (or platoon guide),POST.

(b) FALL OUT AND FALL IN ON THE BASE MAN (or platoon guide).

d. Progression. Progression may be attained by moving from table I to table II or by shortening the quick time marching periods between exercises and performing all exercises a second time.

e. Leadership

(1) To execute the exercises, the Marines continue at quick time while the instructor simultaneously explains and demonstrates the exercise to be performed, and then commands the Marines accordingly. In each instance, the preparatory command will be the name of the exercise and, in all instances, the command of execution will ***MARCH***.. *To* terminate each exercise, the command *is* ***QUICK TIME, MARCH.*** *The* Marines immediately pick up the step as the instructor counts cadence.

(2) Unless specified differently, each exercise should be continued for 20 to 40 seconds depending upon the vigor of the exercise. The leader can determine the duration of each exercise by observing its effect upon the Marines.

(3) To form for double guerrillas, the commands for pairing the Marines (who are in circle formation) are-

(a) PLATOON, HALT.

(b) FROM (designate an individual), BY TWO'S, COUNT OFF(.Example1-2; 1-2; 1-2; etc.)

(c) EVEN NUMBERS MOVE UP BEHIND ODD NUMBERS(At this time, adjust pairs according to height and weight.)

(d) YOU ARE NOW PAIRED UP FOR DOUBLE GUERRILLAS. To change the Marines position, merely command CHANGE.)
(e) FORWARD, MARCH.

f. Place in the Program. Many Marines have not had the opportunity to perform the simple skills involved in guerrilla exercises. The conduct of these exercises is a simple matter since they can be performed easily and quickly in almost any situation. The tables of exercise are applicable to all personnel. The tables can constitute a station within a 1-hour period or be completed within a separate 15-minute period.

3502.GUERRILLA TABLES

a. Table I
(1) Double Time.(See fig. 3-13, A.)Hold arms in the thrust position. Execute a double-time run, maintaining the circle formation and the prescribed distance between your fellow Marines. Duration-1 minute.
(2) All Fours.(See fig. 3-13, B.) Face downward. Support the body with the hands and feet. Walk forward hands first.
(3) Crab Walk.(See fig. 3-13, C.) Get in the sitting position, face upward, and lift the hips. Support the body with the hands and feet. Walk forward feet first.
(4) Squat Walk.(See fig. 3-13, D.) Assume a full knee bend position. Grasp the ankles (left ankle with the left hand, right ankle with the right hand). Walk forward.
(5)Broad Jump.(See fig. 3-13, E.) Jump forward on both feet in a series of broad jumps. Swing the arms vigorously to assist the jumps.
(6)Toe-Touch Walk.(See fig. 3-13. F.) Walk forward, bending at the waist and touching one hand to the toe of the opposite foot while it is on the ground. Raise the trunk to the vertical position between steps. Keep the knees straight.
(7)Bottoms-Up Walk.(See fig. 3-13, G.) Assume the front leaning rest position and move the feet toward the hands in short steps, keeping the knees locked. When the feet are as close to the hands as possible, walk forward on the hands to the front leaning rest position.
(8) Straddle Run.(See fig. 3-13, H.) Run forward, leaping to the right from the left foot and to the left from the right foot.
(9) Fireman's Carry .(See fig. 3-13,1.)
(lO)Single-Shoulder Carry(See fig. 3-13, J.) Two men execute the carries as indicated by the diagram. No. 1 man executes one type; No. 2 man executes the other.

Figure 3-13 is available in the illustration section at the end of the book.

b. Table II
(1) Double Time.(See fig. 3-14, A.)Hold arms at the thrust position. Execute a double-time run, maintaining the circle formation and the prescribed distance between your fellow Marine.Duration-1 minute.
(2) Toe-Grasp Walk.(See fig. 3-14, B.) Bend forward and grasp toes. With knees slightly bent, walk forward.

(3) Hand-Kick Walk..(See fig. 3-14, C.) Walk forward, kicking the moving foot upward on every step. At the same time, lean forward and touch the elevated toe with the hand of the opposite arm.

(4) Pike Juinping.(See fig. 3-14, D.) Jump forward and upward from both feet, keeping the knees straight, and at the same time, swing the legs forward and touch the toes with the hands at the top of each jump.

(5) Squat Jump.(See fig. 3-14, E.) Leap forward from the squatting position, with the hands on the ground and the arms between the legs. Land on the ground with hands and legs extended. Bring up the legs to the squatting position.

(6) Steam Engine..(See fig. 3-14, F.) Lace the fingers behind the neck and walk forward in the following manner: as the left leg moves forward, raise the knee high, bend the trunk forward, and touch the outside of the right elbow to the outside of the knee. Then lower the left leg and step forward on the left foot and raise the right leg. Repeat with the right leg and left elbow.

(7) Knee-Touch Walk.(Scc fig. 3-14, G.) Walk forward, bending the knees and touching the ground on each step. The knees are bent and straightened on each step.

(8)Hobble Hopping.(See fig. 3-14, H.) Hold foot behind back with opposite hand and hop forward. On the command CHARGE, grasp the opposite foot with opposite hand and hop **forward.**

(9) Cross Carry.(See fig. 3-14,1.)(10)Saddle Back Carry. (See fig. 3-14, J.) Two Marines execute the carries as indicated in the diagram. No. 1 Marine executes onetype;No. 2 Marine executes the other.

Figure 3-14 is available in the illustration section at the end of the book.

Section VI. Running and Orienteering

3601. RUNNING

a. General

(l)Description.The general form and technique for all types of running are fairly constant. (See fig. 3-15.)The head is erect, body slightly forward without bending at the waist, and the arms are at a loose thrust position alternating from front to rear in straight planes. A cross-body arm movement wastes energy .The movement of the legs and feet will be discussed in subsequent paragraphs dealing with the different types of running. Of primary importance is the fact that in all types of running, the toes should be pointed straight ahead. Toeing out is a common error in both running and walking and should be an item of individual correction.

(2) Objective. The objective of running is to develop cardiovascular endurance. Despite the fact that Marines have developed their muscle structure and the strength of their muscle tissue, unless they have developed cardiovascular endurance to a satisfactory degree, they are not entirely physically fit or combat ready. Cardiovascular endurance (wind) depends on the efficiency of the lungs and heart. The efficiency of the lungs and heart depend on the amount of oxygen the lungs can absorb with each breath inhaled and the amount of carbon dioxide the lungs can expel. The process of absorbing oxygen and expelling carbon dioxide (cardiopulmonary process) is performed by the blood that circulates through the lungs. The condition of this process will determine the amount of effort a person can exert over a period of time. Running is one of the best activities to develop this vitally important endurance.

b. Running Skills. In the development of running skills, individuals may require instruction to improve their proficiency. Some important skills to consider are-

Figure 3-15 is available in the illustration section at the end of the book.

(1) Action of the Arms. Arm action is important. Check to see that arms are held loosely and that the action is relaxed. The faster the run, the more rapid the arm action.

(2) Breathing. Allow the individual to breathe through the mouth as the body demands a large supply of oxygen. Oxygen can be inhaled in greater quantities through the mouth.

3602. TYPES OF RUNNING

a. Double Time

(1) Description and Objective. Double timing is marching at the rate of 180 steps per minute, each step being 36 inches in length. It takes practice to double time with precision in formation. The Marines should keep in step, placing their feet flat on the ground. This, however, should not be a stamping motion, but should be done with as slight a jolt as possible. Double timing is like a jog, the difference being that in a jog the feet are lifted well off the ground and the running motion is bouncy. In double timing, the feet skim the ground and there is no bounce to the ran. Double timing is a vehicle for teaching proper running form and for the development of the cardiovascular system.

(2) Area and Equipment. This type of running can be completed over a variety of surfaces. Usually a Marine uses a field or road. There is no equipment requirement.

(3)Progression. There is no set standard for alternating quick time and double time in the early conditioning of Marines. A general rule is to begin with enough quick-time marching to ensure a thorough warming up, then double time about 100 paces. Change again to quick time until the individuals have made a reasonable recovery from the running, then double time another 100 paces. The amount of double time can be increased and the quick time decreased from week to week, until the individuals are double timing about 1800 yards. This type of training should be given at least twice a week, but by no means is it adequate as the sole means of conditioning.

(4) Leadership

(a) The instructor should be to one side of the column or group and toward the rear so there is a full view of all Marines. Inexperienced instructors have a tendency to supervise from a position too far forward.

(b) Select an individual who can maintain the proper cadence to act as the guide during double-time running.

(c) There are several ways for the instructor and group to count cadence while double timing. If not contrary to local policy, learn several methods and use them for variety.

b. Wind Sprints

(1) Description and Objective. This type of running involves a series of 30- or 40-yard dashes, usually conducted in successive waves of squads. Each squad is in line and the squad leader on the right flank. Wind sprints assist indeveloping speed and cardiovascular endurance. Any flat and level area may be used which will permit the squad to form a line and run the required distance.

(2) Progression. One or two 30-yard sprints will be adequate at the beginning. As time passes, sprints can be lengthened and up to six or seven sprints may be used.

(3) Leadership

(a) At the command READY (given by squad leader), each runner assumes the sprinter's starting position. At the command GO, the squad sprints approximately 30 yards, takes 10 yards to stop, and lines up immediately with the squad leader who repeats, READY, GO, and again the squad sprints. At the conclusion of the third sprint, the squad waits until all the squads of the platoon have made three sprints. Then they all line up and the squad leaders conduct three more wind sprints in the opposite direction.

(b) Valuable time is gained by having each squad ready to go when the preceding squad has moved off its second sprint mark.

c. Cross-Country Running

(1) Description and Objective. Cross-country running is a distance run conducted on a course laid out along roads, across fields, over hills, through woods, and on any irregular ground. The cross-country ran may be utilized as a conditioner or as a competitive event; the objective is to cover the distance in the shortest possible time. The course should be 2 to 2.5 miles in length and be laid out to avoid heavy vehicular traffic. The course should be marked by directional arrows until the runners know the course. These runs build leg muscles, increase lung capacity, and develop endurance. Any local area of varied terrain is suitable.

(2) Progression. In the mass training of a large group, leaders should be stationed at the head and the rear of the column and should make every effort to keep the individuals together .After determining the abilities of the individuals in cross-country running, it is advisable to divide the unit into three groups. The poorest conditioned group starts first, and the best conditioned group, last. The starting time of the groups should be staggered so that all of them finish about the same time. In preliminary training, the running is similar to ordinary road work in that it begins with rather slow jogging, alternating with walking. The speed and distance of the run is gradually increased. As the condition of the individual improves, occasional sprints may be introduced. At first the distance run is from 1/2 to 1 mile. It is gradually increased to 2 or 2.5 miles. Well-conditioned personnel can run 2 to 2.5 miles within a 15-minute period.

(3) Leadership. Marines should not be required to take part in distance running until they have been through a progressively scheduled training program which requires a considerable amount of running. Cross-country runs should be scheduled occasionally to provide variety in the program. Cross-country running has the advantage of allowing mass participation. Interest can be stimulated by putting the runs on a competitive basis. As a single activity, short cross-country runs can be scheduled once a week, gradually increasing the distance as the physical conditioning improves.

(4)Use With Other Programs. Cross-country running can be combined with other activities such as conditioning exercises.

d. Fartlek Training

(l)Description and Function. Fartlek training is various running exercises conducted along an intense combat-like course. The running is conducted to overwork the lungs, allowing only partial recovery which is followed by another intense period of overload. This sequence is repeated for the duration of the workout. Fartlek training is a useful combat training method and general cardiovascular conditioner. The entire unit must be in uniformly good to excellent condition to qualify for this training.

(2) Progression. The following is an example of one fartlek session:

(a) Warm up by stretching 3-5 minutes.

(b) Warm up running easily 5-10 minutes.

(c) Run at a fast, steady pace for 3/4 to 1 3/4 miles (dependent on terrain).

(d) Walk/jog at a moderate pace for 5 minutes (recovery).

(e) Run easily sprinting 15 to 20 meters occasionally.

(f) Run full speed uphill for 175 to 200 meters.

(g) Warm down by running easily for 1/2 to 1 mile.

(h) Warm down by stretching 3 to 5 minutes.

(3)Leadership. To add variety, try incorporating combat movements, terrain association, a series of exercises or carrying table of equipment weapons (hit and roll). This will take some initiative in setting the course, but it allows for variety. In this manner, the fartlek course can be used as a training session or part of one. It works extremely well with highly motivated, competitive Marines.

3603.ORIENTEERING

Orienteering is land navigation over a prescribed course as a timed, competitive event. It is an excellent way to integrate land navigation training with physical conditioning and requires Marines to be skilled in both areas. Commanders adjust the training as needed to change the degree of difficulty or to emphasize either navigation or conditioning. Taken to the extreme, an orienteering exercise is conducted in full combat gear over rugged terrain against aggressors. Orienteering requires the following physical skills: endurance, speed, strength, and all-around conditioning.

Section VII. Circuit Training

3701. GENERAL

a. Description and Objective. A strength circuit consists of a series of stations where individuals in small groups exercise vigorously for a short period of time and then move (on signal) to the next station where a different form of exercise is conducted. This rotation continues until all groups move through all stations. Strength circuits contain no set or specific types of exercise stations within the circuit. The objective of circuit training is to develop strength. There are three general types of circuits.

(1) Fixed Circuit. This is a circuit in which apparatus of an immovable type (fixed into the ground) is used.

(2) Movable Circuit. This circuit consists of individual exercise apparatus which is portable and can be moved to and from the training area.

(3) Simplified Circuit. This circuit requires no equipment or apparatus.

b. Formation. The exercises are done at will, but rapid, steady, and continuous work is required of all. Each Marine's nervous and muscular system reacts differently to timed vigorous exercises. His performance should be measured on how many movements per exercise he can complete as an individual. For example, one Marine may be able to complete 5 movements, while another may be able to complete 20, and yet each is receiving the maximum benefit. All three circuits contained in this chapter are designed for platoon-sized groups. Expansion beyond this capacity requires a large amount of equipment, as each Marine in the fixed and movable types of circuits must have an item of equipment available for exercise at each station. A group larger than a platoon could be exercised through use of the simplified type of circuit; however, the group would be unwieldy and control could be a problem.

c. Place in the Program. All circuits illustrated can be completed in a 15-minute period. This feature allows the exercise of a platoonor smaller group on the circuit for a single 15-minute period, or the scheduling of the circuit as a 15-minute period within a longer period. A circuit can thus be utilized within a rotating activity system of scheduling. Choice of a circuit by the unit depends upon area, facilities, and other local factors; however, there is a circuit for every need.

3702. FIXED STRENGTH CIRCUIT

a. Description and Objective. The strength circuit is an arrangement of various types of exercise apparatus which are fixed in position. (See fig. 3-16.)Seven basic exercises are used and each exercise requires an apparatus. All apparatus of one type are positioned together to constitute a station. Each station will accommodate 10 Marines. The circuit is designed to be accomplished in 15 minutes when conducted on a time-rotation basis, as normally executed for unit training. For individual use, a Marine may complete the circuit by executing a specific number of repetitions for each exercise. In this case, the time required for completion of the circuit would vary slightly depending on the number of repetitions the Marine completed. The objective of this circuit is to provide a series of exercises which will improve and maintain the strength of the body's major muscle groups.

Figure 3-16 is available in the illustration section at the end of the book.

b. Starting Level. Marines must be thoroughly warmed up prior to participating in the circuit system. If Marines have not engaged in vigorous exercise immediately prior to starting the circuit, then an instructor should execute the following warm up exercises. These exercises should be conducted in the normal formation for set drills. Seven repetitions of each exercise will normally provide sufficient warm up. The exercises are-

(1) High jumper, Conditioning Drill 1.
(2) Bend and reach, Conditioning Drill 1.
(3) Squat bender, Conditioning Drill 1.

c. Progression. The instructor can adjust the circuit system through several methods to accommodate participants of varying physical ability. One method of adjusting an exercise is to change the method of executing the leg lift or to select a heavier weight for the barbell curls. When it becomes apparent during a unit program that the overall fitness of the group has Improved, then the exercise may be made more strenuous in two additional ways. First, the instructor can have one or more of the supplementary stations added. Second, the time spent exercising at each station can be increased in 5-second increments to a maximum of 60 seconds. Also, as Marines learn the circuit, the 45-second period for movement and instruction between stations can be eliminated, thus allowing only enough time to change stations.

d. Starting Positions. The Marine starts on any station, exercises steadily for a certain period (45 seconds initially), then moves on command to the next type of apparatus where he again exercises steadily for an equal period. The Marine continues until the required exercise is completed at each station. If it is desired to expand the number of stations in the circuit to accommodate more participants at one time, the instructor can provide four additional supplementary exercises, requiring no equipment. This will increase the amount of time required to complete the circuit.

e. Leadership. Close control of all Marines is necessary to ensure that a minimum amount of time is spent in moving them to their initial stations and in moving them between stations. One instructor can control the activity on the strength circuit. A stop watch or wrist watch with a second hand is required. When Marines arrive at the strength circuit, the instructor will form them for exercise and conduct the warm up drill. The group is then reassembled and formed into a number of files equal to the number of stations being used in the circuit. Each file is then directed to a station. As soon as all participants have reached an exercise position at a station, the instructor gives the command READY, GO. After 45 seconds of exercise, the instructor gives the command STOP, CHANGE OVER. The instructor allows Marines 45 seconds for moving to the next station and for preparing for the next exercise before the command EXERCISE is again given. In lieu of verbal commands, a whistle may be used to stop and start the exercises. For large groups, a megaphone or loudspeaker is useful.

f. Individual Conditioning Program. For the Marine working alone on the strength circuit, it would be impractical to time the exercise periods. For individual exercise, the Marine should select a number of repetitions of each exercise to complete, then rotate to the next stations after completing these repetitions. The number of repetitions selected should be at or near the maximum that the Marine is capable of doing without halting for rest.

g. Fixed Circuit Stations

(1) Primary Stations (See fig. 3-16.)

(a) Pullups/Chinups. A horizontal bar placed 8 feet above the ground is necessary for each Marine at this station. The Marine will also need a space on the bar that is 45 inches wide. On the command**EXERCISE,**the Marine grasps the bar with both palms facing either forward or to the rear, arms fully extended, and feet free of the ground and executes the exercise as described in MCO 6100.3_.The exercise is repeated as many times as possible until the command **STOP** is given. Then the Marine moves to the next exercise. If a Marine has done his maximum number of pullups/chinups prior to the command **STOP**, he will remain in the "dead" handing position until the command **STOP** and move to the next station.

(b) Twist Grip. The apparatus is a horizontal bar, free to turn, held between uprights placed 30 inches apart. The bar is 52 inches above the ground. A weight of 20 pounds is attached to the center of the bar by a light rope long enough to permit the weight to rest on the ground. The Marine stands at arm's length from the bar and grasps it with his hands on either side of the rope, palms down, thumbs under the bar. On the command **EXERCISE,** the hands are rotated so that the backs of the hands are rotated away from the body, thus winding the rope on the bar. The elbows are kept straight to ensure that the exercise is performed by the hand and forearm. When the weight is drawn up to the bar, the bar is then rotated in the opposite direction to lower the weight to the ground. This exercise is continued until the command **STOP** is given. The Marine then moves to the next station.

(c) Pulley Weights. The apparatus is a T frame with a system of pulleys that suspends a weight of about 90 pounds. The weight is attached to a light steel cable which has a drawbar attached to the other end. The Marine grasps the drawbar and sits down directly under the bar, legs extended to the front and arms extended overhead. The exercise is executed by pulling the drawbar down behind the head, then extending the arms slowly again until they are fully extended overhead. The exercise is repeated as many times as possible until the command STOP is given. The Marine then will move to the next exercise. Upon completion of the exercise, the weight is lowered slowly to the ground.

(d) Barbell Curls. A barbell is necessary for each Marine at this station. The barbell is constructed of 1 1/4-inch pipe 5 feet long, and two concrete-filled No. 10 cans. Each barbell should weigh about 40 pounds. Variance in the weight of the barbells, up to about55 pounds, will allow appropriate overload to be applied to Marines who are above average in strength or weight. The Marine grasps the bar with the palms forward and assumes a standing position with the barbell held in front of the hips, hands approximately shoulder's width apart. On the command **EXERCISE,** the Marine flexes the elbows and draws the barbell up until it touches the upper chest. The elbows remain at the sides. Breath is inhaled with the upward movement and exhaled as the barbell is lowered to the starting position. The exercise is repeated as many times as possible until the command **STOP** is given. The Marine moves to the next apparatus.

(e) Step-Up. The apparatus is a platform or ledge 18 inches high and of such size. At accommodate 10 Marines. The Marine faces the platform and on the command **EXERCISE,** steps up onto the platform, bringing his trailing foot up beside the leading foot. He then steps back down to the original position, stepping down first with the same foot he initially used in stepping up. After 10 repetitions of the exercise, he changes the order of moving the feet to use the opposite leg for stepping up. Repeat this exercise until the command **STOP** is given. The Marine then moves to the next station.

(f) Leg Lift. The apparatus is a horizontal bar constructed as described in (a). To prevent the body from swaying, a horizontal back support is added 40 inches below the horizontal bar. The arms are kept fully extended. On the command **EXERCISE,** the Marine jumps up, grasps the bar with the palms forward and the back support behind him. The exercise is executed by raising the legs to a horizontal position then slowly lowering them to the vertical position. The Marine does not flex his knees. He does not swing the legs to the rear of a vertical position to gain momentum for raising them in the next repetition of the exercise. The movement is repeated until the command **STOP** is given. The Marine then moves to the next exercise. If unable to raise his legs to a horizontal position without flexing his knees, the Marine flexes his knees and draws them up to the chest, then lowers his legs to the vertical position.

(g) Rope Climb. The rope climb is 20 to 30 feet high with five ropes suspended from a horizontal bar which forms the uppermost part of the framework. To prevent the horizontal bar from sagging and to provide safety, only five ropes are attached to it. There are two frameworks per station. The ropes are 6 feet apart. Any method may be used to climb the rope, and the Marines climb as high as possible. Marines who are proficient should climb therope several times during the time allotted. Inexperienced Marines should be cautioned to take care during descent to avoid rope burns on their hands.

(2) Supplementary Stations. The following exercises are designed to expand the basic circuit by being inserted in specific places within the system. For each supplementary station used, there should be adequate room for 10 Marines to exercise.

(a) Bent Knee Situp or Bottoms-Up. These calisthenics are designed to strengthen the abdominal muscles. These exercises will be inserted between the pullup/chinup and twist grip stations. The primary stomach exercise is the situp. In case of inclement weather or other conditions that make ground contact undesirable, the bottoms up exercise is used. In situps on the command **EXERCISE,** the Marine lies on his back (supine position) with knees flexed and both feet flat on the ground and executes the exercise (minus an assistant) as described in MCO 6100.3_.The Marine then returns to the starting position, repeating the exercise until the command **STOP** is given. The Marine then moves to the next station. In bottoms up on the commandEXERCISE,the Marine assumes the front leaning rest and executes the bottoms-up exercise as described in Conditioning Drill 3. He continues this exercise at a moderate cadence until the command **STOP** is given. The Marine then moves to the next station.

(b) Pushup. This exercise is designed to strengthen the arm- and shoulder-girdle muscles. It should be included between the twist grip and pulley weight stations. Upon the command **EXERCISE,** the Marine executes the pushup as described in Conditioning Drill 1. He continues this exercise at a moderate cadence until the command **STOP** is given. The Marine then moves to the next station.

(c) Knee Bender. This exercise is designed to build leg muscles and is included between the pulley weight and barbell curl stations. On the command **EXERCISE,** the Marine executes the knee bender as described in Conditioning Drill 3.He continues this exercise at a moderate cadence until the command **STOP** is given. The Marine then moves to the next station.

(d) Trunk Twister. This exercise strengthens the major muscles of the trunk and is included between the step-up and pullup/chinup stations. On the command EXERCISE, the Marine executes the trunk twister as described in Conditioning Drill 1. He continues this exercise at a moderate cadence until the command **STOP** is given. The Marine then moves to the next station.

3703. MOVEABLE STRENGTH CIRCUIT

a. Description and Objective. The exercises in this circuit are progressive and the course is planned to gain and hold the interest of the participating groups. The circuit consists of a series of stations, with each station designed to develop a particular group of muscles. Along with muscular development, correct posture and deep rhythmic breathing should be stressed on this circuit at all times.(See figs. 3-17 and 3-18.)

Figure 3-17 is available in the illustration section at the end of the book.

b. Equipment. The equipment is set up in files. Six files of 8 stations will accommodate a platoon

c. Formation. The platoon marches to the area where the equipment is positioned and forms a file within each lane of stations, covering down on a piece of equipment. Movements are made on the double, the important factor being that no time is wasted in getting to work.

d. Progression. Initially 40 to 45 seconds per station is adequate. As individuals become stronger the time should be increased in 5-second increments until a minute to a minute and a half is reached.

e. Leadership

(1) The leader stands in front of the barbell station and controls the rotation from this position. The leader supervises the entire group, with the assistance of several instructors who move about in the platoon correcting and encouraging the Marines.

(2) The leader starts each group but does not count cadence nor lead them through the exercises. Each individual exercises rapidly but individually.

(3) As each Marine finishes his repetitions with the barbell, he places the barbells on the ground. The leader calls, READY, followed by the command, FALL OUT ONE. All will double-time to the station directly in front of them, while the individuals on the barbell stations will do an about face to the rear station in their lane.

f. Movable Circuit Stations. The Marine can obtain best results on the movable circuit if the exercises on the various stations are given in the following manner:

(1) Station 1—The Barbell. (See fig. 3-17, 1.)The exercises at this station stress proper posture; deep, rhythmic breathing; and development of the muscles of the arms, shoulders, and upper body. An instructor teaches the proper methods of lifting before the exercises begin. The methods are to lift with the legs, to keep the back straight, and to merely grip with the hands. Two recommended exercises are given below. The instructor will use only one exercise per period .He may use either exercise. At successive periods, the other exercise is used.

(a) Exercise 1--Two-Hands Military Press. (See fig. 3-18, 1.)Grasp the barbell with both hands, knuckles up at shoulder width, and lift to the chest. Steadily press to arm's length overhead; lower to the chest resisting weight all the way. Inhale as the weight is pressed up and exhale as the weight is brought down.

(b) Exercise 2--Two-Hands Regular Curl. (See fig. 3-18, 1.) Lift the weight to the waist, with the palms of the hands out, heels together, stomach in, chest lifted and arched, shoulders back, elbows in close to the sides; inhale deeply and curl the weight to the shoulders, using the arms only, at the same time keeping the elbows close to the sides; exhale Emphasize posture and the use of the arms only. This is a very valuable exercise for the development of the biceps and the grip and should be repeated from 8 to 16 times, depending on the ability of the Marine.

(2) Station 2--The Jump Rope. (See fig. 3-18, 2.) This exercise develops strength and agility in the legs and stamina of the whole body. It makes the Marine agile on his feet and increases his footwork efficiency and timing. Each Marine should progress until able to jump rope at least 3 minutes at top speed.

(3) Station 3--The Twist Grip. (See fig. 3-18, 3.) The twist grip is an excellent exercise for the hands and forearms, and adds greatly to the Marine's ability in hand-to-hand combat. The handle is gripped and twisted, winding the rope until the weighted can is level with the height of the hands, which are held horizontal. The weight is lowered in the same manner; the Marine resists the weight all the way, occasionally stopping the twisting motion and alternately removing first one hand, then the other, from the handle. A variation of the above exercise is to wind the handle with the palms up and the arms bent and with the elbows held close in to the sides. Each Marine should maintain a good posture and keep the stomach muscles taut throughout this exercise.

(4) Station 4--The Incline Plane. (See fig. 3-18, 4,)The use of the incline plane is a very strenuous exercise and well designed for the development of the abdomen. Although 6 to 10 repetitions are sufficient for the beginners, more can be added as ability increases. Marines with hernias or recent operations will be excused from participation at this station.

(5) Station 5--The War Club. (See fig. 3-18, 5.) The war club is a simple and effective means of exercising the principal muscle groups of the body, especially those of the trunk, back, and shoulders. To gain the maximum benefit from this exercise, the Marine must keep both feet flat on the ground at all times. Throughout the exercise period, the weight is swung from arm's length as follows:

(a) As in chopping wood, first on one side, then on the other.

(b) As a batter warming up with a number of bats.

(c) In large circles, first with one hand and then with the other.

(6) Station 6--The Bicycle Ride. (See fig. 3-18, 6.) The bicycle ride is well suited to exercising many of the muscle groups of the body, particularly those of the abdomen. Vary the speed of the exercise, but keep everyone "riding" the entire period. A variation exercise may be performed by placing the legs together, raising them slowly to a height about 2 feet from the ground, and then lowering them slowly to the ground.

(7) Station 7--The Step-up. (See fig. 3-18, 7.) The step-up exercises the legs. The step-up is performed by initially stepping up with the left foot, followed by the right, then stepping down with the left foot followed by the right. Continue for 20 seconds, then change to the right foot as the lead foot for 20 seconds.

(8) Station 8--The Isometric Pull. (See fig. 3-18, 8.) Two trainees work at this station with a cable pull and perform the following exercises:

(a) Initially start with one Marine in the supine position and one Marine sitting. The sitting Marine lowers the upper body to the ground and pulls the partner up to the sitting position. The partner then performs this same action and this is continued for 20 seconds at a rapid rate.

(b) During the last 20 seconds, the same action takes place but, in this case, the Marine in the supine position resists the pull of the partner for approximately 5 seconds before allowing to be pulled up into the sitting position.

3704. CIRCUIT-INTERVAL TABLE

a. Description and Objective. The circuit-interval table is designed to develop strength and endurance within a short period of time, with no equipment requirement, through a rapid and vigorous routine of exercise. Fifteen minutes is an adequate period to execute all exercises and to secure a vigorous workout with the circuit-interval principle.

Figure 3-18 is available in the illustration section at the end of the book.

b. Formation and Starting Positions

(1) A leader forms platoons or smaller groups in an oval or circular formation with 3- to 5-yard intervals between Marines. The Marines face to the right and move forward at quick time and then double time. (See fig. 3-19, A.)After running several platoon circle laps, the leader calls out the name of an arm and shoulder exercise from the list below, orders quick time and commands, for example, PUSHUPS. On this command, all Marines immediately hit the ground and individually and rapidly begin doing pushups. No cadence is counted. (See fig. 3-19, B.) After 30 seconds of exercise, the leader commands. **QN YOUR FEET, FORWARD, MARCH** The platoon resumes the quick time cadence and the leader, when ready, gives the necessary commands for double time. The double time is continued for one or more laps and the leader calls out the name of the next exercise and the process is repeated. This continues, with running between each exercise, until every body part has been exercised.

(2) The instructor controls the running and quick time to observe the effects of the exercise upon the Marines. Cadence, step, and precision are not important to the objective and the instructor should not use them. What is important is speed and the instructor should stress this. After the exercise period is started, the Marines do not stop. This circuit method emphasizes stress and recovery, the recovery occurring during the quick time periods.

c. Activities. The leader can use the following exercises. These exercises can be repeated if necessary, during a second round. The leader can use other calisthenic exercises in the circuit-interval table.

(1) Arms and shoulders-push-ups.
(2) Stomach situps.
(3) Back-squat thrusts.
(4) Legs-bicycle (on back).

d. Progression. lhe progress is controlled by the leader, who must pace the running, quick-time movement, and exercise in such a way that Marines will receive a vigorous workout yet be able to participate throughout the 15-minute period. Marines who are in the initial stages of physical condition will not be able to double time or exercise as long as those who are better conditioned. The idea is to set a pace which can be increased during each workout, thus progressing gradually to a higher level of physical fitness.

e. Leadership. The platoon leader, platoon sergeant, or section leader can lead the group. The leader must execute the exercise with the unit to feel the effects and thereby adjust the pace.

f. Place in the Program. This activity may be scheduled whenever a short period of time is available. The only requirement is that enough space, indoors or out, be available to form the circle.

Figure 3-19 is available in the illustration section at the end of the book.

Section VIII. Basic Physical Skills and Obstacle Courses

3801. GENERAL

The purpose of this section is to list the basic military skills and the methods for their development. Many of these skills are best developed by obstacle courses but other drills are also discussed. The section explains types of obstacle courses, details of construction, and methods of negotiating the various obstacles.

3802. BASIC PHYSICAL SKILLS

a. Objective. The objective is to develop proficiency in the various military physical skills which are essential to personal safety and effective combat operations. In travel by foot over rugged terrain and in the execution of combat duties, Marines must be trained to perform basic skills, such as running, jumping, climbing, and carrying. During training, Marines will develop agility and coordination in these skills .Fast and skillful execution of these skills may mean the difference between success and failure on the battlefield.

b. Place in the Program. These skills are practiced throughout the entire program in many different activities. Many of these skills are best practiced on obstacle courses.(See par. 3803.)

c. Description. These basic skills are the minimum skills required by all Marines. The basic skills are as follows:
(1) Running. Running is used to strengthen the legs and develop cardiovascular endurance. Marines should be exposed to running in various situations: on roads, over rough ground, up and down hills, cross-country, and running over low obstacles.
(2) Jumping. In broad jumping, the take-off foot is planted firmly and the spring comes from the extension of this leg as the other leg reaches for the far side of a ditch or similar obstacle. (See fig. 3-20.) The arms are forcibly raised forward and upward to assist in propelling the body up and forward. Landing may be on one or both feet depending upon the length of the jump. In vertical jumping downward from a height, the jumper should aim his feet at the desired landing spot and he should also jump with the knees slightly bent and feet together, with the trunk inclined slightly forward. As the feet touch the ground, the shock is absorbed by bending the knees into a full squatting position. If the height is too great or the ground too hard to absorb the shock, then the jumper should forward roll or side roll thus eliminating some of the momentum.

Figure 3-20 is available in the illustration section at the end of the book.

(3) Dodging In combat situations, it is often times necessary to change directions quickly. To execute this movement while running, a lead foot is firmly planted, left foot if the direction is to the right and right foot if the direction is to the left. The opposite foot is moved toward the new direction. The knees are slightly flexed during the movement and the center of gravity is low and balanced. (See fig. 3-21.) At the time of the change of direction, the head and trunk are turned quickly in the new direction.

Figure 3-21 is available in the illustration section at the end of the book.

(4) Climbing and Surmounting. All Marines should know how to climb and surmount various types of obstacles. (See fig. 3-22.)

Figure 3-22 is available in the illustration section at the end of the book.

(a) Vertical Climbing, as in Climbing a Rope or Pole. The hands grasp the rope or pole overhead with the palms toward the face. Gripping the object, the body is pulled upward with the arms and shoulders, assisted by the feet which grip the object and assist by pushing downward. If shoulder girdle strength and body coordination are not adequate to permit alternating the hands, the arms act together in pulling upward.

(b) Climbing as in Surmounting a Wall. Ingoing over a wall, the body should be kept as close to the top as possible, since in combat operations it is important to offer as small a target as possible to the enemy. If an individual climbs a wall while carrying a rifle, both hands should be freed by slinging the rifle over the back. There are two methods commonly used for surmounting a wall of moderate height, but only one for dropping from it.The methods are as follows:

1 Run, Jump, and Vault. Approach the wall at a run, jump forward and upward at the wall and place one foot against it as high up as possible. Use the foot in contact with the wall to help push the body upward while grasping the top of the wall with the hands. Pull the body up with the arms, assisted by pressure of the foot against the wall and swing the legs over, propelling the body weight over the wall.

2 Hook and Swing. Approach the wall at a run and jump forward and upward. Hook one elbow over the wall, locking the arm in place by pulling up until the top of the wall is underneath the armpit. Grasp the top of the wall with the other hand. Draw the leg which is closer to the wall up as far toward the top as possible. Then swing the other leg over the top of the wall.The body is then carried over with a rolling motion. A variation of this leg action can be used by Marines who are unable to draw up the leg as described. While hanging with both legs fully extended, start a swinging motion with the legs together. When the legs have enough momentum, swing the outside leg over the top of the wall with a vigorous kick, then follow with the body.

3 **Dropping.** All drops from a height are executed in the same manner regardless of the method used to gain the top. One hand is placed against the far side of the wall while the other hand grasps the top. From this position, the body is rolled over the wall and vaulted away from it with the legs swinging clear. As the body passes over the wall and drops, it should at all times face the wall. This will keep the rifle and other equipment clear. Break the fall by retaining a grasp on the top of the wall as long as possible.

(c) Climbing Ladders and Cargo Nets. Rope ladders, stationary vertical ladders, and cargo nets employ the same general technique. The important element is to grasp the side supports firmly in the hands about shoulder height and place the feet on a rung which would cause the body to be fully extended. In movement upward, one hand is moved upward and a new grasp is secured and, at the same time, the opposite leg moves up a rung. As the knee straightens, the body is elevated. This process is repeated using the opposite arm and leg. Alternation continues in this manner until the climber reaches the objective.

(5) Traversing Horizontal Objects. The traversing of horizontal objects puts heavy stress on the arms and shoulder girdle area as the feet are usually suspended in the air with all of the body weight on the arms and shoulders.(See fig. 3-23.)

(a) Traversing Horizontal Ropes or Pipes. The hands grasp the horizontal support overhead with the palms facing. To propel the body forward, one hand is released and moved forward to secure a new grasp. At the same time, the opposite side of the body is swung forward (some people are able to "walk" in the air, keeping the body to the front and moving the legs in time with the arms as in walking on the ground). The other hand is then released and moved forward; this alternation is continued until the objective is reached.

Figure 3-23 is available in the illustration section at the end of the book.

(b) Traversing Horizontal Ladders. In this situation, the movement is the same as used in traversing a rope or pipe. The hands, however, are placed on the rung with the palms away from the face. Other than this difference the technique is the same.

(6) Crawling. Crawling in combat situations is an often used skill. Crawling may be high or low.(See fig. 3-24.)

Figure 3-24 is available in the illustration section at the end of the book.

(a) High Crawl.

In the high crawl, the Marine moves on hands and knees, moving one hand and the opposite knee and then continuing to move the hands in alternation with the opposite knee following the companion hand.

(b) Low Crawl. The Marine is in the prone position usually with the forearms and palms of the hand on the ground. He propels forward by bending the knee of one leg and pushing with the inside edge of the shoe. At the same time, the opposite arm moves forward and pulls to the rear. The body remains low and movement is continued by bending the opposite knee and pushing, and at the same time sliding the opposite arm forward and pulling. Alternation of hands and legs continues until the objective is reached.

(7) Throwing. Throwing may be executed from the kneeling or standing positions. The object to be thrown is held in the hand, and the throwing arm is bent at the elbow; the hand is then moved to the rear until the hand is behind the ear. The body is turned so that the lead foot and balance arm on the side toward the target are pointing at the target. (See fig. 3-25.) The balance arm is used to sight over and align the throwing hand and the target. When properly aligned, the elbow is moved rapidly forward until it is at a pointjust in front of the body where the arm is straightened and the wrist snapped. This whip motion propels the object to the target. Underhand throws secure momentum by the thrower bending his knees and swinging the throwing arm to the rear. As the knees are straightened, the arm is forcefully swung forward from the shoulder and the object released.

Figure 3-25 is available in the illustration section at the end of the book.

(8) Vaulting. Vaulting is employed to overcome low barriers or fences. (See fig. 3-26.)The object to be surmounted is approached at an angle. The hand on the side next to the obstacle is placed on the top of the obstacle and, with a straight arm, the body weight is pushed upward. At the same time, the leg on the side next to the obstacle is thrown upward and over the top followed by the other leg (side approach). In landing, the weight comes down on the leading leg first followed by regaining the balance on both legs. The free arm serves as a balance. A direct (front) approach can be used at which time both legs go over the object together.

(9) Man Carrying. There are three basic individual means of carrying personnel in combat situations and one of these methods may be used in carrying objects.

Figure 3-26 is available in the illustration section at the end of the book.

(a) Fireman's Carry."A" stands sideways in front of "B", "A" bends his knees and leans forward, placing one arm through "B's" crotch, grasps the wrist of "B's" arm, which is hanging over the shoulder, and then "A" runs forward.

(b) Saddle-Back Carry. With his back toward "B", "A" stands in front of "B". "B" mounts "A's" hips and clasps his arms in front of "A's" chest. "A" grasps "B's" thighs.

(c) Single-Shoulder Carry. "A" stands facing "B"."A" assumes a semi-squatting position. "B" leans forward until "B" lies across "A's" right shoulder. "A" clasps his arms around "B's" legs and straightens up, lifting "B" from the ground. "A" then runs forward. This method may also be used to carry heavy objects.

(10) Balancing. Balancing the body while walking or running on a narrow object when crossing obstacles is a skill which requires practice and confidence. Balance is required in negotiating a log placed across a stream, in crossing a narrow beam or rail, and in similar situations.(See fig. 3-27.) To perform this skill, place the feet on the object to be crossed, hold the arms to the sides at shoulder level, and fix the eyes on the object approximately 5 yards in front of the feet. Generally, it is not a good practice to look down at the feet. Walk the beam by placing first one foot and then the other in the center of the beam, thereby moving forward, using the arms to aid in maintaining balance.

Figure 3-27 is available in the illustration section at the end of the book.

(11) Falling. Injury can be avoided if Marines are taught to fall properly. They should know how to use the body momentum to their advantage during a fall rather than to try resisting that force. (See fig. 3-28.) If enough force is present, such as occurs during a fall while running or in jumping downward from a height, individuals can extend their hands to catch the weight. At the same time, duck the head and roll forward onto the feet. The key to falling without injury from the standing position is relaxation and rolling on the outside of the leg, hip, and buttocks to take the brunt of the fall.

Figure 3-28 is available in the illustration section at the end of the book.

3803. OBSTACLE COURSES

a. General. Obstacle courses are a valuable part of physical readiness training. The challenge presented by the obstacles assists in developing and testing the basic physical skills. In many combat situations, success will depend upon the Marine's ability to perform one or more of these skills, often while carrying field equipment and when fatigued. In this section, two types of obstacle courses will be discussed-Conditioning Obstacle Course and Confidence Obstacle Course.

b. Course Safety Precautions. Commanders and course instructors should take certain precautions to prevent injury on obstacle courses. A few of the precautions are:
(1) Inspect the course for faulty construction of obstacles, protruding nails, rotten logs, condition of the landing pits, and other hazards to safety.
(2) Conduct warm up exercises before the unit runs the course.
(3) Explain and demonstrate the correct techniques for negotiating all the obstacles before allowing the Marines to try them.
(4) Give Marines at least two weeks of conditioning exercises before scheduling the obstacle and/or confidence courses.
(5) Ensure that negotiation of the higher and more difficult obstacles are under the supervision of an instructor.
(6) Do not permit individuals who have neither practiced the basic skills nor run the conditioning obstacle course to participate in the confidence obstacle course.
(7) Weather conditions may cause footing or handhold surfaces to be slippery. If such is the case, postpone training on the course.

3804.CONDITIONING OBSTACLE COURSES

a. Description and Objective. The Conditioning Obstacle Course is commonly known as the Obstacle Course. This course consists of fairly low obstacles which are designed to be negotiated quickly. The obstacles serve to test various basic skills, and running the course is a test of the Marine's physical condition. After receiving instruction and an opportunity to practice the skills, Marines run the course against time.

b. Area and Equipment
(1) Complete standardization of obstacle courses should not be attempted since topographical conditions always vary. Commanders should use ingenuity in constructing a course, making good use of streams, hills, trees, rocks, and other natural obstacles. Since the course is eventually run at high speed, it should not be dangerous.

(2) The course should be wide enough for six or eight men to run simultaneously, encouraging competition. The lanes for the first several obstacles should be wider and the obstacles themselves easier than those that follow. This avoids congestion until the contestants scatter out over the course. The last two or three obstacles should not be too difficult and should not involve high climbing. This prevents injuries and falls resulting from fatigue.

(3) The total distance of the course should range from 300 to 450 yards and include from 15 to 25 obstacles. Normally the obstacles should be 20 to 30 yards apart and arranged so that those which exercise the same groups of muscles are separated.

(4) The obstacles should be substantially built. Peeled logs, 6 to 8 inches in diameter, are ideal for many of the obstacles. Sharp points and corners should be eliminated. Landing pits for jumps or vaults should be filled with sand or sawdust to prevent injuries.

(5) The course should be constructed and marked so that it is not possible to sidestep or detour obstacles. However, it is desirable to provide alternate obstacles of varying degrees of difficulty.

(6) The course should be in the shape of a horseshoe or figure eight so that the finish is close to the start and signs should be placed to indicate the course route.

c. Leadership. Before Marines run an obstacle course, they should be instructed in the proper technique of negotiating each obstacle. In each case this technique should be explained and demonstrated in detail, with emphasis on avoiding injury. Every individual should be given an opportunity to practice on each obstacle until he becomes reasonably proficient at negotiating it. Before the course is run against time, it is advisable for individuals to make several runs at a slower pace. During such practice or trial runs, the instructor should observe the performances and make appropriate corrections. Marines should never be permitted to run the course for time until they have practiced on all obstacles. The best method of timing the runners is to have the timer stand at the finish and call out the minutes and seconds as each individual finishes. If several watches are available, each wave may be timed separately. If only one watch is available, the different waves should be started at regular intervals, such as every 30 seconds. If an individual fails to negotiate an obstacle, a previously determined penalty should be exacted.

d.Types of Obstacles

(1) Jumping-Type Obstacles. These obstacles may be ditches which are cleared with one leap, trenches which the individuals can jump into, heights which require jumping downward, or hurdles which an individual can jump over. (See fig. 3-29.)

Figure 3-29 is available in the illustration section at the end of the book.

(2) Dodging-Type Obstacles. Obstacles of this type are usually mazes consisting of posts set in the ground at irregular intervals. (See fig. 3-30.)The intervals between posts should be rather narrow so that the Marines must pick their way carefully through and around them. Lane guides may be constructed to guide the Marines to dodge and change direction. Obstacles may be put into a maze pattern to cause the Marines to change direction.

Figure 3-30 is available in the illustration section at the end of the book.

(3) Vertical Climbing and Surmounting Type Obstacles. These obstacles may be climbing ropes, either plain or knotted and 1-5 inches in diameter; cargo nets or walls 7 or 8 feet high; or vertical poles 6 to 8 inches in diameter and 38 feet high.(Seefig. 3-31.)

Figure 3-31 is available in the illustration section at the end of the book.

(4) Horizontal Traversing-Type Obstacles. Horizontal obstacles may be pipes, beams, ladders, or ropes. (See fig. 3-32.)

(5) Crawling Type Obstacles. Obstacles which require crawling may be constructed of large pipe sections, low rails, and wire. (See fig. 3-33.)

Figure 3-32 is available in the illustration section at the end of the book.

(6) Vaulting-Type Obstacles. Obstacles of 3 to 3.5 feet in height such as low walls or fences may be used as a vaulting obstacle.(See fig. 3-34.)

(7)Balancing-Type Obstacles. Beams, logs, and planks may be used as balancing-type obstacles.(See fig. 3-35.)These items may be used to span water obstacles and dry ditches, or raised off the ground somewhat to simulate natural depressions.

Figure 3-33 is available in the illustration section at the end of the book.

3805.CONFIDENCE OBSTACLE COURSES

a. Description and Objective. The Confidence Obstacle Course is commonly known as the Confidence Course. This course is composed of higher and more difficult obstacles than those used in the conditioning course. The confidence obstacle course is designed to give Marines confidence in their mental and physical capacities and to cultivate their spirit of daring. They are encouraged but not compelled to negotiate this course and the course is not run against time. The negotiation of a confidence course, however, is strenuous enough to be an excellent physical conditioner. Marines should NEVER attempt to take the obstacles at high speed and should not compete for speed. The obstacles vary from fairly easy to extremely difficult. Some are of considerable height to accustom Marines to climbing such heights without fear. Considerable emphasis is placed on obstacles that train and test an individual's balance.

Figure 3-34 is available in the illustration section at the end of the book.

b. Area and Equipment

(1) The confidence course accommodates four platoons, one platoon at each group of six obstacles. The course should be made up of about 24 obstacles, numbered and marked as follows:! to 6, white numbers on red background; 7 to 12, black numbers on a white background;

13 to 18, white numbers on a blue background; and 19 to 24, white numbers on a black background.

(2)A few simple pieces of equipment should be provided for individuals who do not have the strength, courage, or ability to negotiate the obstacles.

Figure 3-35 is available in the illustration section at the end of the book.

c. Formation. The obstacles should be divided into groups of six, and each group is designated by a different color. Each platoon starts at a different color. Individuals are separated into groups of 8 to 12 at each obstacle. At the starting signal from the company commander, they proceed numerically through. Anyone may skip an obstacle who is afraid to try. Individuals proceed from obstacle to obstacle until time is called, then assemble as ordered.

d. Leadership. If the Marines are new to the confidence course, an instructor will demonstrate or will give a brief orientation at each obstacle, including an explanation and demonstration of a method of negotiating it. Marines are encouraged to try the various obstacles, but they are not compelled to do so. No compulsion is to be used. The manner of negotiating any obstacle is left to the discretion of the Marine. However, the instructor assists anyone who experiences difficulty. Instructors must supervise closely at all times to prevent injuries, as some of the obstacles are quite high. Also, some of the obstacles should not be used when slippery or wet. The example of instructors and especially selected demonstrators will serve to inspire the individuals to greater effort.

e. Negotiating the Obstacles. Although personnel need not conform to any one method of negotiating the obstacles, there should be some uniformity in the approach to them. A general method of negotiating the obstacles is indicated below.

(1) Red Group This group contains the first six obstacles. (See fig. 3-36.)

(a) The Belly Buster Individuals may vault, jump, or climb over. Warn them that the log is not stationary.

(b) Reverse Climb. Climb the reverse incline and go down the other side to the ground.

(c) The Weaver. Move from one end of the obstacle to the other by weaving the body under one bar and over the next.

(d) Hip-Hip. Step over each bar, either alternating legs or using same lead leg each time.

(e) Balancing Logs. Step up on log, retaining the balance, walk or run along log.

(f) Island Hopper. Jump from one log to another until the obstacle is negotiated.

(2)White Group. This group is composed of the second six obstacles. (See fig. 3-37.)

(a) Tough Nut. Step over each X in the lane.

(b) Slide for Life. Climb the tower, grasp the rope firmly, and swing the legs upward. Hold the rope with the legs to distribute the weight between them and the arms. Braking the slide with the feet and legs, proceed down the rope. Warn Marines that there is danger of getting rope burns on their hands. When the rope is slippery or wet, this can be a dangerous obstacle.

(c) Low Belly Over. .Mount the low log and jump onto the high log, both arms grasping over the top of the log, the stomach area in contact with it. Swing the legs over the log and lower the body to the ground.

(d) Belly Crawl Move forward under the wire, belly down, to the end of the obstacle.

(e) The Dirty Name Mount the low log and jump to or reach the higher logs in succession, then jump or drop to the ground. Warn the Marines about the height of the final log.

(f) The Tarzan Mount the lower log and walk the length of it and each successive, higher log until reaching the horizontal ladder. Grasp two rungs of the ladder and swing the body into the air. Negotiate the length of the ladder by releasing one hand at a time and swing forward, grasping a more distant rung.

(3) **Blue Group.** This group is formed by the third group of six obstacles. (See fig. 3-38.)

(a) High Step over. Step over each log, alternating the lead foot or using the same lead foot.

(b) Swinger. Climb onto the swinging log and over to the ground on the opposite side.

(c) Low Wire. Move under the wire on the back, using the hands to raise the wire to clear the body.

(d) Swing, Stop and Jump. Gain momentum with a short run, grasp the rope, and swing the body forward to the top of the wall. Release the rope while standing on the wall and jump to the ground.

(e) Six Vaults. Vault over the logs, using one or both hands.

(f) Easy Balancer. Walk up one inclined log and down the one on the other side to the ground.

(4) Black Group. The last group is formed by the final six obstacles. (See fig. 3- 39.)

(a) Incline Wall. Approach the underside of the wall, jump up and grasp the top and pull the body up and over. Slide or jump down the incline to the ground.

(b) Skyscraper. Jump or climb to the first floor, climb up the corner posts or assist each other to any desired floor. Descend to the ground in any desired manner.

(c) Jump and Land. Climb up the ladder to the platform and jump to the ground.

(d) Confidence Climb. Climb the inclined ladder to the vertical ladder. Go to the top of the vertical ladder, then down the other side to the ground.

(e) Belly Robber. Step on the lower log and assume the prone position on the horizontal logs. Crawl over the logs to the opposite end of the obstacle.

(f) The Tough One Climb the rope or pole on the higher end of the obstacle, then go down the ladder and across the log platform. Climb over or between the logs at the end and go down the rope or pole to the ground. Vault over the final log.

Figure 3-36 is available in the illustration section at the end of the book.

Figure 3-37 is available in the illustration section at the end of the book.

Figure 3-38 is available in the illustration section at the end of the book.

Figure 3-39 is available in the illustration section at the end of the book.

Section IX Individual Exercise Programs

3901.GENERAL

Marines at times will be stationed on independent duty and consequently will be responsible for their own physical fitness program. This section will assist them in understanding the need for exercise and will aid in the planning and execution of an individual exercise program. Exercise activities included are the Bench Conditioner, 6-12 Plan, Weight Training, and Isometric Contraction. Keeping physically fit is a problem that faces every Marine. Even though we are frequently engaged in training that requires some physical effort, in many cases, it is not enough to prepare us to meet the intense physical demands of combat. Attaining a satisfactory level of physical readiness is not an insurmountable objective. Available time appears to be the most difficult obstacle to the development of physical readiness. In most cases, regular physical training programs are centralized, requiring the individual to temporarily leave the work area. The problems involved in setting an hour aside two or three times each are numerous. However, most of us can devote a few minutes each day to physical fitness with little, if any, impact on our daily work schedule, especially if it does not require us to leave our work area.

a. Type of Program. There are many good physical fitness programs available to the individual or group. Regardless of the type or duration, to be effective, the program must contain exercises that are strenuous and are challenging to the individual. Space will not permit the inclusion of all available means of individual exercise. The programs selected for this chapter have met the requirement of minimum space and minimum time.

b. Need to Augment Program. These programs are quite strenuous and will develop a satisfactory level of physical readiness. However, if the Marine desires additional development of endurance, it is recommended that he supplement these programs with a 15- to 30-minute period of wind sprints and double timing on an alternating daily basis.

c. Progressive Training. If Marines are performing duties which require little or no physical activity, they must plan a physical conditioning program that assures a moderate beginning, moderate but steady progression, and sufficient warmup before starting the vigorous exercise. To avoid organic of bodily harm, a Marine activity without adequate warmup. He should conduct conditioning programs on a daily basis over an extended period of time, never on an unduly accelerated or crash basis.

3902.THE BENCH CONDITIONER

a. Description and Objective. The bench conditioning program uses a modified bench to employ both isotonic (moving) and isometric (stationary) exercises are as the nucleus of the program. The exercises are designed to develop strength and endurance in all the major muscle groups of the body. The principles of progression, overload, and balance are employed when the exercises are performed properly.

b. Area and Equipment. The conditioning apparatus can be constructed in any unit motor pool with welding equipment found in most salvage yards.(See figs. 3-40, 3-41.)Additionally, there are available any number of commercially produced apparatus available. The important thing is that in utilizing this equipment the following exercise routine be adhered to.

c. Starting Level and Progression.. The program consists of two tables, each with 10 exercises. The Marine controls progression by required repetitions or, in some cases, by application of maximum effort. Each table can be completed within 15 minutes.

d. Starting Position. To start the program, the Marine begins with the Table 1 and executes each exercise for the required number of repetitions as indicated. The Marine controls the starting level and progression. When the Marine executes the maximum repetitions for Table I within a 15-minute exercise period, he progresses to Table II. To maintain this level of development, the Marine should also execute the maximum repetitions for Table II within a 15-minute exercise period. The Marine should keep substitution of exercises to a minimum. However, it he completes a full 15 minutes of strenuous exercise and exercises all muscles, then there should be no appreciable difference in the overall development.

e. Bench Conditioning, Table I

(1) Exercise 1: Side-Straddle Hop. This is a two-count warmup exercise done at moderate cadence. The starting position is the position of attention. On count ONE jump slightly into the air, swinging the arms out to the sides and up to a vertical position, hands touching.(See fig. 3-40, A.)At the same time, spread the feet wider than shoulder-width apart. On count TWO, using a slight flexing of the knees and ankles, jump slightly into the air and return to the starting position by swinging the arms back down to the sides. Twenty repetitions of this exercise is the standard dosage throughout the program.

(2) Exercise 2: Hand Walk. Remove the lower horizontal bar. Adjust the upper horizontal bar so that it is high enough to permit a "dead hanging" position, with the feet off the ground.(See fig. 3-40, B.j From the "dead hanging" position, release one hand and drop the arm to the side of the body. Then raise that arm and regrasp the horizontal bar. Release the bar with the other hand and drop that arm to the side. Repeat this as many times as possible.

(3) Exercise 3: Situps. Lie down with the fingers interlocked and placed behind the head. Hook the toes under the foot braces. Raise the trunk and upper body to an upright sitting position, twisting it to the left and then forward and downward until the right elbow touches the left knee.(See fig. 3-40, C.)Lower the body to the starting position .Sit up again but twist the body to the opposite direction as before, touching the left elbow to the right knee. Again lower the body to the starting position. The starting dosage is 20 situps. The Marine should continue the progression until he has attained 40 situps.

Figure 3-40 is available in the illustration section at the end of the book.

(4) Exercise 4: Double Step-upStarting at one end of the bench, step up onto the bench, and walk across it. Step down from the other end; turn around and repeat the process to return to the starting point.(See fig. 3-40, D.)Each return to the starting point constitutes a repetition. The starting dosage is 20 repetitions. Maximum dosage is 35 repetitions. This exercise should be done at a rapid cadence.

(5) Exercise 5: Isometric Bar Lift. Adjust the lower bar so that it is slightly higher than the beltline. Placing the feet on the footplates at the base of the frame, grasp the lower bar so that the hands are spread shoulder-width apart. Assume a crouched position and lift with maximum effort using the arms, back and legs.(See fig. 3-40, E.)Starting dosage is 4 repetitions of a stress time of 5 seconds followed by a 5-second rest prior to the next repetition. The Marine obtains progression by lengthening stress periods to 6 and later 7 seconds. Do not increase the number of repetitions.

(6) Exercise 6: Knee Lift. Adjust the upper bar to the same height used in Exercise 2.Adjust the lower bar so that is stops rearward movement of the hips when the "dead hanging" position is assumed.(See fig. 3-40, F.) Keeping the arms extended, flex the legs and raise the knees as high as possible. Hold this position for 5 seconds, then return to the starting position. After 2 seconds in the starting position, raise the knees again. Each return to the starting position constitutes one repetition. The dosage is five repetitions. Progression is the same as in Exercise 5.

(7) Exercise 7: Isometric Pull. Adjust the lower horizontal bar to a position where it is slightly higher than the beltline. Grasp the handles and pull outward.(See fig. 3-40, G.)Apply maximum effort and hold for approximately 5 seconds. Relax for 5 seconds between repetitions; perform four repetitions. Moving the body closer to or farther away from the bar will change the stress from the upper arms to the forearms. Progression is the same as in Exercise 5.

(8) Exercise 8: Isometric Compression. Maintain the position as in exercise 7.(See fig. 3-40, H.)Grasping the handles in the same manner, press in **with** maximum effort and hold for approximately 5 seconds. Relax for 5 seconds between repetitions. Progression is the same as in Exercise 5.

(9) Exercise 9: Isometric Press. Remove the lower horizontal bar. Adjust the upper horizontal bar until it is about 6 inches lower than the extended arms can reach. Stepping on the footplates at the bottom of the frame, grasp the bar with both hands and push up.(See fig. 3-40,1.)Keep both the legs and arms slightly flexed and the back straight. Apply maximum effort for 5 seconds then relax for 5 seconds. Complete four repetitions. Progression is the same as in Exercise 5.

(10) Exercise 10: Pushups. Grasping the foot braces with both hands, assume the front leaning rest position.(See fig. 3-40, J.)Keeping the back and legs straight, lower the body until the chest is lower than the **hands,** then return to the starting position. The Marine should complete the maximum possible number of repetitions.

f. Bench Conditioning, Table II. There is no limit on the maximum number of repetitions attainable in Exercises 3, 4, and 6 of Table II. The only limit imposed is that the entire program of 10 exercises outlinted in either table should not exceed 15 minutes.

(1) Exercise 1: Side-Straddle Hop. This is a two-count warm up exercise done at a moderate cadence. The starting position is the position of attention. On count ONE, jump slightly into the air, swinging the arms out to the sides and up to a vertical position, hands touching.(See fig. 3-41, A.)At the same time, spread the feet wider than shoulder-width apart. On count TWO, using a slight flexing of the knees and ankles, jump slightly into the air and return to the starting position by swinging the arms back down to the sides. Twenty repetitions of this exercise is the standard dosage throughout the program.

(2) Exercise 2: Pull-up. Adjust the horizontal bar so that is is high enough to permit a "dead hanging" position with the feet off the ground. Grasp the bar with both hands, palms facing forward. By flexing the arms, raise the body to a position where the chin is higher than the bar.(See fig. 3-41, B.)Then lower the body to the "dead hanging" position. Repeat as many times as possible.

(3) Exercise 3: Bench situps. Sit on the bench and hook the feet under the foot braces. With the fingers interlocked behind the head, lean back until the head touches the floor.(See fig. 3-41, C.)Return to the starting position. The starting dosage is 15 situps.

(4) Exercise 4: Step-Up. Face the bench and step up on it with one foot, bringing the trailing foot up next to the leading foot. Step back down again, leading with the same foot used first in stepping up.(See fig. 3-41, D.)Perlorm half of the total repetitions, then change the sequence of moving the feet to use the other leg in stepping up, and repeat the same amount of exercise. The starting level is a total of 40 step-up. **This exercise should be done at a rapid cadence.**

(5) Exercise 5: Isometric Bar Lift\djust the lower bar so that it is slightly higher than the beltline. Placing the feet on the footplates at the base of the frame, grasp the lower bar so that the hands are spread shoulder-width apart. Assume a crouched position and lift with maximum effort using the arms, back, and legs.(See fig. 3-41, E.)Starting dosage is four repetitions of a stress time of 8 seconds followed by a 5-second rest prior to the next repetition. Progression is obtained by lengthening stress periods to 10 seconds. Do not increase the number of repetitions.

Figure 3-41 is available in the illustration section at the end of the book.

(6)Exercise 6: Leg Lift. Adjust the bars and assume the starting position as shown in figure 3-41, F. Keeping arms and legs extended, raise the legs to a horizontal position and hold in that position for 2 seconds. Then lower the legs slowly to the starting position. Five repetitions is the starting level.

(7) Exercise 7: Isometric Pull. Adjust the lower horizontal bar so that it is slightly higher than the beltline. Grasp the handles and pull outward.(See fig. 3-41, G.)Apply maximum effort and hold for approximately 8 seconds. Relax for 5 seconds between repetitions; perform four repetitions. Moving the body closer to or farther away from the bar will change the stress from the upper arms to the forearms. The Marine obtains progression by lengthening the stress period to 10 seconds.

(8) Exercise 8: Isometric Compression. (See fig. 3-41, H.)Maintain the position as in exercise 7.Grasping the handles in the same manner, press in with maximum effort and hold for approximately 8 seconds. Relax for 5 seconds between repetitions; perform four repetitions. The Marine obtains progression by lengthening the stress period to 10 seconds.

(9) Exercise 9: Isometric Pres. Remove the lower horizontal bar. Adjust the upper horizontal bar until it is about 6 inches lower than the extended arms can reach. Stepping on the footplates at the bottom of the frame, grasp the bar with both hands and push up.(See fig. 3-41,1.)Keep both the legs and arms slightly flexed and the back straight. Apply maximum effort for 8 seconds, then relax for 5 seconds. Complete four repetitions. The Marine obtains progression by lengthening the stress period to 10 seconds.

(10) Exercise 10: Inclined Pushup. Assume the front leaning rest position with the feet on the bench.(See fig. 3-41, J.)Keeping the back and legs straight, lower the body until the nose touches the ground. By extending the arms, raise the body to the starting position. Repeat as many times as possible.

3903. THE 6-12 PLAN

a. Description and objective. The 6-12 plan of physical fitness has been developed to assist in regulating quantity and progression and to provide a convenient set of exercises. This is a basic program and will take 18 weeks to complete if you follow the moderate progression as to the time prescribed for each level of achievement. The time can be shortened as explained below. This plan consists of six basic exercises a day which can be completed in 12 minutes. There are six tables of six exercises each, thus allowing you to progress from table to table. The plan is progressive, fits any age group, contains balance and variety, and applies the principle of overload in a safe and gradual manner. Begin at Table I, Progression Guide, with the number of repetitions as indicated by age.

b. Formation. If just starting an exercise program, do not rush through the first table. Remember, individuals should remain at each level for about a week before moving upward. The time allotment stated for each exercise at the bottom of the tables is a guide; some people may take more and some less time on the individual exercises. At the end of a one-week period, when the individual can comfortable perform the six exercises in 12 minutes, move on to the next level. To a certain degree, the individual must be the judge of his ability to progress form level to level and table to table. If starting with a certain degree of fitness, some of the beginning tables may present little challenge.

c. Starting Level **and Progression.** There are three levels of achievement for each age group, indicated as A, B, and C. Start at the C level for the appropriate age group. At the end of a one-week period, or when the individual can do all exercises at that level within 12 minutes, progress to the B level. At the end of the second week, or when the individual can accomplish that level within 12 minutes, progress to the A level. At the conclusion of the third week or when the individual can achieve the A level within the time limitation, move on to table II.(See Tables I through VI, Progression Guide.)

d. Maintenance Level. Attempt to work through all six tables. If this proves to be too difficult, then maintain exercise at the—

(1) A-level on Table IV, Progression Guide it in the 45 to 49, 50 to 59, or over 60 age group.

(2) A-level on Table V, Progression Guide if in the 17 to 29, 30 to 39, or 40 to 44 age group.

e. Precautions. To achieve the maximum benefit, perform each exercise exactly as specified. Read the descriptions and study the illustrations. Do not slight the movements. Use a sensible approach and follow these four points as they apply before starting or during your exercise program.

(1) If you have the slightest doubt about your ability to participate in this exercise program, consult a physician.

(2) Stop immediately if you notice unusual breathlessness or chest pain while taking part in these exercises. If these conditions persist, consult a physician.

The following Tables of Progression Guides and Exercises are the progression of the 6-12 Plan program.

Table I. Progression Guide is available in the illustration section at the end of the book.

Table I. Exercises.is available in the illustration section at the end of the book.

Table II. Progression Guide.is available in the illustration section at the end of the book.

Table II. Exercises.is available in the illustration section at the end of the book.

Table III. Progression Guide.is available in the illustration section at the end of the book.

Table III. Exercises.is available in the illustration section at the end of the book.

Table IV. Progression Guide.is available in the illustration section at the end of the book.

Table IV. Exercises.is available in the illustration section at the end of the book.

Table V. Progression Guide.is available in the illustration section at the end of the book.

Table V. Exercises.is available in the illustration section at the end of the book.

Table VI. Progression Guide.is available in the illustration section at the end of the book.

Table VI. Exercises.is available in the illustration section at the end of the book.

(3) Unless you have exercised regularly and know yourself to be in good physical condition, start at table I with the C level appropriate to your age.

(4) If you are out of shape, admit that fact to yourself. Hide your pride; after all, you are in the privacy of your own quarters. Set your goal for the longer, steadier pull toward fitness. Resist the urge to pass over the lower numbered tables to find a table that will test your fitness. You are not trying to test, but rather to develop.

3904.WEIGHT TRAINING

Weight or barbell training should not be confused with the more common types of weight lifting used as a competitive sport. Weight lifting is designed to develop specific muscle groups so that the individual is capable of lifting a large amount during a single lift. In contrast, weight training is the systematic development of all the major muscle groups by the use of calisthenics reinforced with weight to provide resistance.

3905.WEIGHT LIFTING

a. Description and Objective. The weight lifting program is progressive and applies the principle of overload in a safe, gradual manner. The exercises of the table can be completed in 15 minutes. The objective of the exercises is to develop strength and muscular endurance, and muscle tone of the five major muscle groups:legs, arms, back, trunk, and shoulder girdle.(See fig. 3-42.)

b. Warmup. A warm up exercise is important to prepare the body for the more vigorous exercises that are to follow. Ten repetitions of the high jumper exercise are excellent for a warm up period. Also, an exercise which will require fast body movement is needed.To provide such exercise, 3 to 5 minutes of rope skipping is recommended to increase the individual's development potential.

c. Formation. Marines should take care in the completion of these weight lifting exercises. They should ensure that their back is straight during the lifting phase of all exercises. When exercises require assuming the standing position with the weight, they should always grasp the weight while in a squatting position and then rise to a standing position.

d. Progression. Each exercise has a starting number of repetitions and specified pounds of weight. After each 4th or 5th day of exercise, the Marine should increase the repetitions by one until he has reached the maximum of 10 repetitions. At this time, the Marine should increase the weight by 5 pounds. Then he should repeat the process again with the initial number or repetitions.

e. Weight Training

(1) Exercise 1:Squat(See fig. 3-42, A.)Starting level—six repetitions, 50 pounds (commonly called the flatfoot deep knee bend). Place the bar upon the shoulders. Stand with feet about 18 inches apart. Keeping the feet flat, lower the body into the low squat position. Stand erect and repeat. Exhale while lowering into the squat position and inhale while standing. This constitutes one repetition.

(2) Exercise 2:Waist Bender(.See fig. 3-42, B.)Starting level-6 repetitions, 40 pounds. Assume the standing position with the bar across the shoulders and the feet a shoulder-width apart .Bend forward at the waist until the upper body is parallel to the ground; return to the starting position. Each return to the upright position constitutes one repetition.

(3) **Exercise 3:Curl** (See fig. 3-42, C.)Starting level—6 repetitions, 40 pounds. Grasp the barbell with the palms facing to the rear and assume the standing position with the feet a should-width apart. With the barbell held in front of the hips, flex the elbows and lift the weight until the bar touches the upper chest. Lower the barbell back to the hip level position. Inhale deeply with the upward movement and exhale on the downward movement. Each time that the bar touches the chest will constitute one repetition.

(4) **Exercise 4:Side Bender**(.See fig. 3-42, D.)Starting level—6 repetitions per side, 40 pounds. Assume the standing position, with the bar across the shoulders, with feet a shoulder-width apart. Bend to the left as far as possible and return to the starting position. Repeat six times and then execute the same procedure to the right for six repetitions.

(5) **Exercise 5:Standing Press**(See fig. 3-42, E.)Starting level—6 repetitions, 45 pounds. Grasp the bar with the palms facing forward and assume the starting position. Curl the weight to the upper chest position. Inhale deeply and press the bar upward to an overhead position. Exhale while lowering the bar to the chest position. Each time that the bar is pressed upward constitutes one repetition.

(6) **Exercise 6:Upward Row.** .see fig. 3-42, F.)Starting level—6 repetitions, 40 pounds. Grasp the bar, hands close together, palms to the rear, and assume the standing position. Starting with the bar held in front of the hips, flexing the elbows and the shoulder girdle muscles, lift the bar straight up to an overhead position. Inhale deeply while lifting the bar. Exhale while lowering the bar to the hip position. Each time that the bar returns to the hips constitutes one repetition.

(7) **Exercise 7:Shoulder Curl.** See fig. 3-42, G.)Starting level—6 repetitions, 25 pounds. Grasp the bar, palms down, and assume the standing position. Keeping the elbows locked, curl the bar, pivoting the arms at the shoulders until the bar is in an overhead position and as far to the rear as possible. Return the bar in the same manner to the hip position. Each time that the bar returns to the hip position constitutes one repetition.

Figure 3-42 is available in the illustration section at the end of the book.

3906.ISOMETRIC EXERCISING

a. **Description and Objective.** Isometric exercising is the application of maximum effort during an exercise period. The isometric principle is to apply force gradually over a 5- to 10- second period until the maximum effort is applied. Relaxation follows for approximately 5 seconds and then force is applied again. The Marine continues this process at the prescribed level for each exercise for a period of 15 minutes or less. The objective of isometric exercises is to create muscle growth. Isometric exercises are the fastest means of creating muscle growth. They are founded on the fact that a muscle will grow only so fast regardless of the type or duration of the activity. The principle of overload—that the muscles develop commensurate with demand—reinforces that fact. However, isometrics will not develop cardiovascular or muscular endurance. Consequently, Marines who choose to use isometrics in their exercise routine must also include running or some other type of aerobic activity.

b. **Area and Equipment** isometric exercises may be designed to be performed with or without equipment.

c. Use With Other Programs. The isometric exercises presented in this section will provide a variety from which to choose. In addition to the exercises contained in this section, the application of isometric force is used in some of the exercises in section II.

d.Door Frame Exercises. The following exercises are designed for use with a standard door frame found in all offices or barracks.(See fig. 3-43.)

(1) Exercise l:Arm Press. Stand in the doorway with the legs straight, knees locked. Using the arm muscles, press hard upward against the top of the door frame. Repeat for three repetitions applying gradual effort to maximum contraction.

(2) Exercise 2:Leg Press. Stand in the doorway with the hands on the top of the door frame, elbows locked .With knees bent, press hard with the leg muscles .Repeat for three repetitions beginning with a gradual effort and increasing to maximum contraction. A low platform may be necessary to reach the top of the door frame and still maintain a bent knee position.

(3) Exercise 3:Side Press. Extend both arms to the side of the doorway. Palms are shoulder high, facing outward. With both arms, press hard against the sides of the door frame. Repeat for three repetitions. Begin gradually and increase to maximum contraction.

(4) Exercise 4:Lateral Raise. Extend both arms to the sides of the doorway, arms down, palms facing inward. With the back of the hands, press hard against the sides of the door frame .Repeat for three repetitions. Begin with a gradual effort and increase to maximum contraction.

Figure 3-43 is available in the illustration section at the end of the book.

(5) Exercise 5:Neck Press?lace the forehead against the door frame, hands clasped behind the back Using the neck muscles, press hard against the door frame. Repeat for three repetitions, then reverse position so that the back of the head is resting on the door frame. Again do three repetitions .Begin gradually with both exercises and increase to maximum contraction.

(6) Exercise 6:Door Pull Stand facing the edge of the open door and grasp the doorknobs. Pull outward with both arms (if doorknobs are not available grasp the edge of the door).Applying outward pressure, move the body toward and away from the door .Repeat the three repetitions. Begin with gradual effort and increase to maximum contraction.

Chapter 4
COMBAT WATER SURVIVAL
Section I. Marine Corps Water Survival Program

4101.MILITARY SWIMMING

Military swimming emphasizes strokes that result in staying power rather than speed. Marines need to swim easily, quietly, and with adequate vision. Strokes should be used that allow the Marine to carry basic combat equipment and to tow or push a wounded buddy while keeping the face out of the water to allow breathing. The best strokes to use are the side stroke and the breast stroke. Marines must be proficient in floating, drown-proofing, and in using their basic combat equipment to make flotation devices. They must be drilled on using the Kapok and Mae West buoyancy compensators and on emergency egress from helicopters, landing craft, and amphibious assault vehicles.

4102.PROGRAM DEVELOPMENT

Commanders of the Landing Force Training Commands (Atlantic/Pacific) are tasked with developing and maintaining the individual water survival and swimming training program. They are responsible for training all Marine Corps water safety and survival instructors and integrating this training in to their combat readiness training program. All Marines are deployable and as soldiers of the sea must constantly practice the skills necessary to achieve mission accomplishment. Commanders must be imaginative in program development, integrating basic skills development which enhances the total mission accomplishment. An example is teaching waterproofing of the Marine's backpack so that it can be used as a flotation device in streams or rivers. Once learned, this skill would be practiced first in a pool and then in a stream or a river. Routine inspections during deployments would ensure that backpacks could be used as flotation devices.

4103.QUAL STDS AND TRAINING GUIDELINES

The inherent nature of Marine Corps operations and training requires that Marines achieve an ability to survive in water. Water survival and swimming training is designed to reduce fear of water, instill self-confidence, and develop a Marine's ability to survive in water. As a minimum, every Marine should be qualified as swimmer, third-class (S3).This qualification should be met during recruit training or Officer Candidate School. Once qualified, a Marine need not be requalified unless requalification at a specific level (S3 or higher) is required by unit mission, military occupational specialty or duty assignment, or for other reasons determined by the commander. The following classifications and standards apply to Marine Corps water survival and swimming training. Abbreviations will be used for service, medical, and training record entries citing MCO 1510.29A, Individual Water Survival and Swimming Training, as authority.

a. Qualification standards and test procedures are described in paragraph 4104.

b. Swimmer classification and abbreviations are:

(1) Unqualified - UQ.

(2) Swimmer, third-class (S3).Minimum aquatic skill level for all Marines.

(3) Swimmer, second-class (S2)-

(a) Minimum standard for all naval aviators, naval flight officers, and aircrewmen with written waiver from the commanding officer.

(b) Requirements for military occupational specialty 1803 (assault amphibian vehicle officer), 1833 (assault amphibian vehicle crewman), 1302 (engineer officer), 1371 (combat engineer), 1381 (shore party specialist), and 1379 (engineer operations chief).

(4) Swimmer, first-class (S1)-

(a) Naval aviator, naval flight officer, and aircrewman qualification requirement.

(b) Requirements for military occupational specialty 0321 (reconnaissance man), 8652 (reconnaissance man, parachute jump qualified), 8653 (reconnaissance man, self-contained underwater breathing apparatus qualified), 8654 (reconnaissance man, parachute/self-contained underwater breathing apparatus qualified), 9952 (self-contained underwater breathing apparatus Marine (officer/enlisted)), and 9953 (parachutist/self-contained underwater breathing apparatus Marine (officer/enlisted)).

(5) Water Survival Qualified (WSQ).Ultimate water survival/swimming goal for Marines.

c.Instructor and assistant instructor classifications are-

(1) **American Red Cross Advanced Lifesaver.** Individuals who are currently certified as American Red Cross advanced lifesavers may assist qualified instructors during Marine Corps water survival/swimming training and testing. Certification is valid for 3 years.

(2) American Red Cross Water Safety Instruction. Individuals who are currently certified as American Red Cross water safety instructors may conduct swimming training and qualify individuals as 3d, 2d, and 1st class swimmers. Certification is valid for 2 years.

(3) United States Marine Corps Water Safety and Survival Instruction. Those personnel certified as water safety and survival instructors by one of the landing force training commands may conduct Marine Corps water survival and swimming training and American Red Cross swimming and water safety training. These instructors may also participate as assistant instructors during the conduct of landing force training command-sponsored water safety and survival instructor courses. Certification is valid for2 years.

(4) American Red Cross Water Safety Instructor Traine. Individuals qualified who have been certified as instructor trainers by the American Red Cross may train American Red Cross water safety and basic swimming instructors. Certification is valid for 2 years.

(5) United States Marine Corps Water Safety and Survival Instructor Traine. Individuals qualified who are certified as instructor trainers through special training courses conducted jointly by the American Red Cross and the Marine Corps, under the direction of the Commandant of the Marine Corps, may train water safety and survival instructors and American Red Cross water safety and basic swimming instructors. Certification is valid for 2 years.

d. Training Guidelines Instructors engaged in training Marines as swimmers will use the current edition of the American Red Cross swimming and water safety courses instructors' manuals as the authoritative textbooks for swimming procedures, strokes, breaks, and holds.

4104.WATER SURV/SWIM QUAL STD AND TEST PROCS

a. The following qualification standards and test procedures are applicable to Marine Corps water survival and swimming training. The utility uniform without cover, boots/oxfords or socks will be worn for all swimming tests through swimmer, first class. The utility uniform with boots/oxfords will be worn as prescribed for the water survival qualification.

(1) Swimmer, Third-Class (S3) To qualify as a swimmer, third-class, a Marine must enter the water feet first from a minimum height of 5 feet and remain afloat for 5 minutes. During this time the Marine must swim 50 yards using any stroke or combination of strokes. This test should be taken by all Marines as early as possible in their initial training. Those unable to pass this test will be classified as UQ and should be given instruction in fundamental swimming skills. Those who barely meet requirements are swimmers who need help and should receive additional instruction in fundamental swimming skills, whenever possible.

(2) Swimmer, Second-Class (S2.) To qualify, a Marine must enter the water feet first from a minimum height of 10 feet and remain afloat for 10 minutes. The Marine must swim 100 yards and use the three basic survival swimming strokes (side stroke, backstroke, and breast stroke) for a minimum distance of 25 yards each.

(3) Swimmer, First-Class (S1)A prerequisite to qualification as swimmer, first-class, is successful completion of the test for swimmer, second-class. To qualify as a swimmer, first-class, a Marine must be able to do each of the following:

(a) Approach a person of approximately the same size while in the water, demonstrate a release, get that person in a carry position, and tow the "distressed swimmer" 25 yards.

(b) Enter water feet first and immediately swim underwater for 25 yards. Swimmer is to break the surface twice for breathing during this distance at intervals of approximately 25 feet.

(c) Remove trousers in water, inflate for support, and remain motionless for a minimum of 1 minute.

(d) Swim 220 yards using any survival swimming stroke or any combination of survival strokes.

(4) Water Survival Qualification (WSQ)A prerequisite to water survival qualification is successful completion of the test for swimmer, first-class. To be water survival-qualified, a Mannemust:

(a) Enter water by jumping from a height of 10 feet.

(b) Stay afloat in full utility uniform (less cover) for 1 hour. Boots/oxfords will be removed after 5 minutes in the water but will be retained by the swimmer.

(c) Traverse 75 yards in deep water in full utility uniform (less cover) with boots/oxfords.

b. All water survival and swimming training will include instruction in the following:

(1) Employment of safety lines or other comparable expedients to assist in swift stream/river crossings.

(2) Adverse physiological effects caused by cold water (hypothermia) and the precautionary measures to be taken prior to exposure to such an environment.

(3) Employment of a standard and expedient flotation devices in a water survival situation.

(4) Removal of individual field equipment after unscheduled water entry. Entry will be made from a minimum height of 10 feet with a complete set of field equipment, properly packed as an expedient flotation device and waterproofed. The ALICE pack will then be removed, the rifle laid across the top of the pack, the helmet remaining or put back on the individual's head, and the individual will traverse 25 yards to the side of the pool or water's edge.

c. The additional water survival qualification (flight physiology training) and requalification required of personnel assigned to flight status will be conducted in accordance with OPNAVINST 3710.7_,Promulgation of NATOPS General Flight and Operating Instructions. Swimming qualification (first-class, SI) is a prerequisite to water survival (flight physiology) training and is a one-time requirement.

Section II. Water Sports and Competitive Activities

4201.GENERAL

The commander, when developing the combat readiness program, needs to look carefully at the mission. If the mission requires Marines who must perform mission-essential water-related activities of long duration, he may want to look to water sports/competitive activities as a method of developing endurance. Endurance and confidence can only be developed by extended time in the water. While defined training can accomplish this goal, the goal can be enhanced through training which is integrated with the physical conditioning program. This approach will provide variety and enthusiasm. Two activities which will accomplish this goal are water relays and water polo. The important consideration is that the unit achieve total participation by all members of the unit.

4202.WATER RELAYS

Development of water relays will generally conform to the guidelines in chapter 5, section IV. When contemplating relays, the commander should review chapter 5 and adapt the principles to the size of the unit and the facility to be used.

4203.WATER POLO

The official rules provided in this paragraph will need to be adapted to the unit's unique situation, its size, and the size of the facility. Commanders should make every effort to have all members of the unit participate.

a. NCAA Rules This style of play is based on two popular sports: swimming and basketball. Playing is done on the surface of the water by teams of seven players each. The size of the playing area is 25 yards by 14 yards, which is the size of the typical high school or college indoor pool. The deeper the water, the better. If the entire playing area is deep-6 feet or more—this is ideal, although most indoor pools have a shallow end. At each end of the pool is a goal. In deep water, the goal is 10 feet across and 3 feet high. In shallow water, where the goalie can stand on and jump from the bottom, the goal is 10 feet across and 5 feet high. One of the seven players on each team is the goalie. He is the only player on the team who can stand on or jump from the bottom (if it is shallow enough to permit this), catch and pass the ball with both hands at the same time, or hit the ball with a clenched fist. However, the goalie may not swim across or pass the ball across the middle of the pool.

b.Equipment. The goalie on the visiting team must wear a white cap with the number 1 clearly marked on it; his teammates, who are designated as guards and forwards, must wear white caps numbered 2 through 7, with the substitutes wearing white caps numbered 8 through 21.The goalie on the home team must wear a dark-colored cap with the number 22 clearly marked on it; his teammates designated as guards and forwards, must wear similar dark-colored caps numbered 29 through 42.The guards and forwards on each team may swim freely up and down the pool as they see fit, interchanging positions as often as they wish. They may not stand on or jump from the bottom, touch the ball with more than one hand at a time when catching or passing or shooting it, or enter inside the opposing team's 2-yard line unless preceded by the ball. The ball is similar to a soccer ball, except that it is yellow and covered with a waterproofing substance that makes it easier to handle with one hand.

c. Playing Time. In intercollegiate competition the game consists of four 7-minule quarters, the teams changing ends after every period of play. In high school competition the game consists of four-,five-, or six-minute quarters, depending on the ages of the participants. Here again the teams change ends after every period. There is a very brief interval between each quarter. At the start of each period the teams line up at their respective ends, and with ablast of the whistle, the head referee throws the ball into the middle of the pool. The fastest swimmers on each team then race to gain possession.

d. Officials. The head referee walks along one side of the pool watching for infractions. He is aided by the assistant referee, who patrols the opposite side of the pool. The referees are both equipped with whistles and two official's flags. These flags should be 12 inches square and should correspond in color to the caps worn by the respective teams. Whenever a referee sees an infraction, he blows his whistle and signals with his flag indicating that the player nearest to the ball with the appropriate colored flag will be given possession of the ball. Whenever the whistle is blown, all players should react by looking immediately to see which color flag is being waived which will indicate which team should be moving into the defense. Other officials are a timer and a scorer. As in all sports, it is important that the officials be skilled and competent.

e. Technical Fouls. The following are some of the common technical fouls: starting before the referee blows his whistle to open play; holding onto or pushing off from the side of the pool during play; taking hold of the ball underwater when tackled by an opposing player; swimming inside an opponent's 2-yard line unless preceded by the ball; touching the ball with both hands at the same time (goalie excepted); standing, walking on or jumping from the bottom when taking an active part in the game. When a technical foul occurs, thereferee blows the whistle and with his flag awards possession. The player on the team awarded possession nearest the point of infraction then has 5 seconds to put the ball into play by passing to a teammate or by dropping the ball into the water and swimming after it.

f. Personal Fouls. The following are some of the more common personal fouls: committing any of the technical fouls mentioned above for the purpose of scoring or preventing a goal; holding, ducking, pulling, pushing off from, swimming over, or impeding the arm or leg movement of any opponent who is not touching the ball; and splashing water in the face of an opponent. When a personal foul occurs, the referee will follow the same procedure as with a technical foul. While doing this, the referee will also call out clearly and loudly the number of the player who was guilty of the infraction and a personal foul will be marked against that player by the official scorer. When a player accumulates five personal fouls, he fouls out of the game and must be replaced by a substitute.

g. Penalty Shot. A penalty shot can be awarded by either of the officials or the scorer when:

o An offensive player inside the opponents' 4-yard line not touching the ball is held, ducked, pulled-back, kicked, or struck.
o A team has accumulated a total of ten personal fouls. When the former occurs, the head or assistant referee should immediately blow the whistle and by holding a two-flagged stick in a vertical position above his head, signal that a penalty shot has been awarded. When the latter occurs, the scorer should use a buzzer located at the scorer's tabic to signal that one team has accumulated ten personal fouls, thereby entitling the other team to a penalty shot. A penalty shot is taken from the 4-yard line in front of the goal. All players except the defending goalie must leave the 4-yard line until the shot is taken. No player can be within 1 yard of the shooter. After ascertaining that the shooting player is on the 4-yard line and the goalie is onThegoal line, the referee will ask the shooter to lift up the ball. When he docs, the referee will give a sharp blast from the whistle. At the whistle, the shooter must shoot without delay and without any faking at the goal. The goalie may try to block the shot. If he does, or the shot is otherwise missed, it is immediately in play and action continues.

h. When a Goal is Scored. When a goal is scored, either from a shot taken by a player out in the field or by penalty shot, the defending goalie must then pass the ball to the nearest referee. The referee will then pass the ball back to the goalie, who puts it into play by a pass to one of his teammates as soon as possible.

Section III. The Battle Swimming Test

4301.GENERAL

The battle fitness test is the evaluation contained in the Guidance for Basic Warrior Training (BWT) as part of the Basic Warrior Training Concept Plan.

4302.THE BATTLE SWIMMING TEST

The battle swim test consist of the following:

a. Be able to climb a three meter high board, while wearing a field uniform, carrying a M-16, web gear, and four canteens. Jump off blindfolded, recover and swim 100 meters with the M-16.

b. Repeat the same process, without the blindfold, and upon entering the water a second time, recover and retrieve all submerged gear.

c. Remain afloat and tread water for five minutes.

d. Demonstrate a knowledge of how to cross a stream using field expedient techniques and a single strand rope bridge.

e. Successfully complete helicopter egress training.

Chapter 5
COMPETITIVE CONDITIONING ACTIVITIES
Section I. Organization of Competitive Activities

5101.LEADERSHIP OF COMPETITIVE ACTIVITIES

a. Place in the Program. Competitive conditioning activities consist of dual combatives, relays, team contests, and team sports in which individuals or teams compete against an opponent to win. Competition is one of the best ways of maintaining interest in the physical training program during the sustaining stage. Organized competition provides enjoyable, vigorous physical activity that has proved to be one of the best supplements to conditioning drill activities. The benefits of competition are the development of aggressiveness, unit pride and identity, teamwork, and the will to win. These activities help develop cardiovascular and muscular endurance, strength, and coordination. Scheduling competitive activities in an orderly and progressive manner is desirable. The progression should be from relays to dual combatives, to team contests, and finally to team athletics. These activities should be conducted as part of the program after a basic level of conditioning has been developed, usually during the slow-improvement stage of conditioning. Muscles and joints should be strengthened by preconditioning to withstand the strain placed upon them by sudden stops and turns, body contact, bearing of weight, and falls. Competitive activities, however, should not be allowed to dominate the physical training program.

b. Area and Equipment. Some of the competitive activities included in this chapter require specific types of areas and equipment; others do not. The area requirement can usually be satisfied on available training fields. When items of equipment are required or specific courts or field layouts are to be marked off, such information is included.

c. Leadership. The principal factor for success as a leader of competitive activity is an energetic, dynamic, enthusiastic approach. The leader's attitude is reflected by the group, so the commander must carry on the activity in a snappy and vigorous manner. Confidence on the part of the commander will create an impression of decisiveness and certainty. Confidence grows out of experience and a thorough knowledge of the activity. Mastery of subject is the first step in developing confidence, assurance, and poise.

(1) The following suggestions are offered for leaders of competitive combatives, contests, and sports:

(a) Get the activity underway quickly by selecting and teaching only the essentials.

(b) Use rules to add to the enjoyment of the activity and do not allow them to interfere with the spirit of competition.

(c) Stop the activity before interest begins to lag.

(d) In team contests, clearly distinguish sides by attempting to maintain unit identity.

(e) Always insist on fair play, enforcing the rules impartially.

(2) The following procedure is recommended for presenting a competitive activity.

(a) Name the activity.

(b) Briefly explain the objective of the activity and give only the pertinent rules.
(c) Have a demonstration at slow speed and answer questions.
(d) Organize groups into teams and appoint captains.
(e) Arrange teams in the proper starting positions.
(f) Conduct the activity.

d. Competitive Units. Units for competition should be the same organizations in which Marines train: battalions, companies, platoons. In most situations, the unit is the squad.

e. Provisions for Instruction. One of the most effective methods of maintaining interest and participation in competitive activities is to provide instruction in activities with which most Marines are unacquainted. Such instruction can be conducted during regular physical training periods. Careful planning is required to keep all individuals continuously engaged in vigorous activity. The materials in the following sections should be used as a guide for instructional purposes.

f. Officiating and Control. Every effort should be made to provide good officials for all competitive activities. Poor officials quickly cause dissatisfaction among participants in team activities and create a situation which can turn an organized activity into a brawl. Each company should have or develop several competent and qualified officials available for games on company and platoon levels.

5102.MILITARY FIELD MEETS

When units reach the latter part of the slow-improvement and sustaining stage of training, interest in the program may lag. A change in course content can arouse the desire to participate. An event which does not require a high degree of skill yet demands strenuous activity is ideal for the military field meet. This is a series of team contests conducted on a station-to-station basis during a given period of time. Team contests carried on simultaneously provide essential training, vigorous exercise, and stimulating competition in an atmosphere that is enjoyable for all. Because of the healthy rivalry that a field meet arouses, it is an excellent form of inter-unit competition.

a. Objective and Advantages. The objective of the military field meet is to provide activity for everyone in the participating units. Activities are chosen that will develop aggressiveness, teamwork, a will to win, and competitive spirit, and that will stimulate interest and build esprit de corps. The military field meet can **be** included as part of the physical training program or as part of the off-duty recreational program. It is a form of contest that can be conducted in nearly all circumstances because it can be easily modified, requires a minimum amount of equipment, and can be readily organized.

b. Level of Competition. The military field meet is flexible. It can be adjusted to large or small groups. A company-size unit is the most desirable, but it may also be administered within a larger unit. If it is conducted within a company, the participating units will be squads. If it is held within a larger unit, platoons compose the teams. A larger-size unit requires more extensive organization and administration.

c. Selection of Events. In organizing a military field meet, select events that are simple and easy to administer. All rules and regulations should be clearly understood by everyone. No event should require previous practice. In selecting the events, consider the interest and capabilities of the Marines and available equipment and facilities. Select events which will require that all members of the units participate. Events should not be dominated by the athletes; rather, MOS-related events which have significance in producing skills and combat readiness should be used.

d.Equipment and Facilities. The site must be large enough to permit events to be grouped about a central control point. Use available facilities such as parade decks, Softball fields, or tracks. A public address system is desirable at the control point for the initial orientation of teams and for subsequent announcements of time lapses, cumulative scores, and final standings. A tally board is necessary so that all units can constantly monitor how the events are progressing.

e.Administrators. Efficient administration of the military field meet depends on the referees, judges, and scorers. For these positions, choose individuals within the units who have had athletic or officiating experience. Prior to the day of the event, all administrators should be briefed and assigned a specific task in order that they may become familiar with the rules and organization of the contests they will conduct. The following officials should be available:

(1) A primary instructor or supervisor who is in charge of the control center and who is responsible for the successful operation of the athletic carnival. The supervisor must have an assistant to act as a timer and scorer.

(2) One assistant instructor in charge of each event. This person should be-

(a) Familiar with the rules of the event.

(b) Effective as a leader to ensure proper supervision and control over the participating teams.

(c) Enthusiastic to provide proper motivation.

(d) Self-confident of the ability to judge infractions of the rules. The assistant instructor must be fair in judgment and penalize without hesitation when infractions occur.

(3) Runners between event stations and the control point. These individuals collect and deliver scores.

f. Team Organization for Competition. The size of the teams is determined by the level on which the field meet is organized. Maintaining the integrity of the unit promotes esprit de corps, but this does not preclude grouping two squads into one team.

g.Conduct of Events

(1) The assistant instructor at each station takes charge of the group and gives a brief explanation of the major rules of the event. The assistant instructor speaks clearly and distinctly from a position to be seen and heard by all. A short demonstration is desirable if it will help clarify the event.

(2) The assistant instructor should make certain that teams can be clearly distinguished by the use of, for example, T-shirts and fatigue jackets, or caps and no cap, or colored jerseys. The assistant instructor should teach a whistle response (teams stop play immediately upon hearing whistle), get the event started as quickly as possible, and make any necessary corrections as the contest progresses. The rale of good officiating is to use a minimum of calls, yet maintain control of the contest. Penalize when necessary, but refrain from disqualifying contestants or teams.

(3) Keep the activity moving as rapidly as possible. When the central control point sounds the whistle to stop the play, all competition ends immediately. The assistant instructor then assembles the group, forwards the team scores to the central point, and, upon a signal from the central control point, rotates the teams to their next station. It is essential that rotation and orientation of teams be quick and orderly to reduce time spent between contests.

(4) Upon completion of the final event, assistant instructors move their teams to the central control point for the announcement of winning teams, presentation of awards (if any), and final critique.

h. Scoring System. The system for determining the winner of the military field meet should be simple and efficient. At the completion of competition, the scorer totals the points that each team has scored in all contests. The scorer then subtracts the number of points scored against a team (penalty points for nonparticipation,etc.) from the total. The resulting scores are placed in a column with the highest score at the top and the lowest at the bottom. The team with the highest total is the winner. This type of scoring system encourages a team to ensure total unit participation.

Section II.Combatives

5201.DESCRIPTTON AND OBJECTIVE

Combatives are strenuous, short competitive contests in which two individuals attempt to overcome each other in a bout of skill and strength. These contests help to develop Marines' resourcefulness, confidence, strength, agility, coordination, and will to win. Any level ground area can be used. However, extremely hard ground should be avoided as some of the combatives require ground contact. A whistle is needed to control the bouts during the competition since voice commands may go unheeded. The extended rectangular formation is used for dual combatives. Combatives are conducted on an informal basis. Marines are allowed to remain at ease between activities and are allowed to brush themselves off after being on the ground. To get the most out of combatives, individuals must be urged to overcome their opponents as quickly as possible as would be required in combat.

a. Benefits for Marine Training. The possibility of close contact with an enemy in combat faces Marines at all times. They must be trained to react aggressively and violently. Combatives may be used as an introduction to such hand-to-hand contact and should be followed by hand-to-hand combat training. Marines enjoy competition, and this type of activity is a welcome change from formal conditioning activities.

b. Instructor Responsibility

(1) The instructor tells the individuals that all combatives begin and end on the whistle signal. The instructor demonstrates each activity before having it performed, explaining it in simple terms.

(2) After stopping one activity, the instructor gets everyone in place for the next.

(3) The instructor must closely supervise combatives to insure that contestants do not use unfair or unsportsmanlike tactics. To avoid unnecessary injury, instructors must see that the bouts are closely controlled and opponents equally paired. Adjustments should be made in apparent cases of mismatched abilities.

5202.COMBATIVES TABLES

There are three tables of combatives, each of which can be completed in 15 minutes. These tables become progressively more difficult from lower to higher numbers.

a.Combatives Table I

(1) Open-Hand Slap Boxing(See fig. 5-1, A. individuals assume a boxer's stance, palms open, fingers extended and joined. Each contestant tries to slap the opponent about the head and upper body with the open hand. This is a good warm- up activity.

(2) Wrist Tug-of-War(See fig. 5-1, B.)Two individuals sit on the ground with the soles of their feet in contact. Each grasps the opponent's wrists so that the hands are directly over the feet. At the whistle, each individual tries to pull the opponent from sitting to standing position.

(3) Arm-Lock Wrestling.(See fig. 5-1, C.)Two individuals sit back-to-back with legs spread and arms locked at the elbows. Each right arm is inside the opponent's left. Each tries to force the opponent's left arm or shoulder to the ground. The individual who first wins three bouts is the winner.

(4) Bulling(See fig. 5-1, D.)Two individuals assume the Westmoreland wrestling hold, each grasping the opponent's neck with the right hand and the opponent's right elbow with the left hand. Each tries to force the other to move one foot by pushing, pulling, or otherwise manipulating. The individual who first wins two boutsis the winner.

(5) Indian Wrestling(See fig. 5-1, E.)Two individuals lie on the ground, side by side, with their heads in opposite directions. They link right elbows. On the instructor's signal or by mutual agreement, each individual raises the right leg approximately straight and far enough to engage the opponent's heel. To start the contest, each person usually raises the leg three times rhythmically and on the third time engages the opponent's heel. Each tries to roll the other over backward. The right leg is used for three bouts, then the left leg for three bouts.

Figure 5-1 is available in the illustration section at the end of the book.

b. Combatives Table II

(1) Wrist Bending.(See fig. 5-2, A.) Opponents pair off, face each other, raise their arms forward, and, with palms forward, interlock their fingers. At the starting signal, each individual attempts to bend the opponent's wrist. The arms are kept up and forward and are not allowed to swing around and down to the sides. The individual who first wins two bouts is the winner.

(2) Back-to-Back Push.(See fig. 5-2, B.)Two individuals stand back-to-back with elbows locked. Each right arm is inside the opponent's left arm. At the starting signal, each pushes backward, trying to move the opponent. Opponents are not allowed to lift and carry each other; only pushing is permitted. The one who pushes the opponent the farthest wins the bout. The individual who first wins two bouts is the winner.

(3) Hop and Pull Hand.(See fig. 5-2, C.) The individuals are matched in pairs. Each grasps the opponent's right hand and, hopping on the right foot, attempts to pull the opponent off balance. Contestants automatically lose if they touch their free hand or their lifted foot to the ground. For successive bouts, they alternate hands and feet.

(4) Westmoreland Wrestling. (See fig. 5-2, D.)Each contestant grasps the back of the opponent's neck with the right hand and the opponent's right elbow with the left hand. In this position, each attempts to pull, push, or force the opponent to touch the ground with any part of the body other than the feet. The individual who first wins two bouts is the winner.

(5) Crab Fight.(See fig. 5-2, E.)Two individuals sit on the ground facing in opposite directions with their hands on the ground behind them. At the whistle, they raise their hips and push with their shoulders and bodies, each trying to make the other's hips touch the ground. The individual who first wins two bouts is the winner.

Figure 5-2 is available in the illustration section at the end of the book.

c. Combatives Table III

(1) Hand Wrestling.(See fig. 5-3, A.) Two individuals stand facing each other. Their right feet are forward and braced side by side. They grasp right hands for the first bout, left hands for the second bout. Each pulls, pushes, makes sideward movements, and otherwise maneuvers to force the opponent to move one or both feet from the original position. The contestant who first wins two bouts is the winner.

(2) Back-to-Back Tug.(See fig, 5-3, B.) Two individuals stand back-to-back with both arms linked at the elbows. Each has the right arm inside the opponent's left arm. At the starting signal, each attempts to pull the opponent. Lifting and carrying are permitted. The contestants must maintain their original direction and keep their arms linked. After a predetermined time, the player pulled or carried the farthest is the loser.

(3) Wrestling to Lift off Feet.(See fig. 5-3, C.) Contestants face each other. Each places the right arm under the left arm of the opponent and around the body. The left arm is over the opponent's right shoulder. Each tries to lift the other off the ground. The individual who first wins two bouts is the winner.

(4) Arm Pull Between Legs.(See fig. 5-3, D.) Two individuals are paired off, back-to-back. Each bends forward and, extending the right arm between the legs, grasps the opponent's right wrist. At the starting signal, each person attempts to pull the opponent After a predetermined time, the player who has pulled the opponent the farthest is the winner of the bout. The one who first wins two bouts is the winner. Repeat with the left hand and then both hands.

(5) Rooster Fight.(See fig.5-3, E.) Each contestant grasps the left foot with the right hand from behind, and the right arm with left hand. Each hops on the right foot, and, by shoulder-butting the opponent or by feinting and sudden evasions, forces the opponent to let go of the foot or arm. The contestant who first wins two bouts is winner.

Figure 5-3 is available in the illustration section at the end of the book.

Section III .Relays

5301. DESCRIPTION AND OBJECTIVE

Relays are races in which each member of a team runs one leg of the race. The team effort decides the winner. Relays provide stimulating competition and contribute to the conditioning of personnel. They also develop aggressiveness, team spirit, and the will to win. Relays should be dispersed throughout the program for short periods of time to provide a change of activity.

a. Team Organization. Relays are conducted most efficiently in platoon-size groups. Teams of equal size must be organized. Competitive spirit is encouraged and teams are organized better by basing teams on units such as squads, crews, or sections. Team captains should be designated. Extra Marines may be used as officials. The number of individuals on a relay team should be limited to squad size. If larger teams are used, the runners will spend too much time awaiting their turns and too little time actually participating. Two to six teams are ideal for relay competition. It is difficult to keep track of winners when more teams compete.

b. Administration of Relays

(1) The time spent on any one relay should be relatively short. If one team achieves a substantial lead in a long relay, the competitive spirit and enthusiastic participation of the other teams may decrease. Several short relays are generally better than one long relay.

(2) To maintain competitive spirit throughout a number of relays, determine the teams that win, place, and show in each relay and their total points for all relays. This can be done by awarding points to all teams on the basis of position at the finish of each relay. The team with the greatest number of points is the winner of the entire set of relays.

(3) Difficulties commonly encountered in conducting relays may be avoided by the following procedures:

(a) The last player in a relay race should be conspicuously identified. For example, the last runner can tie a handkerchief around the head or arm, take off the shirt, put on a hat or take it off, or use some other means.

(b) Another way to keep track of the progress of the race is to have each player sit or squat as soon as each is finished.

(c) Judges at the starting line can keep the runners from starting too soon.

(d) To prevent contestants from turning before they run the full distance, they should be required to run around a peg, pole, or assistant instructor.

(e) Batons, handkerchiefs, tent pegs, or other objects should be passed from one runner to the next when relays are run on a circular track.

(4) Before a unit's first participation in a relay, inform participants of the rules and scoring system. Violation of the rules should not result in disqualification. Instead, impose point penalties. A point penalty is imposed by subtracting a point from the team total at the conclusion of the relay.

(5) Careful administration will prevent most violations. For efficient conduct of relays, follow this procedure:

(a) Announce the name of the relay.

(b) Form the individuals in relay position.

(c) Briefly explain the relay and the rules for running it.
(d) Demonstrate.
(e) Have a definite finish line, and ensure that all know where it is.
(f) Answer questions.
(g) Conduct the relay.
(h) Determine winner and award points.

5302. RELAY TABLES

a. Events. The following relays are grouped into a table of activities. Each relay table can be completed in 15 minutes. Thus, relays can be used as a sole activity or as a part of a longer period. Each table is planned for a platoon-size group (30-60Marines). Adequate warmup for participants is provided by conducting several repetitions of exercises 1 and 2 of a conditioning drill. The recommended relay tables require an area 40 by 60 meters in size. Each table provides a variety of activity. The tables are progressive in the overload applied and should be scheduled in numerical order although not necessarily on successive days.

b. Relay Table I
(1) **60-Meter Lane Relay** (See fig. 5-4, A.) Each team is assembled in single file behind the starting line. On signal, the first individual of each team runs to the turn-around line 30 meters away, then runs back and touches the next teammate waiting at the starting line. The winning team is the first team to get its last member across the finish line. If an individual starts before being touched by the preceding runner, the team may be penalized.
(2) **Wheelbarrow Race**.(See fig. 5-4, B.) The players of each team pair off and line up in single file behind the starting line. The first individual walks on the hands while the partner grasps the ankles. They advance to the distance line (25 meters) behind which they exchange positions and return to the starting line. The rear individual must always hold the partner's ankles. After the first pair returns across the starting line, the next pair starts.

Figure 5-4 is available in the illustration section at the end of the book.

(3) **Squad Front Relay** .(See fig. 5-4, C.) The teams form in a line along the starting line with a 10-foot interval between teams. The members of each team lock elbows so that they are linked together. At the starting signal, the teams run to the distance line (20meters) where the left flank individual acts as a pivot. The team swings around on the pivot and returns to the base line. If a team breaks its links, it must reform before continuing. The first team to completely cross the base line intact is the winner.
(4) **Crab-Walk Race**.(See fig. 5-4, D.) The players of each team line up in single file. The first person of each team assumes the crab-walk position with feet forward on the starting line, hands on the ground behind it. At the starting signal, contestants move forward to the distance line (10 meters). They touch the line with their feet and then return to the starting line in the reverse position with the head and hands leading. The second person may not start until the first individual touches the finish line.

c.Relay Table II

(1) 100-Meter Lane Relay.(See fig. 5-5, A.)This relay is conducted exactly as the first relay of table 1 except that the start and turn-around line are 50 meters apart. This relay provides progression in sprinting.

(2) Frog-Jump Relay. (See fig. 5-5, B.) Each team lines up in single file. The first individual assumes a squatting position on the starting line. At the starting signal, the contestant progresses to the distance line (15 meters) and back by leaping forward, catching the weight on the hands, and bringing up the legs to the squat position for the next leap forward.

(3) Simple Relay. (See fig. 5-5, C.)Each team lines up in single file. Place a marker on the distance line (20 meters) in front of each team. Each team member, behind the first one, grasps the belt of the individual ahead. At the starting signal, each team runs as a unit to the marker, circles it, and returns to the starting line. The first team to completely cross the starting line intact is thewinner.

(4) Fireman's Carry Relay.(See fig. 5-5, D.)The players of each team line up in pairs, one behind the other. One individual in each pair carries the partner to the distance line (30 meters), using the fireman's carry. At the distance line, individuals exchange places and return to the starting line. As a variation, the person to be carried lies on the ground. The one carrying lifts the partner to the proper position. This relay may be performed with the other carries described in guerrilla exercises in chapter 3.

d.Relay Table III

(1) 200-Meter Circle Relay .(See fig. 5-6, A.)A course is laid out in a circular, rectangular,or oval pattern that is 200 meters around. Each team provides one runner on the starting line. On signal, the runner races around the 200-meter track and touches the next teammate, waiting at the starting line, who runs the same course. Each team member runs one lap of the course.

Figure 5-5 is available in the illustration section at the end of the book.

(2) Bear-and-Crab Race.(See fig. 5-6, B.) Each team lines up in single file. At the signal to start, the first contestant in each column assumes the bear-walk position (hands on ground in front), walks to the distance line (15 meters), and then runs back to the starting line. The returning contestant touches off the second individual and goes to the rear of the line. The second contestant assumes the crab-walk position (hands on ground behind) and crab-walks with feet leading to the distance line. This contestant also runs back to the starting line and touches off the next individual, who walks bear fashion. The rest of the members of each team alternate in this manner. The relay ends when the first person is back at the head of the line.

(3) Pilot Relay.(See fig. 5-6, C.) The players are grouped in threes, arms interlocked at the elbows, and outside players facing backwards. The middle person runs forward; the two outside individuals run backward. They run to the turning point (15meters), where they start back, this time with the middle person running backward and the two outside individuals running forward. The next set of three players starts when the first set crosses the starting line.

(4) Saddle-Back Relay.(See fig. 5-6, D.) Mark two parallel lines 15 meters apart. Each team selects a rider. The remaining members of each team count off. The even-numbered players from each team form in single files behind one line and the odd-numbered players from each team form in single files behind the other line directly across from their teammates. At the starting signal, the rider mounts the back of the first player of the team who carries the rider across the other line. There the rider changes mounts to the first player in the second line without touching the ground. This person carries the rider to the next player waiting in the first line. The relay continues until all of the mounts have carried the rider. A rider who falls off must mount again at the point of the fall. A rider who falls in changing mounts must get back on the original mount before making the change.

e. Relay Table IV

(1) 100-Meter Circle Relay (See fig. 5-7, A.)A course is laid out in either a circular, rectangular, or oval pattern that is 200 meters around. Each team is divided in half with each half positioned at starting lines on opposite sides of the track. Each runner races halfway around the track and touches a teammate who completes the lap. Each runner then waits in file at the halfway point until touched by the runner and then completes the second half of the lap. The first team to return all runners to their original starting line is the winner.

Figure 5-6 is available in the illustration section at the end of the book.

(2) In-and-Out Relay.(See fig. 5-7, B.) Each team lines up in a file with players 2 meters apart. At the starting signal, the first player runs back through the column in a zigzag fashion. He alternates, going to the right of one teammate and to the left of the next. Upon completing the run, the player lines up 2 meters behind the last one in the column. As soon as the first runner has passed the second person, the latter starts to run. This continues until all the players have realigned their original order. The team that finishes first is the winner. It may be desirable to have this relay continue until all individuals have run through their entire team two or three times in succession.

(3) Circle Race.(See fig. 5-7, **C.)** Each team forms a circle and holds hands with all individuals facing out except one who faces in and is the "driver."At the starting signal, the teams race to the distance line (20 meters) and back, keeping the circle intact. All the individuals in the circle must completely cross the distance line. The "driver" gives directions and orders. When the circle breaks, it must be reformed before it can continue. The first team completely over the starting line is the winner.

(4) Horse-and-Rider Relay.(See fig. 5-7, D.)Each team lines up in a single file. At the signal to start, the second player in each column leaps upon the back of the first who carries the rider across the distance line (30 meters). At the distance line, the rider dismounts and runs back to the starting line. There the rider picks up the third individual in the column, and carries this player to the distance line where the first player has remained. This continues until the last person is carried across the distance line.

Figure 5-7 is available in the illustration section at the end of the book.

Section IV. Team Contests and Athletics

5401. DESCRIPTION AND OBJECTIVE

Team contests are competitive activities in which Marines as a team compete with another team to win. They are guided by simple rules and organization. The function of team contests is to provide competition and an opportunity for body contact and to contribute to the development of physical readiness. In competing and working together as a team, individuals develop aggressiveness, the will to win, and teamwork. Team athletics deserve a prominent place in the physical training program because they contribute to increased combat efficiency. Because of the competitive nature of athletics and their natural appeal, individuals take part with enthusiasm. Athletic teams formed at intramural and higher levels are a strong unifying influence and provide one of the best means of developing esprit de corps.

a. Area and Equipment A level training field is sometimes the only area required. Many contests need no equipment. In contests requiring equipment, the need is for standard items such as logs, balls, nets, goals, and similar types of equipment. Specific requirements for area and equipment are listed with each contest description.

b. Progression. Team contests of a strenuous nature should be introduced after a basic period of conditioning has been completed and individuals are in the slow improvement stage of conditioning. Progression can take place from the less active to the more vigorous contests, and then from the non-contact to the contact or combative-type contests.

c. Necessity for Preconditioning Individuals must undergo conditioning prior to participation in athletics. Muscles, organs, joints, and ligaments not accustomed to stress and strain from sudden stops and starts, falls, body contact, rapid turns, prolonged running, and other rigors of athletic competition are subject to injury. Although athletics should not be introduced until players are physically prepared, there is still opportunity to engage in competition through lead-up contests. Individuals learn many of the skills required for athletics while participating in team contests.

d. Benefits. Athletics are beneficial primarily to sustain interest in the program and to maintain an achieved level of physical fitness. Athletics are a supplement, not a substitute for other types of conditioning activities which should continue. All of the desirable traits of physical fitness cannot be developed through athletics, yet their contribution is significant. For athletics to make a proper contribution to physical conditioning, the selected sports must bevigorous. The team contests recommended here are carefully selected for their simplicity, aggressiveness and applicability to squad- teaching teamwork. They are generally strenuous and involve the whole team throughout the entire duration of the contest.

5402. TEAM CONTESTS

a. Pushball

(1) **Players.** 10 to 50 players on a side. (See fig. 5-8.)

(2) Equipment A large pushball 5 to 6 feet in diameter.

(3) Area. A field 240 to 300 feet long, 120 to 150 feet wide. Mark a center line in the middle of the field parallel to the end lines. Mark a line 45 feet on either side of this center line and parallel to it, extending it across the width of the field. Mark another parallel line 15 feet from each end line, extending it across the width of the field.

(4) Game. Four 10-minute quarters are played. Give 2-minute rests between quarters and 5-minute rests between halves. The object of the game is to propel the ball over the opponent's goal line by pushing, rolling, passing, carrying, or any other way except kicking.

(a) The ball is placed on the center line with the opposing captains 3 feet from the ball. The rest of the players are 45 feet from the ball, on their half of the field. On the referee's starting whistle, the captains immediately play the ball with their respective teams coming to their assistance.

(b) At quarter time, the ball remains dead for 2 minutes at the spot where it was when the quarter ended. At half time, the teams exchange goals. The play is then started as it was in the beginning.

Figure 5-8 is available in the illustration section at the end of the book.

(c) Players may use any means of interfering with an opponent's progress except striking and clipping. Clipping is throwing the body across the back of an opponent's leg, while the opponent is running or standing. Legal use of force may be applied to all opponents whether or not they are playing the ball. For striking an opponent, the offender is removed from the game. The team penalty is half the distance to their goal. The penalty for clipping is the same.

(d) When any part of the ball goes out of bounds, it is a dead ball. The teams line up at right angles to the side lines and 3 feet apart at the point where the ball went out. The referee then tosses the ball between the teams.

(e) When, for any reason, the ball remains in one spot for more than 10 seconds, the referee declares the ball dead. The ball is then put into play as it is for an out-of-bounds situation.

(5) Scoring. A goal is scored when the ball, or any part of it, is propelled across the opponent's end line. A goal counts 5 points. The team scoring a goal has the privilege of trying for a point after the goal. To try for this extra point, the ball is placed on the opponent's 5-yard (or 15-foot) line. The teams line up on either side of this line separated by the width of the ball. Before the whistle blows, one player on each team may place hands on the ball. On the referee's signal, the ball is put into play for 1 minute. If any part of the ball is driven across the goal line in this 1-minute period, the offensive team scores 1 point. The defending team may not score during the opponent's try for the extra point.

b. Line Rush

(1) Players. Any number up to 50 on each side. (See fig. 5-9.)

(2) Equipment. None.

(3) Area. A field, 75 by 100 feet.

(4) Game. One team lines up behind one goal line and the other in midfield. On the starting signal, the team standing behind the goal line seeks to cross the goal line at the opposite end of the field within 30 seconds. The team in the center seeks to prevent it by catching and holding the runners. At the end of 30 seconds, the teams change.

(5) Scoring. Count the number of individuals who have crossed the far goal at the end of 30 seconds. After each team has had from three to five tries, the scores are added and the winner declared. A player scores 1 point when any part of the body is across the goal line.

Figure 5-9 is available in the illustration section at the end of the book.

c. Human Tug-of-War

(1) Players. 10 to 20 on a team.(See fig. 5-10.)

(2) Equipment. None.

(3) Area.40 to 60 feet.

(4) Game. Draw a line in the center of the area. Divide the players into two equal teams. Place them in single file on opposite sides of the center line facing each other. Each individual places the arms around the waist of the teammate in front. The two leaders of the opposing teams grasp each other around the waist. On signal, the teams try to pull each other over the center line within 30 seconds.

(5) Scoring. The team pulled across the center line loses. If neither team is pulled over the center line, but one team breaks its file, that team loses the match.

(6) Variation. Use a 3/4- or 1-inch rope. Space the leading players on each team 10 feet apart, each holding one end of the rope. The team pulled across the center line loses.

Figure 5-10 is available in the illustration section at the end of the book.

d. Master of the Ring

(1) Players. Any number. (See fig. 5-11.)

(2) Equipment. None.

(3) Area. A clearly marked circle large enough to contain all the players.

(4) Game. All the players stand inside the circle. At the signal, all players attempt to throw each other out of the circle. All tactics are fair except unnecessary roughness. When any part of the body touches across the line, the player is out and leaves the circle at once. Several officials are needed to spot the players who cross the line.

(5) Scoring The player who remains in the circle when all the others are out is the master of the ring.

(6) Variation 1. The players are divided into two equal teams. Each team is clearly marked. On signal, each team tries to throw the opponents out of the circle. The winning team is the team that eliminates all opponents from the circle.

(7) Variation 2. The players are divided into two equal teams. Each team sends only one individual into the circle. When one person has been forced out of the circle, the losing side sends in only one player. The team which eliminates all opponents is the winner.

(8) Variation 3.A pit, approximately 4-feet deep, is used rather than a circle on level ground. The contest may then be an individual or team activity.

Figure 5-11 is available in the illustration section at the end of the book.

e. Log Pivot Circle. Space teams far enough apart so each team can pivot in a circle without colliding with any other team. Each team holds a log in the bend of the arms in front of the chest. (See fig.5-12.) At the command **CIRCLE RIGHT, MOVE**,the left-flank individual holds the pivot while the log is carried around 360 degrees in a counter- clockwise motion, back to the original position. This movement may also be performed to the left by facing in the other direction and pivoting clockwise (**CIRCLE RIGHT, MOVE**).Other commands may be used such as: **CIRCLE HALF-RIGHT, CIRCLE HALF- LEFT** and so on. The first team to complete the prescribed movement is the winner.

Figure 5-12 is available in the illustration section at the end of the book.

f. Log-Rolling Race. Each team tries to roll its log a measured distance by pushing it with the hands and driving the body forward with the legs. (See fig. 5-13.)The first team to get the entire length of the log across the finish line wins.

Figure 5-13 is available in the illustration section at the end of the book.

g. Prone Pushing Contest.. Two teams lie prone, facing each other with a log between them. (See fig. 5-14.)Both teams place their hands against the log, keeping their arms straight. Then, by driving with the legs, each team attempts to push the other a measured distance to the rear.

Figure 5-14 is available in the illustration section at the end of the book.

h. Shuttle Relay Race. Each team in this race divides into two parts, A and B. Members of part A run 50 yards with a log held under their right arms. (See fig 5-15.)At the distance line, they give the log to their teammates in part B who bring it back to the starting line. The team pair finishing first is the winner.

Figure 5-15 is available in the illustration section at the end of the book.

5403.CROSS-COUNTRY AND DISTANCE RUNNING

a. Value of Running. Long-distance running gives some benefits that cannot be obtained in the same degree from any other sport. It builds powerful leg muscles, increases lung capacity, and develops endurance. For these reasons, cross-country and distance running should be included in the physical training program. These sports require only a few miles of open space which is generally available. Short cross-country runs and middle-distance runs can be used to supplement other activities, particularly team sports or the sports that develop precision or agility rather than endurance. Short cross-country runs can be scheduled once a week, gradually increasing the distance as the physical condition of the runners improves.

b. Cross-Country Runs. A cross-country run is a distance run held on a course laid out along roads, across fields, over hills, through woods, and on any irregular ground. A flat cinder or dirt track is not a suitable surface for cross-country running. Some runs are as long as 6 miles while others are as short as 3 or 4 miles. If cross-country running is being used to supplement other activities, the 3-mile course is long enough for most. Cross-country runs should be used only after individuals reach the sustaining stage of conditioning. This type of running should then be scheduled occasionally to provide variety. Cross-country running has the advantage of allowing mass participation. Interest can be stimulated by putting the runs on a competitive basis.

c. Practice Methods. Conditioning is essential to distance and cross-country running. Championship distance running depends on stamina, which can be developed only through constant training. An individual of only average ability can become an outstanding distance runner by steady and careful training. Hiking is an excellent method for getting into condition before the season opens. Long walks build up leg muscles. During the first month of the season, training should be gradual, starting with short distances, and increasing day by day. At first the legs will become stiff, but the stiffness gradually disappears if running is practiced for a while every day. To prevent strain, it is essential to limber up thoroughly each day before running.

(1) In training a large group, leaders should be stationed at the head and the rear of the column. They should make every effort to keep the runners together. After determining the abilities of the unit in cross-country running, it is advisable to divide them into three groups. The poorest-conditioned group is started first, the best-conditioned group, last. The starting time of the groups should be staggered so that all of them come in at about the same time.

(2) In preliminary training, running is similar to ordinary road work in that it begins with rather slow jogging, alternating with walking. The speed and distance of the run is gradually increased. As conditioning improves, occasional sprints may be introduced. Atfirst, the distance run is from one-half to 1 mile. It gradually increases to 2 or 3 miles. On completing runs, require runners to walk for 3 or 4 minutes before stopping to permit a gradual cooling off.

d. Facilities and Equipment. There should be at least one timer with a stopwatch (preferably three) for timing the runners. A course 3 or 5 miles long should be measured and marked by one of the three following methods:

(1) Fasten directional arrows to the top of tall posts and place them at every point where the course turns. Such signs should also be placed at every other point where there may be doubt as to the direction of travel.

(2) Place a lime line on the ground over the entire course.

(3) Place flags, clearly visible to the runners.

- **(a)** A red flag indicates a left turn.
- **(b)** A white flag indicates a right turn.
- **(c)** A blue flag indicates the course is straight ahead.

e. Rules/Scoring. A cross-country team consists of seven individuals, unless otherwise agreed. In dual meets, a maximum of 12 on one team may be entered, but 7 or fewer are scored. First place scores 1 point, second place 2, third place 3, and so on. All who finish the course are ranked and tallied in this manner. The team score is then determined by totaling the points scored by the first five of each team to finish. The team scoring the least number of points is the winner. If fewer than five (or the number determined prior to the race) finish, the places of all members of that team are disregarded. If two or more teams score the same number of points, the event is called a tie.

5404.TEAM ATHLETICS

Team athletics add variety to the combat readiness program. At no time should they be allowed to replace intramural activities or to provide practice time for intramural or base teams. If team athletics are integrated into the program, the goal should be total participation and competition. Therefore, the games selected can be played by the skilled or the unskilled. They require limited equipment, yet will teach competitiveness, aggressiveness, and teamwork while they improve conditioning.

a.Soccer. Soccer is one of the best athletic activities for developing endurance, agility, leg strength, and great skill fulness in using the legs. The most popular sport in the world, soccer is the national game of many European, Asian, and Central and South American countries. In recent years, it has become popular in the United States. A soccer ball is the only equipment needed for the game, and Marines can learn to play it easily. Players do not need much skill to participate, but the amount of skill they can develop is almost unlimited.
(1) Place in the Program. Soccer should be introduced during the latter part of the slow improvement stage of physical training. It can be used as a competitive activity in the sustaining stage. It is primarily a spring or fall sport. Any level field is suitable for competition. The boundaries for the soccer field are similar to the dimensions for a football field. (See fig. 5-16.) Goal posts are essential, but they are easily constructed and usually temporary so that they may be removed when not in use.

Figure 5-16 is available in the illustration section at the end of the book.

(2) Basic Skills

(a) Passing. Passing with the feet is the primary means of moving the ball. Short passes are easier to control. Emphasis should be continually placed on the skill of passing.

(b) Dribbling. The ball is dribbled by a series of kicks with the inside or outside of the foot. Do not kick with the toe. Keep the head over the ball when kicking and propel it only a short distance at a time. Keep it close to the feet. If a player lets a ball move far from the feet, an opposing player can easily take it away.

(c) Instep Kicking. The instep kick is the basic soccer kick. The toe does not come in contact with the ball. The toe is pointed downward, and the instep (the area under the shoe laces) is applied to the ball with a vigorous snap from the knee.

(d) Inside-of-the-Foot Kicking. The ball is kicked with the inside of the foot, and the leg is swung from the hip. The toe is turned outward, and the sole of the foot is parallel with the ground as the foot strikes the ball. This kick is used for short passes and for dribbling.

(e) Foot Trapping. The foot trap is the method of stopping the ball by trapping it between the ground and the foot. Place the sole of the foot on top of the ball at the instant it touches the ground, keeping the foot relaxed. This is an effective way to stop a fast-moving ball.

(f) Shin Trapping. .The shin trap is a method of stopping the ball with the shins. Stand just forward of the spot where the ball should strike the ground and allow it to strike the shins in flight or on the bounce. Use either one or both legs from the knee down so as to absorb the bounce.

(g) Body Trapping. .Intercept the ball with any part of the upper body except the arms and hands. Keep the body relaxed and inclined toward the ball. To keep the ball from bouncing, move backwards from it as it strikes the body. This drops the ball at the feet in position for dribbling or passing.

(h) Heading. Heading is a technique for changing the direction of the flight of a ball by butting it with the head. Tense the neck muscles and jump up to meet the ball. Butt the ball with the forehead at about the hairline to reverse its direction. Use the side of the head to deflect it to the side.

(3) Offensive and Defensive Positions. Forwards usually play on the offensive half of the field. Fullbacks usually play on the defensive half of the field. Halfbacks are the backbone of the team. They move forward on the offense and back on defense. The goalkeeper almost always remains within a few feet of the goal.

(4) Abridged Rules

(a) A soccer team is composed of 11 players. (See fig. 5-16.)

(b) The player propels the ball by kicking it with the feet or any part of the legs, by butting it with the head, and by hitting it with any portion of the body except the arms or hands.

(c) The goalkeeper is the only individual allowed to place hands on the ball, but only in the goalkeeper's area. The term "hands" includes the whole arm from the point of the shoulder down.

(d) A goal is made by causing the ball to cross completely the section of the goal line lying between the uprights and under the cross bar.

(e) Each goal scores 1 point for the team scoring the goal.

(f) The penalty for a foul committed anywhere on the playing field (except by the defensive team in its penalty area) is a free kick awarded to the opposing team.

(g) All opponents must be at least 10 yards from the ball when a free kick is taken.

(h) The penalty for a foul committed by the defensive team in its penalty area is a penalty kick.

(i) A penalty kick is a free kick at the goal from the spot 12 yards directly in front of the goal. The only players allowed within the penalty area at the time of the kick are the kicker and the defending goalkeeper.

(j) An official game consists of two 30-minute halves,

(k) Teams change goals at the end of every quarter.

(l) In the event of a tie, an extra quarter may played. This may be followed by a sudden death period. If the tie still exists, the game can be decided by penalty kick.

(m) After a team has propelled the ball across a side line and out of play, the ball is put back into play by a throw-in from the side line by a member of the opposing team. The ball is thrown in from the point where it crossed the side line as it went out of bounds. The ball must be thrown in with both hands using an over-the-head motion, keeping both feet on the ground.

(n) When the offensive team propels the ball across the defensive team's goal line but not in the goal, the defensive team is awarded a goal kick-a free kick taken from within its own goal area. The ball must come out of the penalty area to be in play.

(o) When the defensive team causes the ball to go behind its own goal line, excluding the portion between the goal posts, the offensive team is awarded a corner kick—a free kick taken by a member of the offensive team at the quarter circle, at the cornerflag post nearest to where the ball went behind the goal line. The flag post must not be removed.

(p) The game is started and, after a goal has been scored, is resumed by placing the ball in the center of the midfield line. Players must be on their side of the line until the ball is kicked. The ball must be kicked forward and must move at least 2 feet to be legal. The first kicker may not touch the ball twice in succession at the kick off. The opposing team must be 10 yards from the ball until it moves.

b.Speedball. Speedball offers vigorous and varied action with plenty of scoring opportunities. It is easy to learn and provides spontaneous fun. Little equipment is needed-a ball is all that is absolutely necessary. Speedball combines the kicking, trapping, and intercepting elements of soccer; the passing game of basketball; and the punting, drop kicking, and scoring pass of football.

(1) Place in the Program.Speedball, like soccer, should be introduced into the physical training program during the latter part of the toughening stage and used as a competitive activity in the sustaining stage. Speedball may be played any time weather permits but is primarily a spring or fall activity.

(2) Method of Play.The game is played as follows:

(a) Two teams of 11 each play the game under official rules, but any number of players may successfully constitute a team.An inflated leather ball, usually a soccer ball, is used. The playing field is a football field with a football goal post at each end. (See fig. 5-17.)

(b) The game starts with a soccer-type kickoff. The kicking team tries to retain possession of the ball and advance it toward the opposite goal by passing or kicking it. Running with the ball is not allowed, with the result that there is no tackling or interference. After the ball touches the ground, it cannot be picked up with the hands or caught on the bounce but must be played as in soccer until it is raised into the air directly from a kick. Then the hands may be used.

(c) When the ball goes out of bounds over the sidelines, it is given to a player of the opposing team and is put into play with a basketball throw- in. When it goes over the end line without a score, it is given to a player of the opposing team who may either pass or kick it onto the field.

(d) When two opposing players are contesting possession of a held ball, the official tosses the ball up between them as in basketball.

(e) Points are scored by kicking the ball under the crossbar of the goal post, drop kicking the ball over the crossbar, or completing a forward pass into the end zone for a touchdown.

Figure 5-17 is available in the illustration section at the end of the book.

(3) Basic Skills

(a) Shared Skills. Skills include kicking, passing, heading, and trapping (from soccer); punting, dropkicking and forward passing (from football); and passing, receiving, and pivoting (from basketball).

(b) Kickups. The kickup is a play in which a player lifts the ball into the air with the feet so as to legally play the ball with the hands. The kickup generally makes the transition from ground play to aerial play. The technique of making the play depends upon whether the ball is rolling or stationary. If the ball is rolling or bouncing toward the player, the foot is held on the ground with the toe down until the ball rolls onto the foot. Then the foot is raised, projecting the ball upward. If the ball is stationary, the player rolls it backward with one foot. Then the player places a foot where the ball will roll onto it. The player then raises the foot, projecting the ball as before. If a ball is rolling away, the player should stop it with a foot and play it as a stationary ball. The player can also raise the ball by standing over it with a foot on either side. The player presses the feet against the ball and jumps into the air, propelling the ball into the hands.

(4) Offensive and Defensive Play. The positions of players in speedball are much the same as in soccer. However, some of the positions have different names. Each team has 11 players. The forward line is composed of five players, the right end, right forward, center, left forward, and left end. The second line consists of right halfback, fullback, and left halfback. In the next line is the right guard and left guard. The player who defends the goal is the goal guard. The strategy employed in speedball during offensive play is very similar to that of soccer. There are two types of defensive formations in speedball: man-for-man and position defense. Man-for-man defense is recommended for beginning players.

(5) Abridged Rules

(a) Field. A speedball field is 100 yards long with additional 10-yard end zones on each end and 55 yards wide, the same as a regulation football field.

(b) Players. Eleven players are on a team. The goal guard has no special privileges.

(c) Time. Ten-minute quarters. Two minutes between quarters. Ten minutes between halves. Five minutes for extra overtime periods.(Begin the first overtime by a jump ball at center, same goals; change goals in the event of a second overtime period.)

(d) Winner of Toss. The winner of the toss has the choice of kicking, receiving, or defending a specific goal.

(e) Kickoff. A kick off starts the first and second half of a game and starts play after a score. The kick off is made from any place on the 50-yard line. The kicking team must be behind the ball when it is kicked. The receiving team must stay back of its restraining line (10 yards distance) until the ball is kicked. The ball must go toward the opposing team's goal before it can be played. Kick off out of bounds goes to the opponent at that spot. A kick off touched by the receiving team before going out of bounds, with no impetus added, still belongs to the receiving team. A kick off, in possession and control of the receiving team and then fumbled out of bounds, belongs to the kicking team at that spot.

(f) Starting Second and Fourth Quarters. The ball is given to the team that had possession at the end of the previous quarter, from out of bounds, as in basketball.

(g) Second Half. The first-half receiving team kicks off at the beginning of the second half.

(h) Fly and Ground Ball. The most characteristic playing rule of speedball is the difference between a fly ball (or aerial ball) and a ground ball. A player is not permitted to touch a ground ball with the hands and must play it as in soccer. A fly ball is one that has risen into the air directly from the foot of a player (for example, a punt, drop kick, place kick, or kickup).Such a ball may be caught with the hands provided the catch is made before the ball strikes the ground again.

(i) Kickup and Overhead Dribble. A kickup is a ball that is so kicked by a player that he can catch it himself. A bounce from the ground may not be touched with the hand because it has touched the ground since being kicked. This rule prohibits the ordinary basketball dribble, but one overhead dribble (throwing the ball into the air and advancing to catch it before it hits the ground) is permitted.

(j) Out of Bounds. If a team causes the ball to go out of bounds over the sidelines, a free throw-in (any style) is given to the opposing team. When the ball goes over the end line without scoring, it is given to the opponents who may pass or kick from out of bounds at that point.

(k) Tie Ball. In case two players are contesting the possession of a held ball, even in the end zone, a tie ball is declared. The ball is tossed up between them.

(6) Scoring

(a) Field Goal(3 points).A soccer-type kick, in which a ground ball is kicked under the crossbar and between the goal post from the field of play or end zone, is a field goal. (A punt going straight through is not a field goal for it is not a ground ball. The ball must hit the ground first.) A drop kick from the field of play that goes under a crossbar does not count as a field goal. A drop kick from the end zone that goes under the crossbar counts as a field goal;if it goes over the crossbar, it is ruled as a touch back, and the other team takes possession of the ball.

(b) Dropkick(2 points). A scoring drop kick must be made from the field of play and go over the crossbar and between the uprights. The ball must hit the ground before it is kicked (usually with the instep).

(c) End goal(1 point). This is a ground ball which receives its impetus (kicked or legally propelled by the body) from any player, offensive or defensive, in the end zone and which passes over the end line but not between the goal posts.

(d) Penalty Kick(1 point).A ball kicked from the penalty mark that goes between the goal posts and under the crossbar is a penalty kick. The penalty mark is placed directly in front of the goal at the center of the goal line.

(e) Touchdown(1 point). A touchdown is a forward pass from the field of play completed in the end zone. The player must be entirely in the end zone. If the player is on the goal line or has one foot in the field of play and the other in the end zone, the ball is declared out of bounds. If a forward pass is missed, the ball continues in play, but it must be returned to the field of play before another forward pass or drop kick may be made.

(7) Substitutions. Substitutions may be made any time when the ball is not in play. A player who is withdrawn may not return during that same period.

(8) Timeout. Three legal timeouts of 2 minutes each are permitted each team during the game.

(9) Fouls

(a) Personal Foul. Kicking, tripping, charging, pushing, holding, blocking, and unnecessary roughness of any kind, such as running into an opponent from behind, are personal fouls. Kicking at a fly ball and thereby kicking an opponent is a personal foul. Four personal fouls disqualify a player from the game.

(b) Technical Foul. Illegal substitution, more than three timeouts in a game, unsportsmanlike conduct, and unnecessarily delaying the game are technical fouls.

(c) Violation. Traveling with the ball, touching a ground ball with the hands or arms, double overhead dribble, violating tie ball, and kicking or kneeing a fly ball before catching it are violations.

(d) Penalties. The offended player shall attempt the kick.

(e) Officiation of Fouls

1 Fouls in the field of play allow no follow-up while fouls in the end zone always allow follow-up.

2 On penalty kicks, with no follow-up, only the kicker and goalie are involved.

3 On penalty kicks, with a follow-up, the kicking side is behind the ball and the defending side behind the end line or in the field of play. No one is allowed in the end zone or between the goal posts except the goal guard. The kicker must make an actual attempt at goal and cannot play the ball again until after another player plays it.

c. Volleyball. Volleyball is a popular sport. The game entails much physical activity, yet it is not strenuous. It is, therefore, a game for young and old alike, for beginners and for skilled players. It may be played indoors or outdoors on any type of terrain. As an informal activity, volleyball can be played by any number; as an organized activity, it provides, as few other sports do, a game for 12 to play in a limited area. While volleyball requires no great skill to play, it permits a high degree of proficiency. Players naturally get more enjoyment when they know the game and play it well. For this reason, instruction in the basic skills should be provided.

(1) Place in the Program Volleyball may be used occasionally as a competitive activity during the sustaining stage. Though it is a year- round sport, it should be included in the physical training program only when a more strenuous activity is impractical. Volleyball is an excellent off-duty activity.

(2) Instruction. Usually during the first three or four classes, a 10- to 15- minute period of instruction, followed by scrimmage, is enough to teach the basic skills, rules, and techniques of volleyball. More time, if available, can be given to teaching basic skills, but emphasize competitive play rather than formal instruction. Divide the unit into 6-person teams, organized on the basis of ability .Teams should be as nearly equal as possible.

(3) Basic Skills

(a) Passing the High Ball. The chest pass is the most effective method of playing the ball. To receive the ball, the knees are fixed, and the body is tilted forward. The elbows are raised sideward to a point in line with the shoulders. The wrists are extended in line with the forearm. The arms, wrists, and hands are rotated inward. To pass the ball, the hands are chest-high, thumbs pointing inward. The fingers are flexed, forming a cup, allowing them to contact the ball. On contact with the ball, the wrists are snapped while the fingers and elbows are pushed upward, sending the ball upward. A high ball is much easier to handle than a low one.

(b) Passing the Low Ball. A ball that is lower than the waist is one of the easiest to hit, but it is also a frequent cause of fouls (holding or carrying the ball). The best position for handling a low ball is to have the knees flexed and arms flexed at the elbows and rotated so the thumbs are pointing outward, the palms up. When the fingers contact the ball, the entire body reacts in a lifting motion. The arms and hands swing upward in a scooping action. It is important that the fingers, not the palms, contact the ball, and that the ball is batted, not thrown.

(c) Underhand Serve. Take a position behind the back line facing the net, left foot forward, holding the ball in the palm of the left hand. The left knee is flexed; the right knee is straight. Swing the right arm back and at the same time move the left hand (holding the ball) across the body in line with the right hip. Then swing the right arm forward hitting the ball off of the left hand with the palm of the right hand, raising the hips and arching the back in the same motion. Be certain to swing the right arm in a straight line, or the ball will be difficult to control. When the opposition is in formation, the server should try to place the ball in the right or left back area, and not near the net.

(d) Setup. A setup is a ball hit into the air near the net by one player, so that a teammate may hit or "spike" it sharply downward into the opponent's court. The chest pass is the best pass to use. The ball is sent approximately 10 feet into the air toward the spiker so it will descend from 4 to 20 inches from the net.

(e) Spiking. The spike is a leap into the air and a sharp downward hitting of the ball into the opponent's court. A spiker must be able to spring easily from the floor, judge the movement of the ball, and strike it with a downward movement of the arm. To jump from the floor, step off with one foot and jump with the other. Stand with the right or left side to the net, facing the setup player. Much depends upon the setup player to place the ball in the proper position. The spiker jumps into the air and strikes the ball above its center so as to drive it downward. A snapping movement of the arm and wrist will drive the ball forward and downward with power and control. Aim for a weak spot in the opponent's defense.

(f) Blocking. The block is a defense technique used to prevent a spiker from driving the ball across the net. It is an attempt by one or more defensive players at the net to block a hard-hit shot by using the force of the ball to send it immediately back into the opponent's court. In an effective block, forwards on the defensive team spring into the air at the time of the spike, placing both hands and arms in the expected path of the ball. An effective block tends to upset the offense and presents another element for the spiker to worry about. To be effective, the blocker must anticipate the path of the ball and time the block with the spike.

(4) Abridged Rules

(a) The volleyball court is 30 feet wide by 60 feet long. (See fig. 5-18.)

(b) The top of the net is 8 feet high.

(c) A volleyball team consists of 6 players.

(d) A match consists of the best two out of three games.

(e) The first team scoring 15 points wins the game, provided that they have 2 points more than their opponents.

(f) A deuce game is a game in which both teams score 14 points. The game is continued until one team obtains a 2-point advantage over the other.

(g) Only the serving team can score. If the serving team commits a fault, it loses the serve to the opposing team.

(h) The team members who receive the served ball rotate one position in a clockwise direction.

(i) The ball is put into play by serving from behind the back line.

(j) A served ball touching the net results in the loss of the serve. At any other time during play, a ball touching the net is still in play.

Figure 5-18 is available in the illustration section at the end of the book.

(k) The ball is out of play when it touches the ground or goes outside one of the boundary lines.

(l) All line balls are good.

(m) The players must hit or bat the ball; they may not throw, lift, or grasp it.

(n) A player may not touch the ball with any part of the body below the knees.

(o) A player may not play the ball twice in succession. In receiving a hard-driven spike, a defensive player may make several contacts with the ball even if they are not simultaneous. All such contacts, however, must constitute one continuous play, and all must be above the knees.

(p) The ball may be touched no more than three times on one side of the net before being returned across the net to the opposing team.

(q) A player must not touch or reach across the net.

(r) A player may touch the line under the net, but may not cross it.

Chapter 6
EVALUATION OF PERFORMANCE DURING TRAINING

6001.OBJECTIVES OF COMBAT READINESS

The objectives of the Marine Corps physical fitness program are-

a. To contribute to the health and well-being of every Marine through regular exercise and health education.

b. To develop Marines who are physically capable of performing their duties in garrison and in combat.

c. To develop Marines with a reserve level of physical fitness that will help them win in combat.

d. To provide a medium for developing the self-confidence of the individual Marine and thereby enhancing overall discipline, morale, esprit de corps, unit efficiency, and the desire to excel within the Marine Corps.

6002. THE PHYSICAL FITNESS TEST

The physical fitness test (PFT) has distracted attention from the proper goal of the physical fitness program. Simply stated, the goal of the physical fitness program is the success of Marines in combat. Unit commanders must not allow the PFT to become an end in itself. The PFT should be used within the commander's program of combat readiness training as a standard, an index of the physical fitness of individual Marines at a given time and place. The PFT is not difficult, and it must be kept in perspective. All athletes should be first class, and first class is attainable for all Marines. The conduct of the PFT is described in Marine Corps Order 6100.3_.

6003. OTHER METHODS OF EVALUATION

The commander should use other ways of evaluation to provide variety and imagination. Intangible objectives, such as confidence and aggressiveness, are hard to measure. The use of either inspection or observation is necessary to evaluate objectives of this nature. The commander has several other methods available when evaluating the physical condition of the command. The additional methods of inspection, observation, medical examination, and testing are summarized as follows:

a. Formal Inspections. Formal inspections, using inspection officers and standardized rating criteria, may help evaluate unit physical fitness.

b. Daily Observation. Routine observation of physical performance and appearance can serve as an indicator of an individual's or unit's physical readiness. However, mere observation is not a totally reliable or accurate means of evaluation.

c. **Physical Examination.** Medical examination may detect any individual disability or detrimental physical condition. It may guide in application of remedial, therapeutic, or limited exercise programs.

d. Other Physical Fitness Tests. Some other tests are described in detail in this manual for information purposes. Commanders are cautioned, however, not to place too great an emphasis on testing.

6004. RESPONSIBILITY

The commander is responsible for the physical fitness of the command and for the measurement and evaluation of its physical readiness. The goal of combat readiness training remains to prepare Marines and units for the physical demands of combat. The bulk of the commander's combat readiness training program should be such activities as obstacle and confidence courses, endurance courses, and progressive load-bearing marches which will prepare Marines for combat. Group activities which are competitive and combative in nature are advantageous and add imagination and enthusiasm to the program. Activities such as orienteering, which is not only physically demanding but also competitive and which teaches essential subject skills, are excellent and motivate all concerned. A commander's goal should be to provide an integrated training program that is so aggressive and demanding of Marines and leaders alike that a PFT administered at any time will reflect that each and every Marine is both physically and mentally prepared for the demands of combat.

6005. COMMANDER'S PHYSICAL FITNESS PROGRAM

The Commandant of the Marine Corps has directed that the physical fitness test shall be the universal measure of adequate individual physical fitness. The focus of a commander's physical fitness training program, however, should not be to prepare Marines to pass the physical fitness test, but to prepare them for the physical rigors of combat.

a. Administration. The physical fitness program requires each Marine to spend a minimum of 3 hours per week in physical fitness training, to be tested semiannually, and to obtain a minimum standard of third class. The attainment of a higher score is a laudable individual goal which should be encouraged but should not become a unit objective. Overemphasis could be detrimental to training required to develop the complete Marine. All commanders should exercise sound judgment as well as positive, aggressive leadership in striving toward the training objective.

b. Variety of Activities. The program must include the following elements:
(1) Physical Readiness Training. physical readiness training is a complete physical training program which develops and maintains the strength, endurance, and physical skills needed to sustain the individual during combat.
(2) Remedial Physical Conditioning. Remedial physical conditioning is a process by which physically substandard Marines are conditioned to meet prescribed standards.
(3) Competitive Conditioning Activities. Competitive conditioning activities consist of teams or individuals competing against an opponent to win. This involves a combination of sports and military skills designed to foster competitive attitudes and develop unit pride and esprit de corps.
(4) Occupational Conditioning. Marching long distances, dry net training, etc., should not be ignored as a means to fulfill the time requirement.

c.Time. In order to develop the desired level of physical fitness, unit commanders are advised that a minimum program of 3 hours per week is mandatory. The program must encompass at least three exercise periods weekly. Each period should include calisthenics, running, or other forms of vigorous activity .The minimum desirable length for each exercise period is 60 minutes. Shorter periods may be authorized by the unit commander if dictated by the local training situation. Unit commanders are encouraged to use the normal working day to satisfy this training requirement. However, commanders are authorized to conduct required physical fitness training during off-duty hours when the mission, workload, personnel status or other significant factors preclude fulfilling it during duty hours.

d.Action. Marine Corps Order 6100.3_ requires that commanders-

o Establish and maintain an effective physical fitness program which ensures that all Marines maintain an acceptable level of physical fitness.

o Ensure that all individuals are medically qualified prior to participating in the physical fitness program.

o Establish a minimum physical fitness program of at least 3 hours per week. Physical training may be authorized on an individual basis at the discretion of the unit commander.

o Ensure that all Marines participate in physical fitness conditioning activities commensurate with their medical qualifications and limitations.

o Conduct physical fitness testing for all Marines.

o Place all Marines who fail to pass the PFT on a daily command-supervised remedial physical conditioning program until they pass the test.

o Ensure that results of physical fitness testing are entered on section A of the fitness report. Comments are placed in section C in connection with attainment of superior physical performance or a medical excuse from the PFT.

6006. OTHER FORMAL TESTS

Commanders must realize that the best index of combat readiness is personal observation of performance on conditioning marches of 15 miles while under load. This manual presents three additional indicators of unit fitness. These indices are the physical readiness test, which was formerly the Marine Corps standard and is now routinely administered at Officer Candidate School; the airborne trainee physical fitness test which many Marines who attend U.S. Army formal schools are required to take; and the battle fitness test, the ultimate test of the combat readiness of Marines. Commanders may wish to include these tests or portions thereof within their combat readiness programs.

6007. PHYSICAL READINESS TEST

a. Purpose. The purpose of this test is to measure the Marine's ability to meet the minimum standard of physical readiness for duty in the field. This test is currently administered to all male officer candidates at Officer Candidate School, MCCDC, Quantico, VA. Female candidates do not take this test.

b. Test Area. Administration of this test requires an area suitable for crawling, jumping, and running 160 yards to include a trench that is 8 feet wide.

c. Equipment. The equipment required for this test includes—
o A platform, rail on a ledge, or a box 18 inches high,
o A 20-foot climbing rope,
o A stopwatch.

d. Conduct of Test. Adequate timers and other supervisory personnel must be provided by the company staff. All events should be conducted in a single session of one morning or afternoon. The events may be run in any sequence.

e. Uniform. The uniform for the physical readiness test is as follows: full utilities, cartridge belt, belt suspender straps, two canteens (full of water), rifle, and helmet. Depending on the heat condition, the PT shirt may be worn in place of the utility jacket. All equipment will remain secure throughout the test. Except for emergencies, canteens are not to be emptied.

f. Events

(1) Event 1: Climbing Uphill (Step-ups)This event simulates marching uphill at a rapid and steady pace. The Marine stands in front of the step. On the command GO, the Marine places either foot on the step and steps up. Hand pressure on the knee may be used if desired. The Marine then stands erect and steps backward and down, on one foot at a time. The same leg may be used for each repetition or the legs may be alternated. For 100 points, Marines will complete 60 repetitions in the maximum allotted time of 1 minute and 20 seconds (80 seconds).

*(2) Event 2: Rope Climb. T*his event resembles entering and leaving a hovering helicopter, using ropes in house-to-house fighting, and ascending and descending landing nets. The Marine assumes a sitting position on the ground at the bottom of the rope with hands at the highest point that can be reached. On the command GO, the Marine jumps up and climbs to the top. The Marine must touch the 20-foot mark with one hand, then descend without sliding or dropping. Marines must reach the 20-foot mark in the maximum allotted time of 30 seconds or less.

(3) Event 3: Evacuation. This event simulates reaching and evacuating a wounded Marine under fire. The Marine assumes the prone position at a distance of 50 yards from a casualty who is of approximately equal weight. On the command GO, the Marine springs up and covers the distance in a sprint. The Marine lifts the casualty to a fireman's carry and returns to the starting point. Assistance by the casualty is permitted. The tested individual must carry all equipment; the casualty has none. Marines must negotiate the course in a maximum time of 47 seconds or less.

(4) Event 4.'Advance by Fire and Maneuven. This event simulates advancing and assaulting during an attack. The Marine assumes the prone position at the starting point. On the command GO, the Marine then alternately crawls, runs in a zigzag manner, rolls, jumps, and assumes different firing positions for the next 130yards. Before reaching the objective, the Marine must "hit the deck," roll over, and assume the firing position three times. While assuming the firing position, the Marine must pull back the charging handle, take aim, and squeeze the trigger before rising from the ground. After rising from the ground for the third time,the Marine continues running and jumps to clear the 8-foot trench in one leap. The Marine then finishes the course in

the maximum allotted time of 2 minutes. The Marine is allowed multiple attempts only within the specified time limit.

(5) Event 5: Forced March. This event is a combat run of 3 miles without halts and against time. The Marine must double-time. Marines must complete the 3-mile course in a maximum allotted time of 30 minutes or less.

g. Event Standards

(1) To successfully pass the physical readiness test, the Marine must pass each event with a minimum of 80 points for each. Points are then assessed for each event in accordance with the scoring tables. (See fig. 6-1.)

(2) If a Marine fails one event, the highest possible score that the Marine may receive for the entire test is 79 percent. (See example 1 in fig. 6-2.) If two events are failed, the highest possible score for the entire test is 69 percent, regardless of the total points accumulated. (See example 2 in fig. 6-2.) If three events are failed, the highest possible score is 59 percent. Failure in four or more events will result in a score of zero for the entire test.

(3) To convert the total points for all five satisfactorily passed events to a final percentage score, total the accumulated points for all five events and divide by five. (See example 3 in fig. 6-2.)

Figure 6-1 is available in the illustration section at the end of the book.

Figure 6-2 is available in the illustration section at the end of the book.

6008. AIRBORNE TRAINEE PHYSICAL FITNESS TEST

a. Use and Composition of the Test.

(1) Use. The airborne trainee physical fitness test is a means of determining the physical ability of the applicant for acceptance to and retention in the airborne course of instruction.

(2) Test Events The test battery consists of five events as follows: chinups, knee bender, pushups, situps, and an endurance run.

b. Method of Scoring and Standards

(1) Scoring. The Marine will be scored by a trained scorer who is thoroughly familiar with the minimum standards for the test events. The Marine will be scored on a pass or fail basis. The performance on each event may be recorded on the reverse side of DA Form 705 (Physical Fitness Testing Record).The examinee records the personal information on the face of the card as specified on the first, second, and third lines. The face of the card is identified by writing diagonally across the lower half of the card "Airborne Trainee PPT—See reverse side."Using line 7 on the reverse side of the scorecard, the scorer enters the test title in the block entitled "(Other (Specify))" and completes the required information on the remainder of that line. The "Remarks" section can be used to record the test event titles and the applicant's performance on each test event.

(2) Standards. To successfully pass the test, the examinee must reach the standard in each test event. The standards follow:

(a) Chinups

(b) Knee bender 80(2-minute period)

(c) Pushups 45(2-minute period)

(d) Situps45(2-minute period)

(e) Endurance runs:2 miles in 15:54 minutes or less in athletic gear and 4 miles in 32 minutes in utilities and running shoes.

c. Uniform for Testing

(1) Examinees. The prescribed uniform for test participation is boots and the work uniform of the season. No headdress is worn. When climatic conditions permit, jackets or outer shirts may be removed.

(2) Officials. Scorers and other test officials should be uniformly and distinctively dressed for contrast with Marines being tested.

d. Test Administration Procedure

(1) Preparation for Test. The administration of the test to a large group makes it mandatory that the test effort be organized and efficiently operated. All testing is not completed with large test groups; on certain occasions, individuals and small groups are tested. Care must be exercised to administer the test uniformly and to standardize the conduct of all elements of the test. Regardless of the size of the test group, the following elements of sound test administration should be included:

(a) An orientation to include the purpose, method of administration, scoring of the test, preparation of the score card, and required standards.

(b) A correct demonstration of each event to ensure that there is no misunderstanding of the proper form and required standards.

(c) Completion of all five test events in one test period with all Marines taking the events in the same sequence.

(d) Adequate rest periods between the test events to allow for recovery before the next event.

(2) Method of Administration with a Large Group. With a minimum of 12 lanes per test event, 14 officials can administer the test battery to 150 or 200 Marines in 2 hours. The officials are designated as follows: one officer in charge, one demonstrator,and 12 scorers. If more or fewer Marines are to be tested, a greater or lesser number of officials will be required. The following procedure is recommended:

(a) Conduct an orientation and ensure the examinees have properly completed their score cards.

(b) Assign Marines to lanes and caution them to remain in the same lane order throughout the test.

(c) Explain and demonstrate the chinup event, administer it, and score it. Then proceed to the knee bender and pushup events, and administer them in the same manner.

(d) Grant a 5- to 10-minute rest period after the pushup event. Advise against excessive consumption of water during the break period.

(e) Explain and demonstrate the situp event, administer it, and score it as prescribed. Then move to the run area, explain the running event, and complete it.

(f) Retain the scorecards at the completion of the running event.

(3) Method of Administration with a Small Group. A similar procedure is followed for the testing of individuals and small groups. The informality usually associated with small groups must not conflict with sound test administration. With fewer examinees, fewer officials are required.

e. Description and Explanation of Test Events

(1) Test Event 1:Chinups

(a) Purpose. This event is devised to test arm and shoulder flexor strength.

(b)Equipment There is one horizontal bar per lane, made of plumber's pipe or a gymnasium horizontal bar 1 1/2 inches in outside diameter. The bar must be rigidly supported at a height of 8 feet above the ground. The upright supports must be 5 feet apart. There must be a movable stand at each bar for short Marines to stand on to reach the bar.

(c) Officials There is one scorer per lane.

(d) Organization. The examinees, holding their scorecards, stand in order behind the restraining line in their respective lanes. The scorers take each scorecard when the Marine is called forward for the test.

(e) Starting Position The bar is grasped with the palms turned toward the face, the thumbs underneath the bar. The body is fully extended in a "dead" hanging position with the arms straight and the feet above the ground.

(f) Movement. Pull the body directly upward until the chin is placed over the bar. Lower the body until the elbows are completely straight and the body is again in the "dead" hanging position. Repeat as many times as required.

(g) Instructions. Explain and demonstrate the fully extended "dead" hanging position with the proper grasp. Show that the chin is placed over the bar at the top of the movement and that the arms are fully extended, the elbows completely straight, at the bottom of the movement (the hanging position). Explain that the body must be kept from swinging and that it is permissible to raise the legs and flex the hips when pulling up, but that any kicking, bicycling, or jerking motion with the trunk or legs is not acceptable. Inform the examinees that no penalty is exacted for hanging on the bar to rest in the bottom position but that this is not to their advantage. Tell them that half-completed chinups are not counted, and that the scorer will repeat the number of the last correct chinup when incorrect execution is detected.

(h) Administration and Scoring. Caution the examinees to assume the "dead" hanging position and wait for the scorer's command to begin. The scorer is at the examinee's left with a clear view of the bar. If the examinee begins to swing widely, the scorer should stop the swinging by extending the left arm across the front of the examinee's body, being sure not to hinder the execution of the chinup. The scorer counts aloud the number of chinups correctly executed. When a chinup is not correctly executed, the scorer repeats the number of the last correct one. The scorer records the number of correct chinups on the scorecard and returns the card to the examinee.

(2) Test Event 2:Knee Bender

(a) Purpose. This event measures the strength and endurance of the leg muscles.

(b) Equipment. None.

(c) Officials There is one scorer per lane.

(d) Organization. Marines stand in numerical order behind the restraining line in their respective lanes. The scorer takes each scorecard when the Marine is called forward for the test.

(e) Starting Position. The feet are spread less than shoulder width apart, hands on hips, thumbs in the small of the back, elbows back.

(f) Movement. Do a knee bend and at the same time bend slightly forward at the waist and thrust the arms between the legs until the extended fingers touch the ground. The hands are about 6 inches apart. The bend is approximately a three-quarters bend. From this knee bend position, recover to the starting position by moving the body upward, straightening the knees, and returning the hands to the hips. Repeat as many times as required.

(g) Instructions. Explain and demonstrate the correct starting position. Be certain examinees understand the correct knee bend and that only the tips of the fingers touch the ground. Tell them the scorer will repeat the number of the last correct knee bender when incorrect execution is detected. Some of the common errors are failure to correctly bend the knees, failure to touch the ground, and failure to assume the completely erect position after the bend has been executed.

(h) Administration and Scoring. The scorer stands to one side so as to see that the knees are properly bent and the fingers touch the ground as prescribed. From this position, the scorer can view the examinee to see that a properly erect position is assumed after each knee bend. The scorer counts aloud the number of correctly executed knee bends. When a knee bend is done incorrectly, the scorer repeats the number of the last correct one. The scorer records the number of correct knee bends on the scorecard and returns it to the examinee.

(3) Test Event 3:Pushups

(a) Purpose. Pushups measure arm and shoulder extensor strength.

(b) Equipment. None.

(c) Official. There is one scorer per lane.

(d) Organization. Marines stand behind the restraining line in their respective lanes until the scorer calls on them to perform. The scorer takes each scorecard when the Marine comes forward.

(e) Starting Position. The front-leaning rest position is the starting position. The body is straight from head to heels, palms are flat on the ground directly underneath the shoulders, and elbows are straight and locked. The body weight is supported on the hands and toes throughout the event.

(f) Movement. Bending only the elbows, lower the body in one straight plane until the chest touches the scorer's hand. Straightening and locking the elbows, raise the body in one straight plane, returning to the original front-leaning rest position. Repeat as many times as required, keeping the body in a straight line from head to heels.

(g) Instructions. Explain and demonstrate that the arms are straight at the beginning and completion of the movement and that the chest must touch the scorer's hand, but the stomach and thighs must not touch the ground. Also explain that the whole body must be maintained in a straight line as it is lowered and raised; that is, there is to be no breaking at the hips or shoulders so that any body part is lowered or raised in advance of the other or as a separate segment .Dipping or rolling through the shoulders is illegal, as is lowering or raising the body with one arm or shoulder at any time. Resting is not permitted during repetitions. Instruct

Marines that the scorer will repeat the number of the last correct pushup when incorrect execution is detected.

(h) Administration and Scoring. It is recommended that Marines assume a prone position while placing their feet and hands in the proper positions. This permits them to rest while the scorer gets into position and, at the same time, provides a feel of the body in a straight plane from head to heels. The scorer lies on the right hip and side to the right of the examinee. This gives the scorer a clear view of the examinee's body to see any errors. The palm of the scorer's right hand rests flat on the ground underneath the lowest part of the examinee's chest. By keeping the right forearm flat on the ground at an angle in front of the examinee's right arm, the scorer's position will not prevent the examinee from lowering the body completely. The scorer's left hand is free to test the straightening of the elbow at the completion of the movements and to point out body segments being lowered or raised separately. When in position and ready, the scorer has the examinee assume the starting position and begin doing pushups. The scorer counts aloud the repetitions done correctly and repeats the number of the last correct pushup if one is done incorrectly. There is no penalty if the contour of the examinee's body causes the hips to protrude slightly out of line, provided that the whole body is raised and lowered simultaneously. The scorer enters the number of repetitions on the scorecard and returns it to the examinee.

(4) Test Event 4: Situps

(a) Purpose. This event primarily measures abdominal strength.

(b) Equipment. None.

(c) Official. There is one scorer per lane.

(d) Organization Marines stand behind the restraining line in their respective lanes until the scorer calls on them to perform. The scorer takes each scorecard when the Marine comes forward.

(e) Starting Position. The examinee lies flat on the back with knees flexed, both feet flat on the ground. The correct angle of the thighs to the ground is 45 degrees. If the heels are too near the buttocks, the applicant will not be able to sit up. Legs are spread shoulder width apart. The examinee interlaces fingers and places them behind the head in contact with the ground. The feet are not held by another person.

(f) Movement. Bend forward at the waist and raise the upper body until the head is directly over the knees. Heels are not to leave the ground. Elbows remain in the same plane to the head and body throughout the event. The upper body is slowly lowered to the starting position until the head touches the ground. Repetitions are done at a slow cadence with no rest periods.

(g) Instructions. Explain and demonstrate the correct starting position and the proper execution of the situps to be sure that Marines understand the movement. Warn them that their knees must remain flexed during each situp, the heels cannot leave the ground at any time, and they may not roll up on one side and push up with one elbow. Tell them they must do the repetitions at a slow cadence with no rest periods. Instruct Marines that the scorer will repeat the number of the last correct situp when incorrect execution is detected.

(h) Administration and Scoring. When the performer is in position and ready, the scorer has the performer assume the starting position and begin doing situps. The scorer counts aloud the correct executions. When a situp is improperly done, the scorer repeats the number of the last correct one. No situp is credited if the hands are unclasped from behind the head, if the back is used to bounce up from the ground (which means the shoulders would not touch the ground), or if one shoulder or elbow is used to push up. The examinee is not penalized if the heels slide forward slightly so long as the knees remain flexed and the heels maintain contact with the ground. The scorer enters the number of repetitions on the scorecard and returns it to the examinee.

(5) Test Event 5:Endurance Run

(a) Purpose. This event measures cardiovascular endurance.

(b) Equipment One stopwatch or watch with a sweep second hand.

(c) Area. A large training field on which a quarter-mile track has been staked out or a level road over flat terrain may be used as a running surface. A 1-mile route is designated with wooden stakes marking the start point, finish point, and one-quarter mile intervals.

(d) Officials For large groups, a scorer times the event and controls the conduct of the run, and a guide runs with the group and sets the pace.

(e) Organization The run is conducted with groups of Marines in a column formation. Company-sized units may run at the same time with platoons serving as running groups. The scorer issues the command to assume double-time.

(f) Starting Position. Marines are assembled in proper column formation (column of twos, threes, or fours, as appropriate to the size of the group) with short men to the front. When all is ready, the column is moved forward a short distance before the running period is started.

(g) Movement. At the command DOUBLE-TIME, MARCH, the examinees retain their places in the column formation and execute the command. Length of steps is about 40 inches. The scorer has the group execute the run. The formation is maintained during therun.

(h) Instructions. Marines are instructed to maintain formation while running and are informed that the guide will set the proper pace. They will be instructed in the command used to control the column in the execution of the test. The scorer should announce the 4-minute, 2-minute, 1-minute, and 30-second remaining time intervals.

(i) Administration and Scoring. The event may be administered as previously prescribed to a large group, to several Marines, or to an individual. An individual examinee usually does not require a guide or pacer. If the event is administered on a training field, the scorer may stand in the center as the runner(s) circles about the field. This method of administration relieves the scorer of running with each group to be tested. Scoring is based on successful completion of the run as prescribed. The scorer should announce the remaining times as prescribed in (h).

6009. BATTLE FITNESS TEST

The battle fitness test is a physical fitness evaluation which is currently under development as part of the basic warrior training concept plan.

Chapter 7
THE HUMAN BODY

7001.GENERAL

To implement an effective physical conditioning program, a leader must understand how the human body functions and how exercise affects the body .This chapter provides only an introductory discussion of anatomy and functioning.

a. Body Functioning During the Stages of Conditioning. For more on the stages of conditioning, see chapter 1.

(1) During the toughening stage, the waste products of muscle activity (lactic acids) collect more rapidly than the blood can remove them. This acid waste builds up in the muscle tissue and irritates the nerves in the muscle fiber, causing pain and stiffness. As the exercise program continues, more blood is carried through the muscle, removing the waste materials more rapidly and eventually causing the soreness to disappear.

(2) During the slow improvement stage, the blood circulation in the muscles increases, and the body as a whole becomes a more efficient machine. The improvement is rapid in the first few weeks, but as a higher level of skill and conditioning is reached, the improvement becomes less noticeable. The body reaches its maximum level of performance after 6 to 10 weeks and should then be maintained at this peak.

(3) During the sustaining stage, it is possible to maintain this state of conditioning through 15 to 20 minutes of exercise a day, but the exercise must be quite strenuous.

b. Diet. Regular exercise increases the appetite. If the desire for greater amounts of food is satisfied by a balanced diet, the body benefits.

(1) There are two main types of foods: body-building and energy- producing.

(a) Body-building foods consist of proteins, which build up tissue and repair wear and tear.

(b) Energy-producing foods include carbohydrates and fats. Carbohydrates provide a quick source of energy, while fats act as a reserve store of energy.

(2) In addition, food contains vitamins, mineral salts, and water. During hot weather and strenuous training periods, the body requires greater fluid intake.

(3) Diet should be supplemented with proper rest to allow the digestive system to digest the lood.

(4) Occasionally, especially during early stages of conditioning, strenuous exercise may cause vomiting. Although vomiting is not a frequent occurrence, it is not usually a cause for concern.

7002. SYSTEMS OF THE BODY

The systems of the body include the skeletal, muscular, circulatory, respiratory, endocrine, digestive, genitourinary, and nervous systems. Each has a different function, but all must work in cooperation with one another to insure a sound body. Of these systems, the first four are the most affected by exercise and are discussed in succeeding paragraphs.

7003. THE SKELETON

a. Bones. The skeleton (see fig. 7-1) is composed of about 206 bones, which-
o Provide structure for the body and support for the attachment of muscles,
o Protect vital organs, such as the brain, lungs, and heart,
o Manufacture red blood cells, which carry oxygen through the body.

Figure 7-1 is available in the illustration section at the end of the book.

b. Joints. A joint is a place of union between two or more bones. Joints can be-
o Immovable, such as in the face and head.
o Slightly movable, such as in the vertebrae or sternum.
o Freely movable, such as in the shoulder, hip, knee, ankle, and elbow. The bones in a freely movable joint are held in place by ligaments. Freely movable joints are of the greatest importance in physical training because exercise improves their mobility and stability.

c. Cartilage and Ligaments. The joints in the body, where bones connect, are supported by cartilage and ligaments.
o Cartilage is a tough, elastic, translucent tissue that acts as a shock absorber or buffer between bones. Examples are the discs between the vertebrae, the tissue attaching the ribs to the sternum, and buffers in the knee joints. Damaged cartilage does not heal.
o Ligaments are connective tissue that bind bones together. They contract and extend but are not elastic. Consequently, ligaments can be torn or strained. Damaged ligaments will heal to some extent.

7004. THE MUSCLES

a. General. Muscles are tissue, or an organ composed of tissue, which contract or extend to produce movement. Muscles are of three classifications: involuntary, voluntary, and cardiac. The voluntary and cardiac muscles are the most affected by physical training; physical training has little effect on involuntary muscles.(Seefigs. 7-2, 7-3, 7-4.)
o People have no control over the involuntary muscles, an example of which is the diaphragm,
o Voluntary muscles are the larger skeletal muscles which cause movement and which the individual can control.
o Cardiac muscle is found only in the heart and for all practical purposes is an involuntary muscle. Physical training has a significant effect on the fitness of cardiac muscles.

b. Muscle Structure. Muscle is composed of cells which are grouped into bundles called fibers. Groups of fibers form muscles. These bundles of fibers are held in place by sarcolemma, a thin, sheath-like material that surrounds the muscle bundles and secretes a fluid that lubricates the muscle tissue. The fused ends of the sarcolemma form the tendons which attach the muscles to the bones.

c. Attachment of Muscles. The arrangement of muscles on the skeleton provides the proper angle of pull to make movement possible. Voluntary muscles are usually attached to the skeleton in two places.

d.Action of Muscles. To produce motion and to do work a muscle usually shortens its fibers. The movement may be flexion such as bending the arm at the elbow, or extension, such as the straightening the arm. Some muscles raise the arms or legs, others depress the raised limb. Some muscles have the primary function of rotating the trunk from side to side. In addition, muscles act as stabilizers as well as prime movers.

Figure 7-2 is available in the illustration section at the end of the book.

Figure 7-3 is available in the illustration section at the end of the book.

Figure 7-4 is available in the illustration section at the end of the book.

7005. FUNCTIONING OF THE SKELETON AND MUSCLES

a. General. The skeleton and muscles function in close coordination. Muscles move the body, and bones support the muscles.

b. Effect of Exercise on Bones. Continuous exercise, particularly among younger people, usually strengthens the bones, allowing them to withstand greater stress and strain. Bones which are not used lose much of their strength, a factor which should be considered when individuals return to physical training after a prolonged period of inactivity. People in this category should refrain from activities which might result in bone breakage before the bone is exercised back to normal condition. The condition known as "march fracture" is usually a result of this condition.

c. Muscular Strength. When exercised vigorously to improve strength, muscle grows in size. In general, the larger the muscle, the stronger the muscle. Furthermore, trained muscle functions more efficiently than untrained muscle. Trained muscles contract more vigorously and with less effort. Regular and strenuous exercise also toughens muscle, making it firmer and able to stand more strain.

d. **Muscular Endurance.** Muscles naturally become fatigued during continuous, repeated work. Through specific exercises, such as chinups or situps, local muscle groups become exhausted long before an individual fatigues. Training to develop muscular endurance enables people to continue a relatively heavy load of exercise over a long period of time. Lengthened exercise periods improve muscular endurance.

e. Muscular Coordination. Improved speed and strength result in part from improved muscular coordination. An unskilled individual may use irrelevant muscles to perform a particular activity, increasing the amount of work without increasing the mechanical output. This increase in skill is highly desirable, but it should be offset by greater effort in duration or intensity to compensate for the loss in overload due to increased skill.

f. Muscular Fatigue. When the rate of work is increased, the energy required is proportionately much greater than the increase in rate. For example, if an individual doubles running speed, the amount of power demanded to do this is increased eight times.

g. Circulation in Muscles Regular, strenuous exercise causes the creation of new capillaries and the opening of inactive, latent capillaries, increasing blood circulation within the muscle by as much as 400 percent. This increases the supply of food materials and oxygen to the muscle, improving its endurance. This process takes about 8 to 12 weeks of regular conditioning in young adults and longer as age increases.

7006. THE CIRCULATORY AND RESPIRATORY SYSTEMS

a. The Circulatory System. The circulatory system transports blood to all parts of the body, removes waste products for disposal, and delivers protecting and repairing substances where needed. The circulatory system includes the heart and blood vessels.

(1) The Heart. The heart is a pump which forces blood through the blood vessels to the parts of the body. The heart is a little larger than a fist and is located in the left center of the chest between the lungs.

(2) Blood Vessels. The vessels carrying blood away from the heart are the arteries. They eventually divide into capillaries, the very small vessels through which diffusion and osmosis take place. The capillaries gradually increase in size, forming the veins which carry blood back to the heart.

b. The Respiratory System. The respiratory system performs the function of breathing, whereby oxygen is inhaled and carbon dioxide is exhaled. The respiratory system consists of the mouth, nose, trachea, lungs, and diaphragm.

o Trachea. The trachea, or windpipe, is a hollow, tube-like structure that carries air from the mouth to the lungs.

o Lungs. The lungs are elastic bags in the chest where the exchange of oxygen and carbon dioxide takes place.

o Diaphragm. The diaphragm is a thin, sheet-like muscle just below the lungs. During inhalation, the diaphragm flattens out and lowers, allowing the lungs to expand and fill with air. During exhalation, the diaphragm raises into a dome shape, helping to expel air from the lungs.

7007. CARDIOVASCULAR FUNCTIONING

a. General. Cardiovascular functioning is the combined functioning of the circulatory and respiratory systems. The chief organs of these systems, the heart and lungs, function together to provide oxygen to the body.(See fig. 7-5.)

Figure 7-5 is available in the illustration section at the end of the book.

b. Heart Action. The heart, the organ which propels blood through the blood vessels, is the chief organ of cardiovascular endurance. A person tires quickly if the heart lacks the capacity to circulate the blood through the body. The heart is a muscular organ and is developed by exercise just like any other muscle. The heart cannot be exercised alone; any exercise that puts a sufficient load on the heart, such as running or cycling, will also exercise other body parts. A conditioned heart tends to beat more slowly but pump more blood when at rest. This is known as an increase in "stroke volume,"a desirable condition because it enables the heart to pump more blood with a slower contraction rate.

c. Functions of the Lungs. Strenuous and regular exercise can improve the efficiency with which the lungs transmit oxygen to the blood by as much as 25 percent In a poorly conditioned person, some of the alveoli (air sacs) within the lungs are closed or collapsed. The forced breathing created by exercise over a period of weeks causes the air sacs to slowly expand, increasing the ability to absorb oxygen.

d. Relationship of Heart and Lungs. The heart pumps carbon dioxide-laden blood through the pulmonary artery to the lungs. In the lungs the carbon dioxide is exchanged for oxygen and the purified blood is returned to the heart by way of the pulmonary vein. The heart then pumps the blood through the aorta for circulation throughout the body. Blood moving into the muscles exchanges oxygen for carbon dioxide, after which time it is ready for the return trip to the heart.

e. Cardiovascular Functioning in High Altitudes. Marines to be employed in high altitudes should be acclimated in a similar area for 10 to 14 days prior to employment. Persons not accustomed to the rarefied air of higher altitudes tire more quickly and may collapse after rapid physical exertion. Air is much less dense at high altitudes than at sea level. Persons accustomed to sea level inhale only about 80 percent of the oxygen they are used to no matter how hard they breathe. Further, people accustomed to sea level or moderate altitude simply do not have enough red corpuscles in the blood to fulfill their needs at high altitudes. The red blood cells will increase over time.

f. Symptoms of Cardiovascular Malfunctioning. The brain is the first organ to react to a lack of oxygen. Unconsciousness results when the brain is denied sufficient oxygen. "Blacking out" is actually a defense mechanism to keep the body alive.

APPENDIX A
INSTRUCTOR HINTS AND INSTRUCTOR TRAINING

1. General. This appendix informs instructors on administration of exercise periods, commands, the extended rectangular and circle formations, methods of instruction and conduct of exercise activities, explanation of basic positions associated with the various activities, and instructor training.

2. Aim of the Physical Fitness Program. The aim of the physical fitness program is to prepare Marines physically for war. This can be conveniently split into two phases.

a. Preparation Phase

(1) Development of all-around physical fitness.

(2) Development of purposeful physical skills.

(3) Development of mental alertness.

(4) Development of character and leadership.

b. Maintenance Phase. Once the preparation is completed, the program must progress to a maintenance program. This phase revolves around the need to maintain the trained Marine's physical fitness for a particular role. This will include—

(1) Maintenance of a high standard of all-around physical fitness.

(2) Application of purposeful physical skills to combat skills.

(3) Development of leadership and self-confidence.

(4) Development of mental toughness and endurance.

3. Physical and Mental Development

a. Marines need to be developed physically and mentally. The physical training (PT) instructor must ensure that the program accomplishes this.

b. The system of the body functions best in response to progressive exercise. This means that Marines should master the easier exercises before they attempt the more difficult ones.

c. The body and mind are linked inseparably and must be considered as one in physical fitness training. Physical exercises assist in developing the mind.

d. The confidence in oneself and in one's comrades attained by physical achievement helps to promote the indomitable will to win. The Marine must be taught to realize this. It will help during arduous training, which at all times must be imaginative, realistic, and filled with enthusiasm. All physical fitness training must be associated with developing the qualities required of a Marine.

4. Planning and Preparation of Programs The following factors must be considered when compiling a program (see also ch. 2)~

a. Aim or desired skill level.

b. Duration of the complete training program and number of physical training lessons.

c. Purpose of included exercises,drills, or practical applications.

d. Beginning/inventory fitness levels.

e. Medical facilities, medical personnel, and Marine Corps or local command regulations governing their proximity to the training.
f. Number of Marines to be trained.
g. Availability of instructors, assistants, and facilities.
h. Apparatus and equipment available.
i. Alternative training areas for varying weather conditions,
j. Stages of progression and tests of achievement.

5. Five Basic Principles of Program Design

a. Regularity. Regularity of exercise is far more important than the amount of exercise performed on a given day. A Marine should exercise a minimum of three times per week. Human muscle begins to atrophy after 3 days without exercise.

b. Balance. The program must be balanced in terms of total physical fitness. Imbalance is created when, for example, an individual merely runs. While the cardiovascular system is improved, general upper body strength may be lacking. Conversely, a person who just does weight training may develop strength, but not cardiovascular endurance.

c. Overload. If total body fitness is to be developed, then the muscular and cardiovascular systems must be made to perform more work than previously. This is done by increasing the frequency, intensity, and duration of the exercise. The body adapts to the amount of stress placed upon it and becomes stronger and more efficient.

d. Progression. To be successful, any type of training program must be progressive. This is an area where untrained persons conducting PT can cause injury to Marines resulting in failure of the PT program. Close supervision must be made of untrained instructors. The result of progressing too fast is that the class begins to reject PT because it is no longer enjoyable or motivating. The program should start at a slow pace, and the pace should increase as proficiency improves. Average persons will encounter three stages of progressiom the toughening stage, the slow improvement stage, and the sustaining stage. (See ch. 1.)

e. Variety. Some programs fail because the training becomes boring. Variety is essential and perhaps presents the greatest challenge to the commander and staff.

6. Grouping. Providing for different levels of physical fitness is particularly recommended in the early stages of conditioning. Older Marines and those in poorer physical condition should be expected to attain a group level of fitness, but they should be given more time to do it.

a. Homogeneous Grouping One simple method of providing for the difference in levels is to group them according to their condition. A two-group classification would divide Marines into highly conditioned and average groups. A finer classification could be obtained by dividing them into three groups—a highly conditioned, a moderately conditioned, and an unconditioned group.

b. Performance or Age Grouping. The segregation of Marines into different exercise groups should be based on physical fitness test scores or on the level of fitness they demonstrate. They may also be grouped at first according to age. A common classification by ages is under 30, 30 to 34, and 35 and above.

7. Preparatory Commands and Commands of Execution. The prepreparatory command describes and specifies what is required. The command of execution calls into action what has been prescribed. All preparatory commands are given with a rising inflection. The interval between commands is long enough to permit the average Marine to understand the first one before the second one is given.

8. Extended Rectangular Formation. The traditional formation for carrying on many physical training activities is the extended rectangular formation (fig. A-1). This formation is more compact than any other. It is the best type to use for large numbers of Marines because it is simple and easy to assume.

a. For the formation of oneplatoon, the base man paces off five paces from the stand, faces left and moves five paces, halts, and again faces left. With the base man positioned facing the stand, the platoon leader then commands. **FALL OUT AND FALL IN ON THE BASE MAN**. At this command, all Marines run to the designated area and reform. This procedure is preferred to marching the unit into position. If more control is desired, the unit may march at double time to the vicinity of the base man and then be directed to fall out and fall in on him. Much valuable time is wasted in the field by needless maneuvering of Marines at quick time in an effort to position the platoon or unit on the exact spot for the exercises.

b. A company-size unit assumes the extended rectangular formation from a column of threes or fours at normal intervals between squads. This extension can also be executed from a company mass without interval between platoons. In extending either a platoon or company-size unit, take your place at the head of the column and command:

(1) TAKE INTERVAL TO THE LEFT, MARCH. At this command, the Marines in the right flank file stand fast with arms extended sideward. All others turn to the left and run forward at double time. After taking a sufficient number of steps, all Marines face the front with both arms extended sideward. The distance between finger-tips is about 12 inches, and dress is right.

Figure A-1 is available in the illustration section at the end of the book.

(2) ARMS DOWNWARD, MOVE. At this command, the arms are lowered smartly to the sides.

(3) LEFT, FACE

(4) EXTEND TO THE LEFT, MARCH. At this command, those in the right flank file stand fast with arms extended sideward. All others turn to the left and run forward at double time. Spacing is the same as in (1) above, and dress is right.

(5) ARMS DOWNWARD, MOVE. Same as in (2) above.

(6) RIGHT, FACE.

(7) FROM FRONT TO REAR, COUNT OFF. At this command, the leading Marine in each column turns the head to the right rear, calls of **ONE** and faces the front. Successive Marines in each column call off in turn **TWO, THREE, FOUR, FIVE** in the same manner.

(8) EVEN NUMBERS TO THE LEFT, UNCOVER. At this command, each even-numbered Marine stridejumps sideward to the left, squarely in the center of the interval. In doing this, each one swings the left leg sideward, jumps from the right foot to the left foot, and smartly brings the right into position against the left.

c. To assemble the unit, you commandASSEMBLE TO THE RIGHT, MARCHAt this command, all return to their original position in the column at double time and reform on the base man.

d.It is recommended that the area for grounding equipment and arms be at the edge of, or well away from, the area to be used for exercising. To conserve time and ensure proper position of the unit, the base man or, if the unit is composed of several platoon-sizegroups, the various base men may precede the unit and establish their positions in relation to the instructor's stand.

9. Circle Formation.The circle formation is effective for conducting various exercise activities (fig. A-2). This formation has an advantage in that supervision of all Marines is facilitated and a moving formation is available which provides control. Guerrilla exercises, grass drills, and some forms of running are examples of activities which are more easily conducted in the circle formation than in the extended rectangular formation.

a. It is not advisable to have more than 60 Marines in a circle. When more must be accommodated, separate circles should be used.Concentric circles may be employed to reduce the size of the circle or to accommodate more persons. If concentric circles are formed, the different circles are made by designating squads for each circle.Each additional circle requires more Marines than the one inside it.For example, one squad of a platoon may form the inner circle and theremaimng three squads the outer circle.

b. When a platoon is to form a circle, the commands ai€IRCLEFORMATION, MARCH, FOLLOW MEUpon this command, the right flank squad of the column moves forward at double time with the leader of the platoon group gradually forming a circle in a counterclockwise direction. Each succeeding file falls in behind that on the right. After the rough outline of the circle is formed, the leader commandsPICK UP A 5-YARD INTERVAL.

Figure A-2 is available in the illustration section at the end of the book.

c.The group may be halted and faced toward the center, or, if instruction is not necessary, the exercise activity may be executed without stopping the platoon.

lO.Leadership Techniques

a. Unless you experience all the exercises, you cannot appreciate how arduous they are, what movements are most strenuous and difficult, where the errors in performance are likely to occur, and what the proper cadence should be.

b. You must give everyone careful supervision and participate in the exercises to show that you can do them. When you participate, your assistant instructors should supervise because it is difficult for you to supervise and exercise at the same time.

c. Marines should never be kept too long in one position, especially a constrained one. They should never have to perform an exercise more times than they can do it without losing the proper form. You should insist upon proper form in the execution of all exercise activities.Even slight deviations from the proper form reduce the value of the exercise.

d. Avoid long explanations.As a rule, it should be necessary to give a full explanation of new exercises only. Explain the most essential features of an exercise first; add details later. Too many details at one time are more likely to confuse than to assist them.Minor corrections should be made to the entire class while the exercise is in progress (for exampHEADS **UP,** KNEES STRAIGHT). If necessary, follow this direction by the name of the person who is particularly at fault. If a Marine requires special attention, givethat person separate instruction by an assistant instructor to avoidwasting the time of the group.

e. The heavy demand on your voice can be lightened by training assistant instructors to assume some of the instruction. Using mass cadence is also an effective method of lessening the demand on your voice.

“Illustrations”

DAY	TIME	ACTIVITY
MONDAY	60 MIN	RUNNING ACTIVITY
TUESDAY	60 MIN	COMBATIVES
WEDNESDAY	60 MIN	CONDITIONING & LOG DRILLS
THURSDAY	60 MIN	CONDITIONING & RELAYS
FRIDAY	60 MIN	RUNNING & CONDITIONING
SATURDAY	60 MIN	RUNNING ACTIVITY

Figure 2-1. Sample A- -Single Activity.

DAY	TIME	ACTIVITY	REMARKS
MONDAY	20 MIN 20 MIN 20 MIN	RIFLE DRILL COMBATIVES RUNNING	PROGRESSIVE ACTIVITY SYSTEM
TUESDAY	30 MIN	GUERRILLA EXERCISES TABLE I	SINGLE ACTIVITY
WEDNESDAY	30 MIN	RIFLE DRILL	SINGLE ACTIVITY
THURSDAY	30 MIN	CIRCUIT INTERVAL TRAINING	SINGLE ACTIVITY
FRIDAY	20 MIN 20 MIN 20 MIN	STRENGTH CIRCUIT CONDITIONING DRILL CROSS-COUNTRY RUN	ROTATION ACTIVITY
SATURDAY	60 MIN	WARM UP RUN AND GRASS DRILL	SINGLE ACTIVITY

Figure 2-2. Sample B- -Various Time Periods, Activities, and Systems.

DAY	TIME	ACTIVITY
MONDAY	45 MIN	RUNNING – CROSS-COUNTRY
TUESDAY	75 MIN	TEAM ATHLETICS – TOUCH FOOTBALL
WEDNESDAY	45 MIN	RUNNING – CROSS-COUNTRY
THURSDAY	20 MIN 40 MIN	LOG DRILL DUAL COMBATIVES
FRIDAY	90 MIN	SPEED MARCH
SATURDAY	90 MIN	TEAM CONTESTS

Figure 2-3. Sample C--Sustaining Activities.

Route	Kilometers from SP	Recommended rate of march (kmph)	Remarks
SP: RJ 8th Div Rd—Superhighway	......	4	Bridge; hard surface; two guides.
RJ Jamestown—Hersey Rd	2.25	4	Traffic heavy; two guides; Jamestown Rd bears to right.
RJ Jamestown—Yankee Rd	7.10	4	Two guides
RJ Jamestown—Lightning Rd	10.50	4	Two guides
RJ Jamestown—Sunshine Rd	12.40	4	Light traffic; one guide Sunshine Rd; gravel, poor traction when wet.
RP; RJ Sunshine—Sedan Rd	14.80	4	Two guides

Figure 3-1.Example of Route Reconnaissance Report.

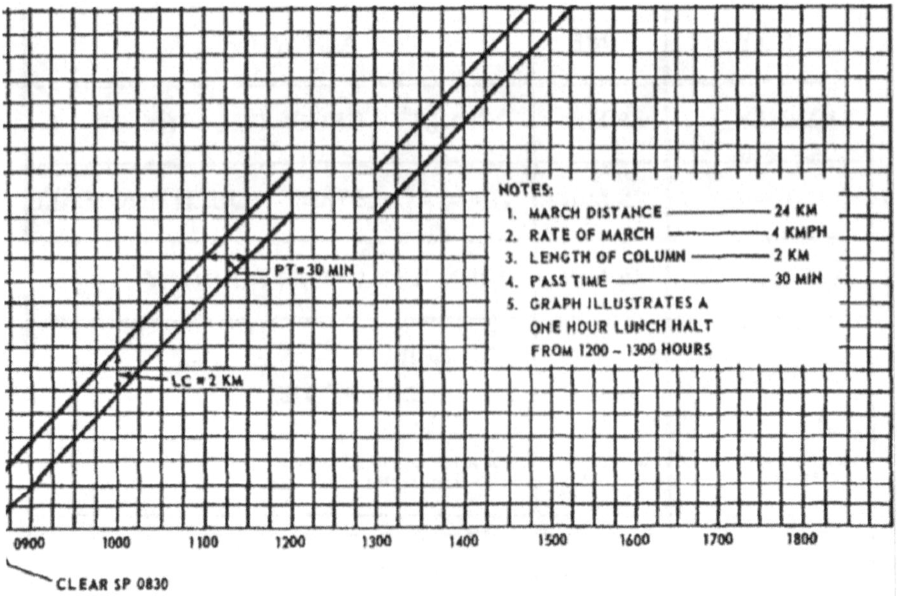

Figure 3-2.Road Movement Graph.

Figure 3-3.The Pace-Setter.

	SINGLE FILE	COLUMN OF TWO
2 m/MAN APART	24	12
5 m/MAN APART	54	27

GAP DISTANCES			
SERIAL:	(3-1) × 100 m	=	200 m
MARCH UNIT:	(12-1-2-) × 50 m	=	450 m
	TOTAL COLUMN GAP		650 m

Figure 3-4.Conditioning Drill One.

Figure 3-5.Conditioning Drill Two.

Figure 3-6.Conditioning Drill Three.

Figure 3-7. Rifle Drill.

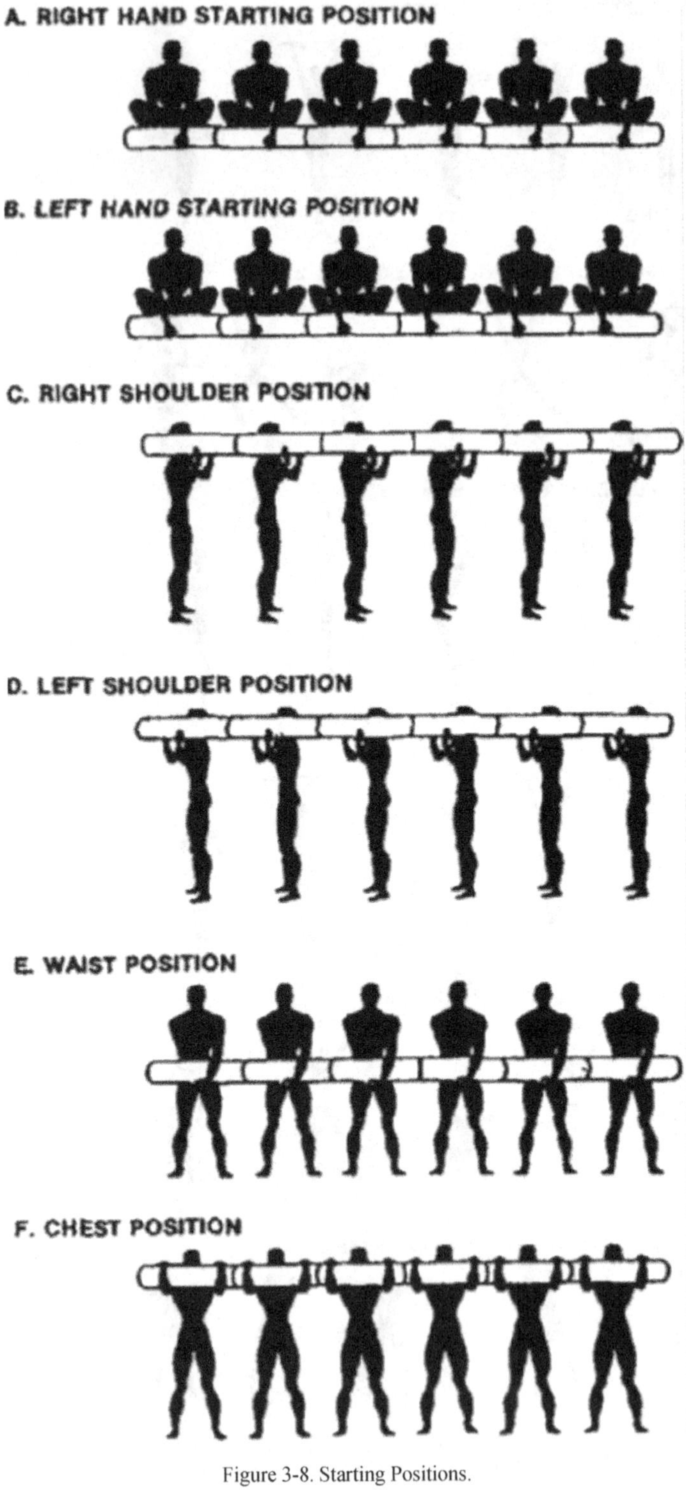

Figure 3-8. Starting Positions.

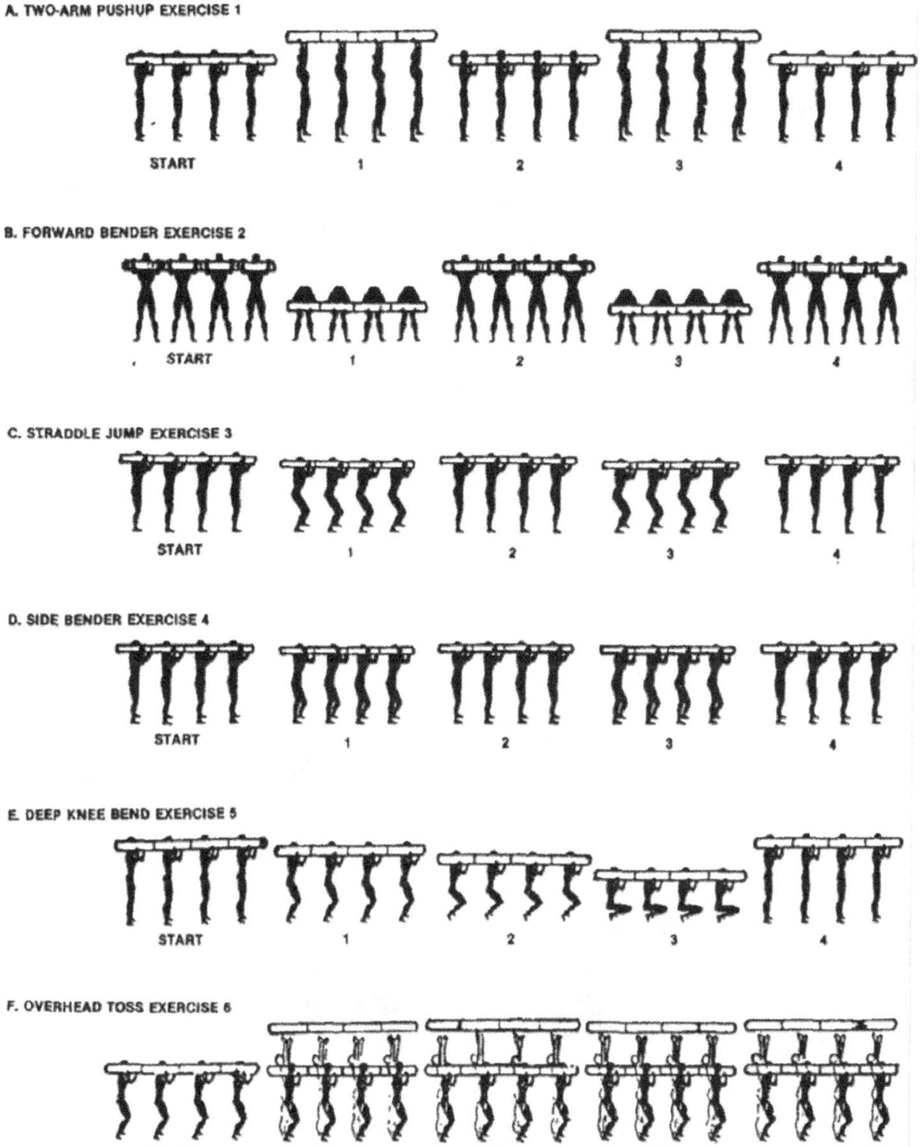

Figure 3-9. Log Drill

Figure 3-10.Basic Positions for Grass Drill.

Figure 3-11.Grass Drill One.

Figure 3-12.Grass Drill Two.

Figure 3-13.Guerrilla Table I.

Figure 3-14.Guerrilla Table II.

Figure 3-15.Proper Running Form.

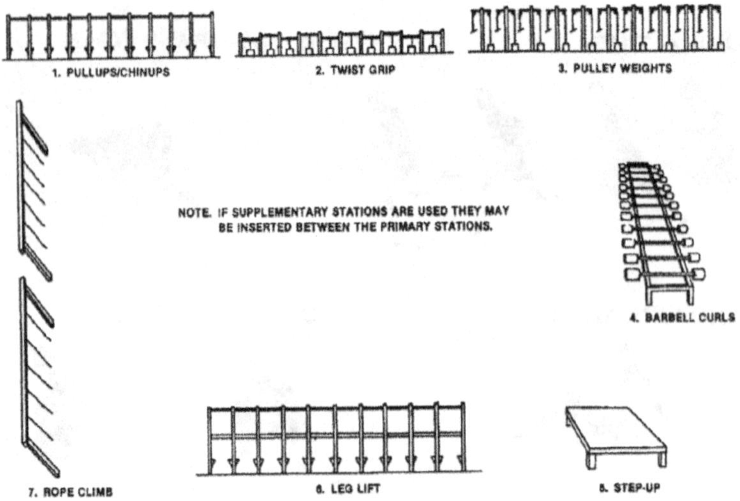

Figure 3-16. Fixed Strength Circuit.

STATION	ITEM	NO.	SPECIFICATION
1	BARBELL		1¼-INCH PIPE 5 FEET LONG WITH CONCRETE FILLED NO. 10 CANS
2	JUMP ROPE		¼- OR ⅜-INCH ROPE, 10 FEET LONG
3	TWIST GRIP		HANDLE 12 INCHES LONG, ROPE 4 FEET LONG, NO. 10 CAN CONCRETE FILLED.
4	INCLINE PLANE		¾-INCH PLYWOOD PLATFORM 2 FEET WIDE AND 6 FEET, 6 INCHES LONG, ELEVATED 10 INCHES AT ONE END. STRAP TO HOLD FEET DOWN.
5	WAR CLUB		HEAD IS 6 BY 12 INCHES, HANDLE IS 14 INCHES LONG BY 1¼ INCHES IN DIAMETER, ABOUT 20 POUNDS.
6	BICYCLE RIDE		PLYWOOD BOARD OR PLATFORM 2 BY 3 FEET WITH 2 BY 2 RUNNERS.
7	STEP—UP		A BOX OR STURDY PLATFORM 18 INCHES HIGH, 18 INCHES WIDE, 24 INCHES LONG.
8	ISOMETRIC PULL		TWO HANDLES 12 INCHES LONG WITH 4 FEET (BETWEEN HANDLES) OR LIGHT WIRE CABLE OR ¼-INCH ROPE

Figure 3-17.Movable Strength Circuit.

Figure. 3-18.Stations of Movable Strength Circuit.

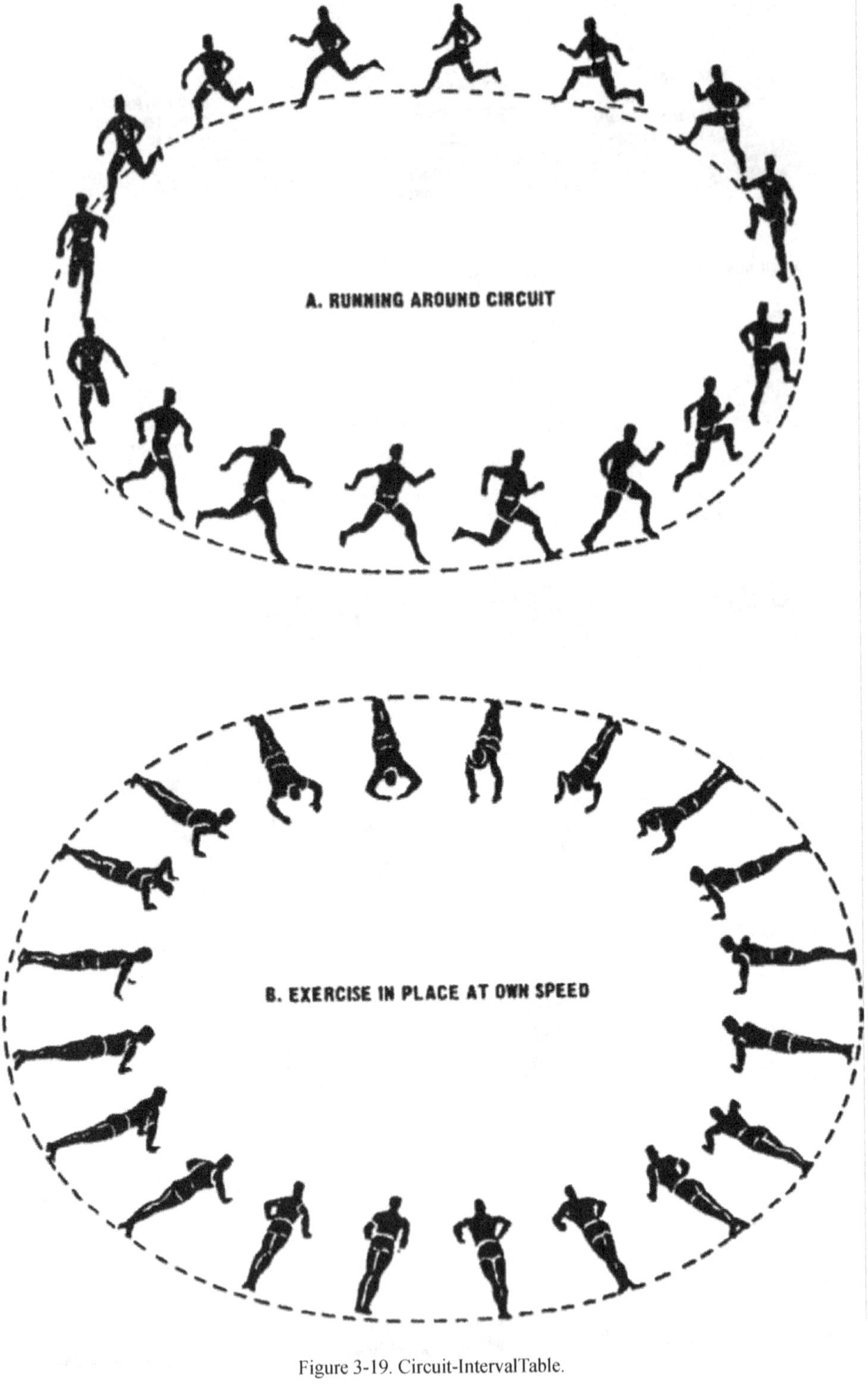

Figure 3-19. Circuit-IntervalTable.

Figure 3-20. Jumping.

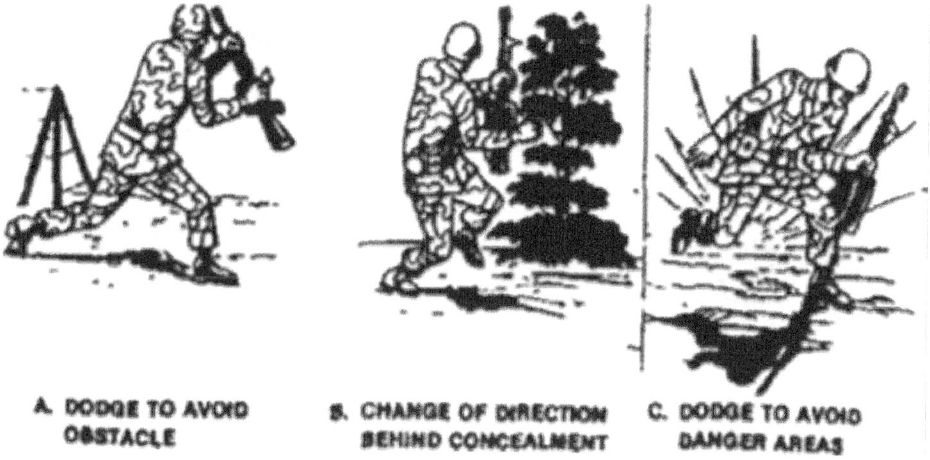

Figure 3-21.Dodging.

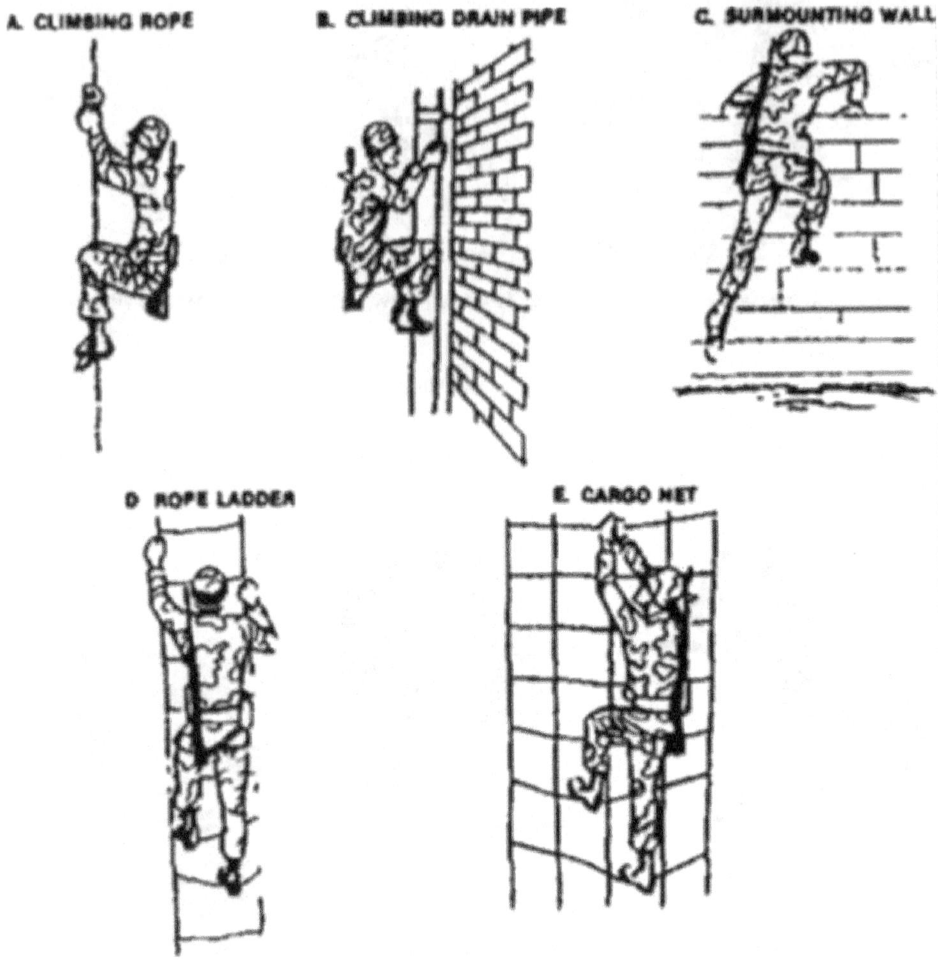

Figure 3-22.Climbing and Surmounting Vertical Objects.

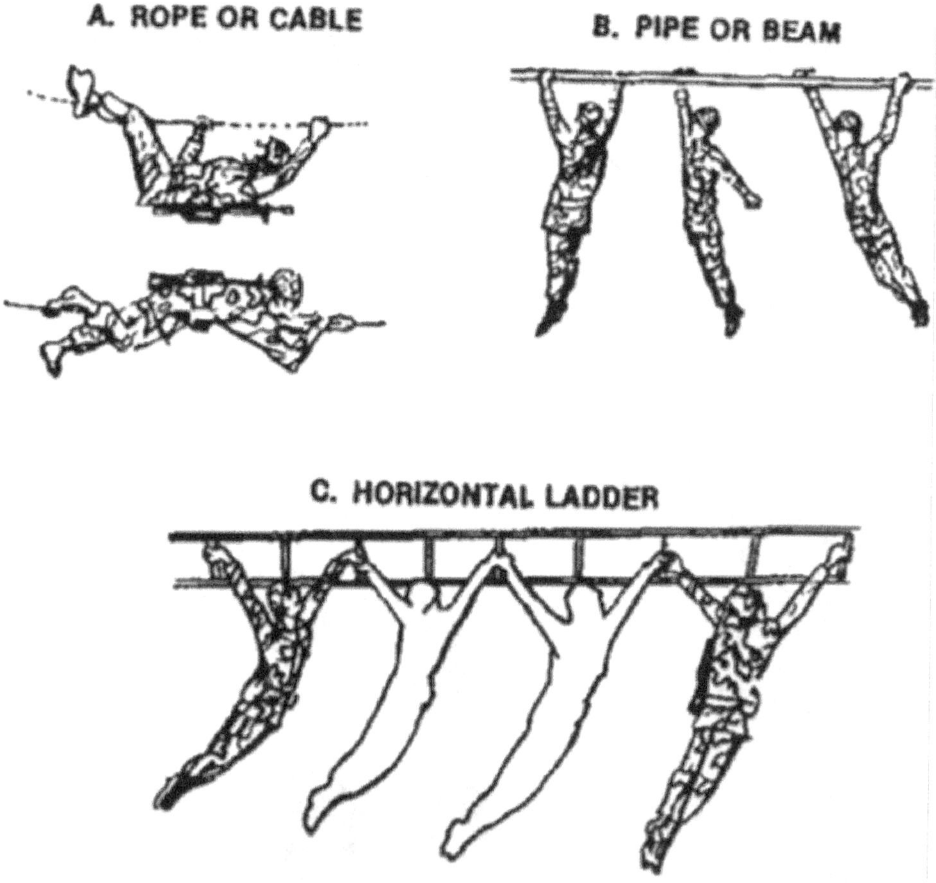

Figure 3-23. Traversing Horizontal Objects.

Figure 3-24.Crawling.

Figure 3-25.Throwing.

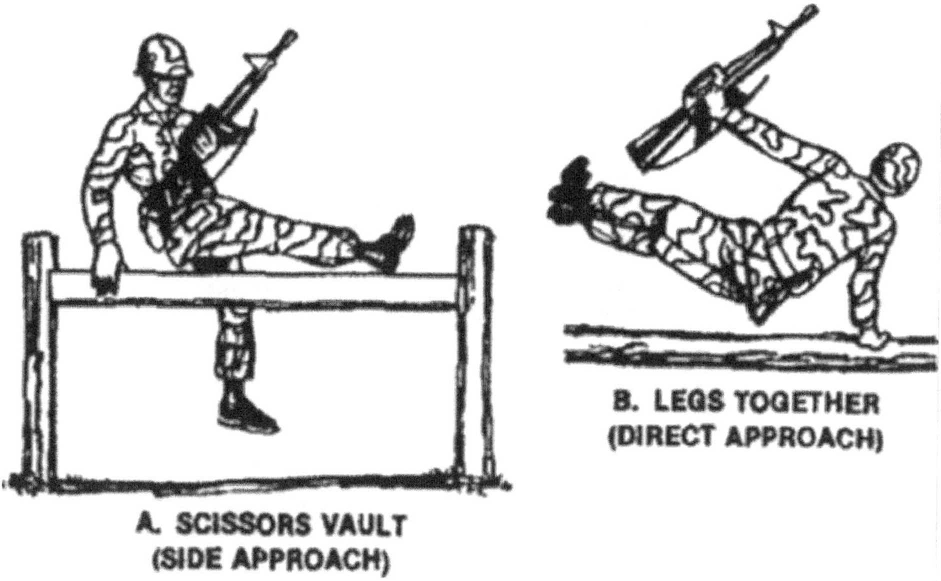

Figure 3-26. Vaulting

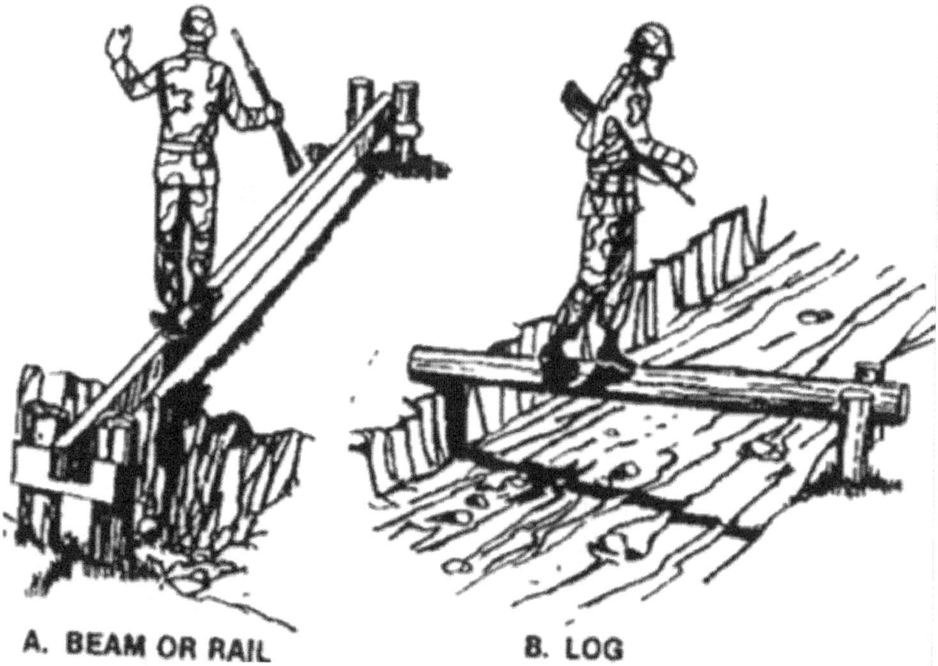

Figure 3-27.Balancing.

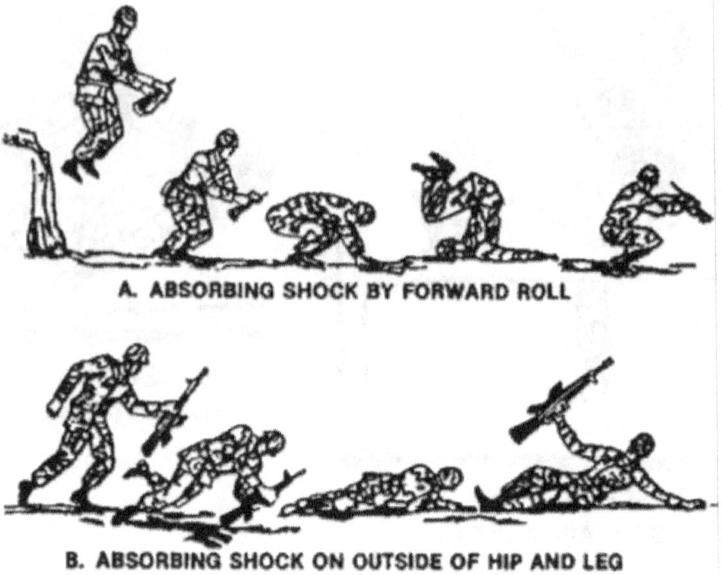

Figure 3-28.Falling.

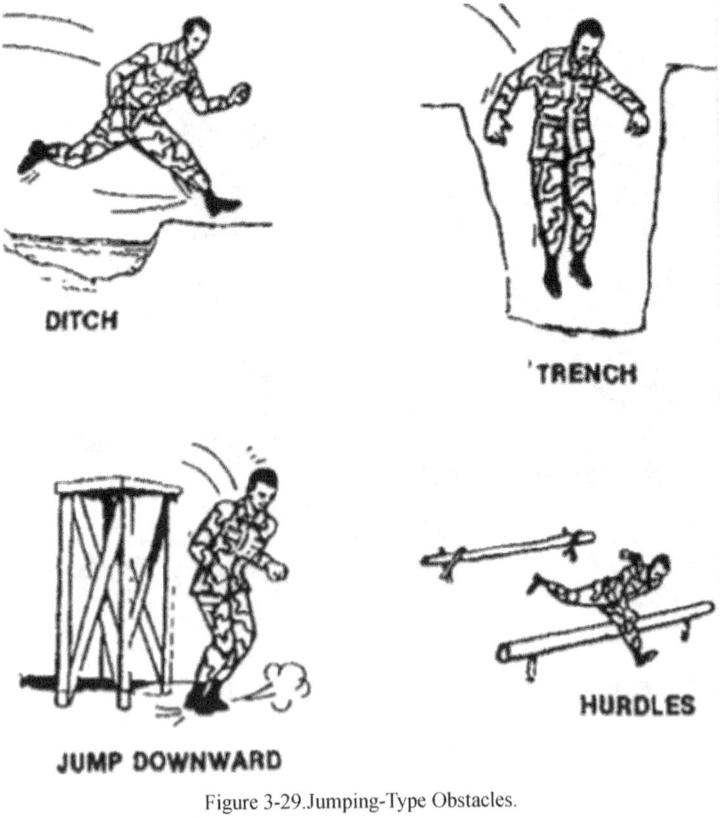

Figure 3-29.Jumping-Type Obstacles.

LANES TO GUIDE CHANGE OF DIRECTION

MAZES TO CAUSE CHANGE OF DIRECTION

Figure 3-30.Dodging-Type Obstacles.

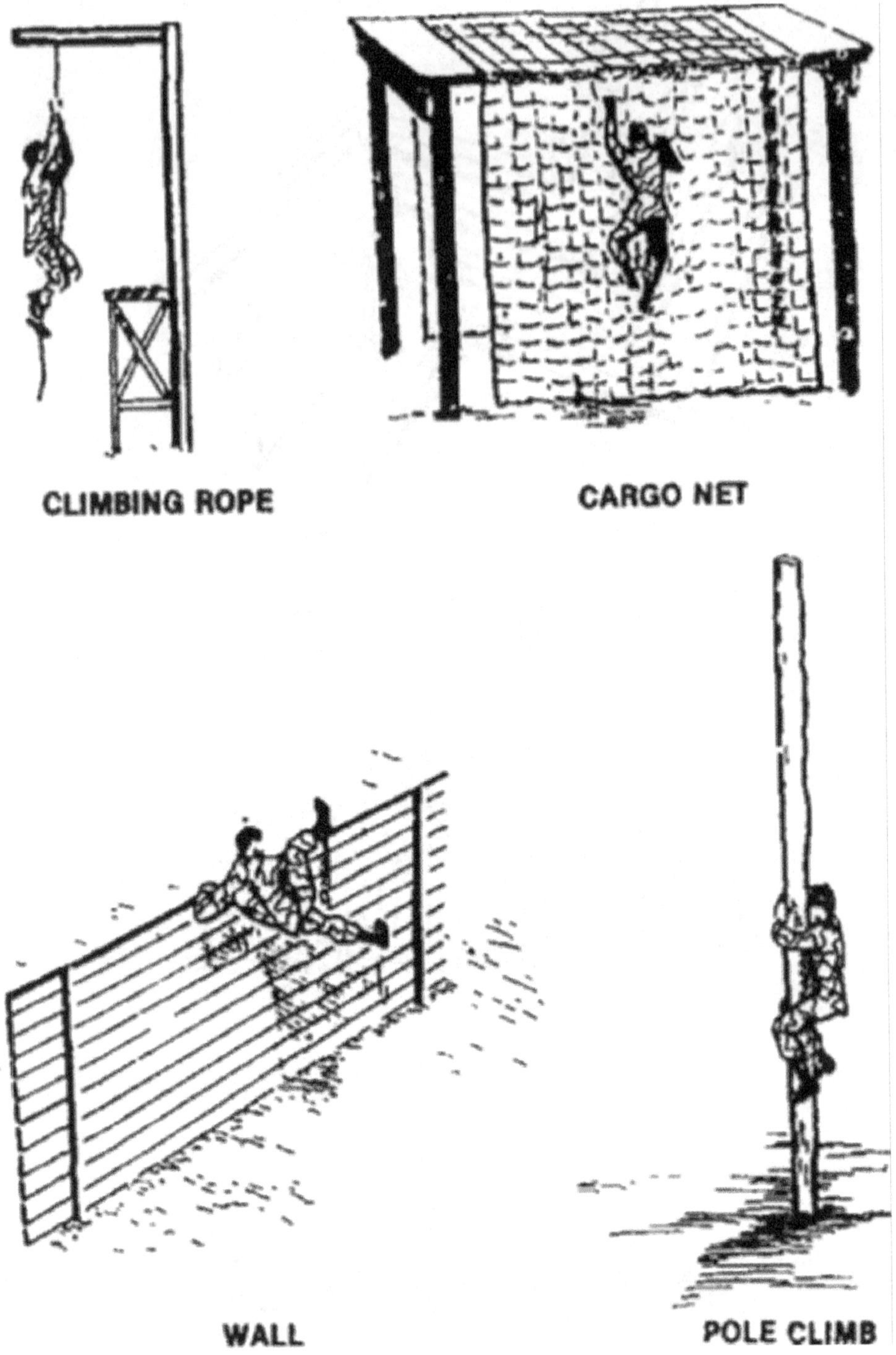

Figure 3-31. Vertical Climbing Obstacles.

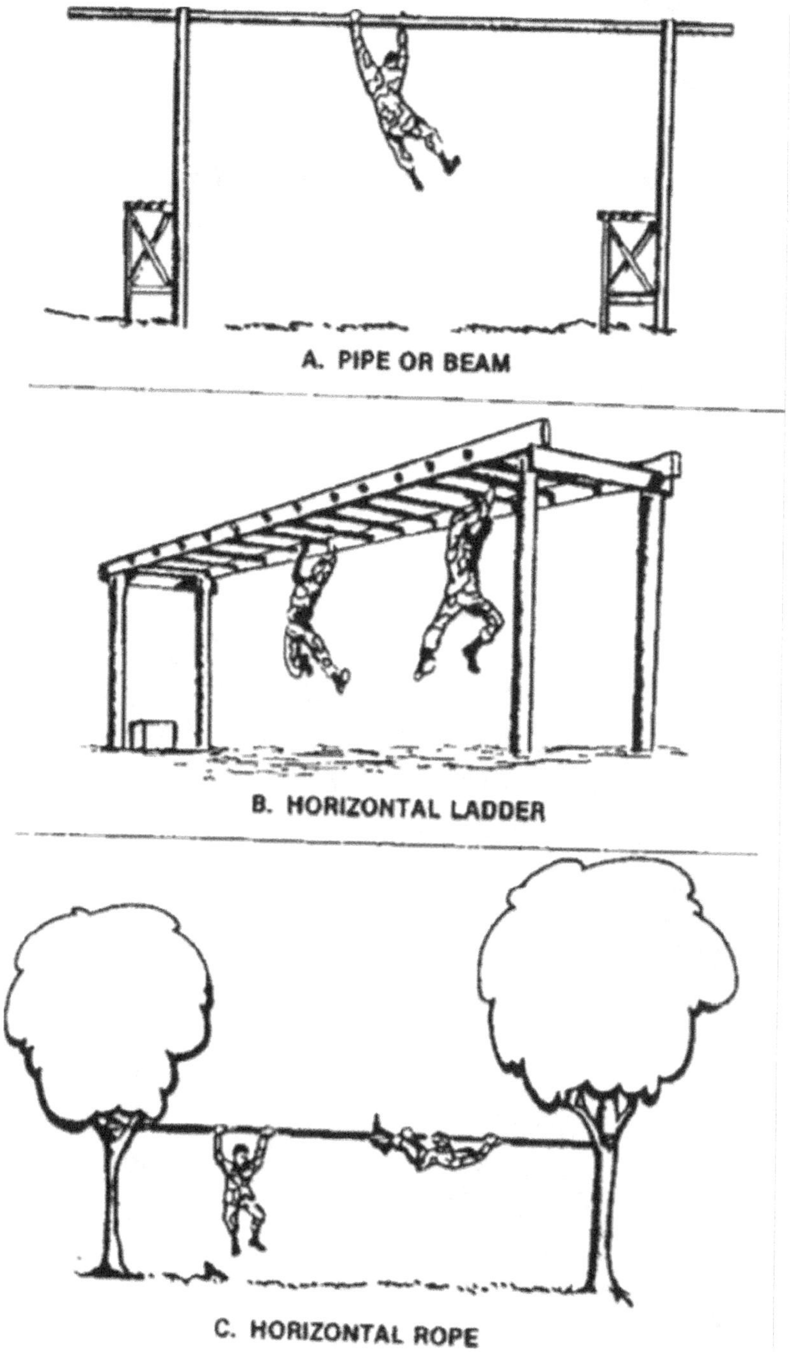

Figure 3-32.Horizontal Type Obstacles.

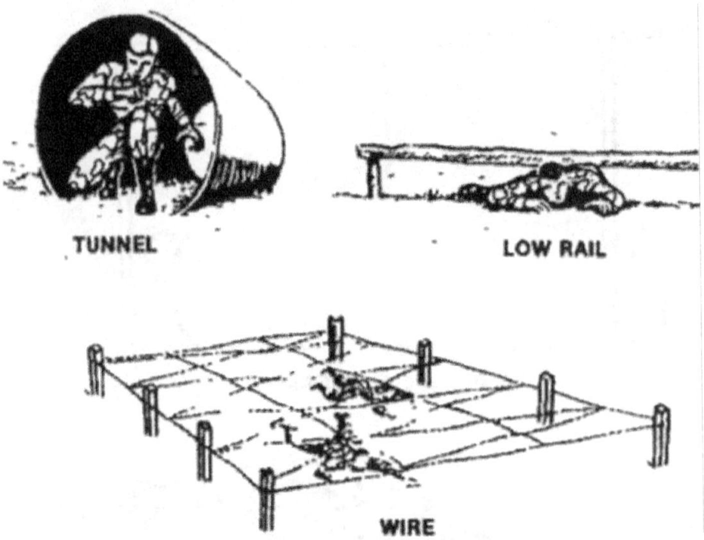

Figure 3-33.Crawling-Type Obstacles.

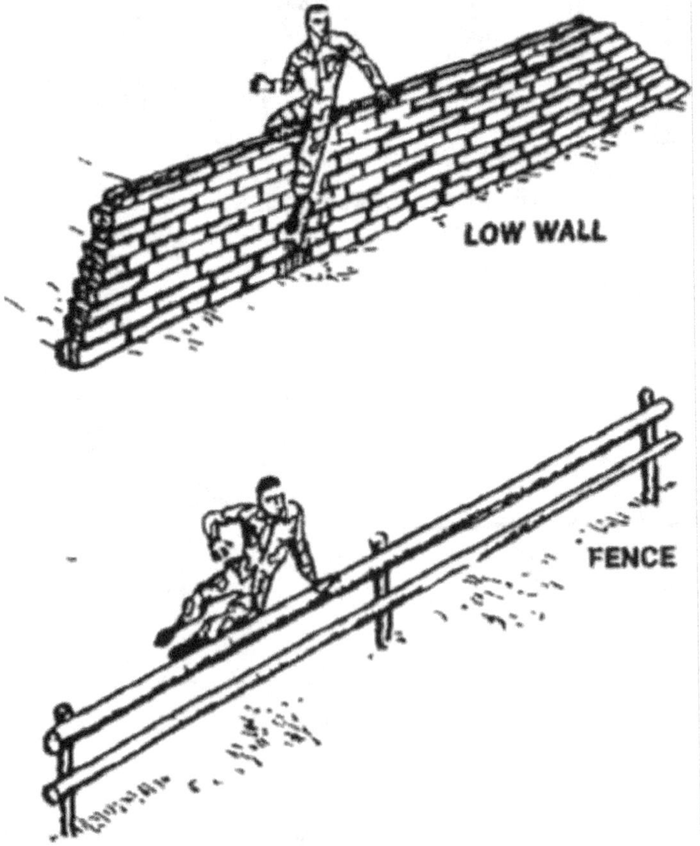

Figure 3-34.Vaulting Obstacles.

Figure 3-35.Balance-Type Obstacles.

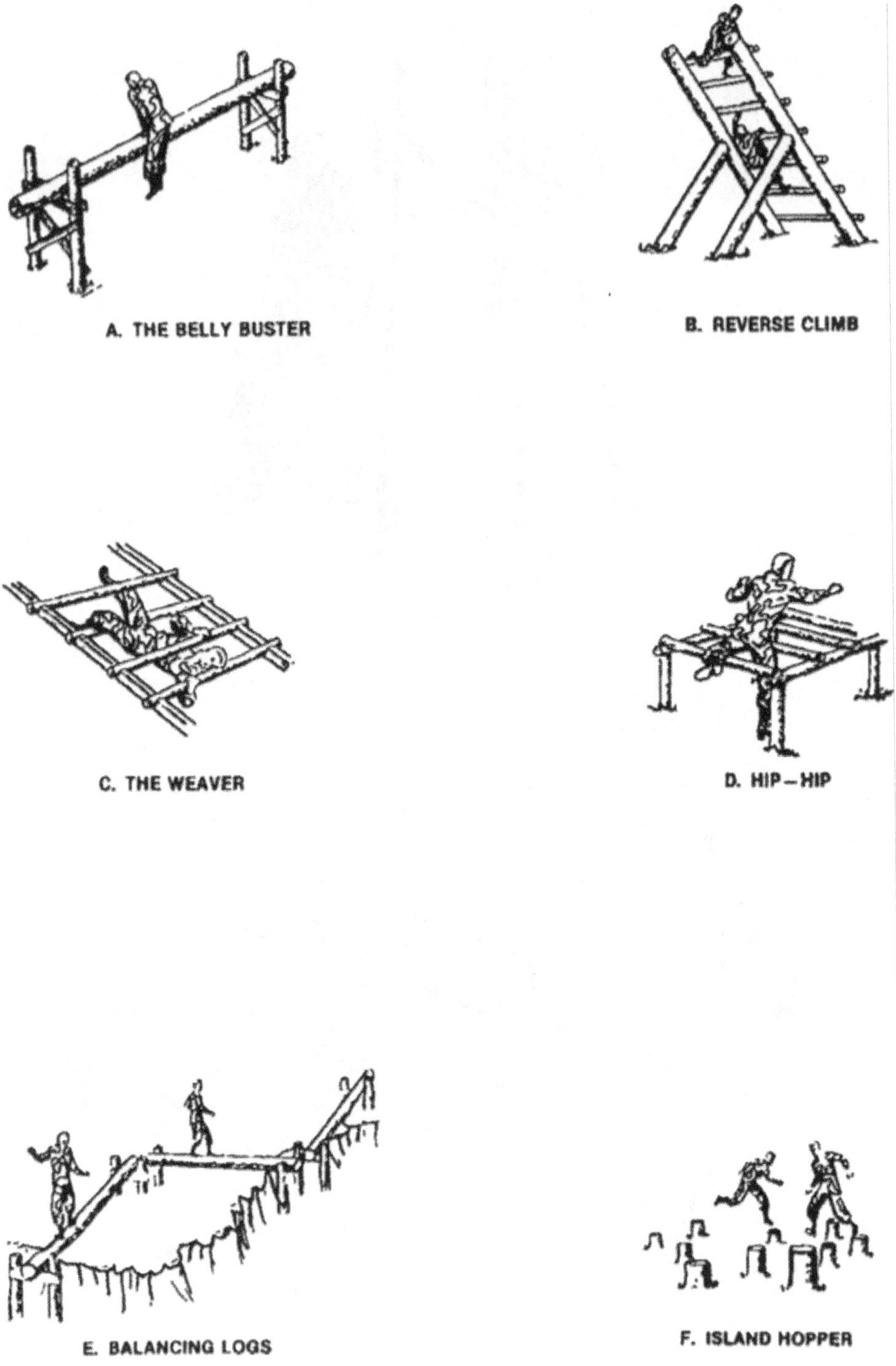

Figure 3-36.Red Group.

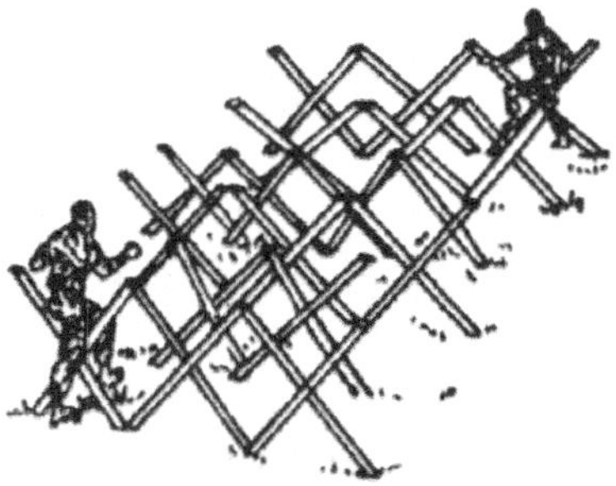

A. THE TOUGH NUT

B. SLIDE FOR LIFE

C. LOW BELLY OVER

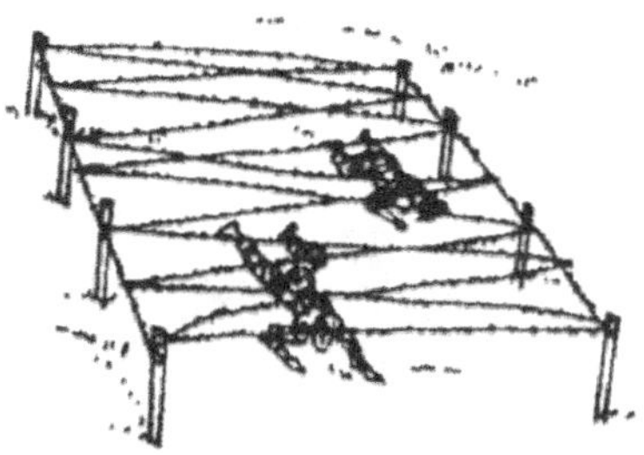

D. BELLY CRAWL

E. THE DIRTY NAME

F. THE TARZAN

Figure 3-37.White Group.

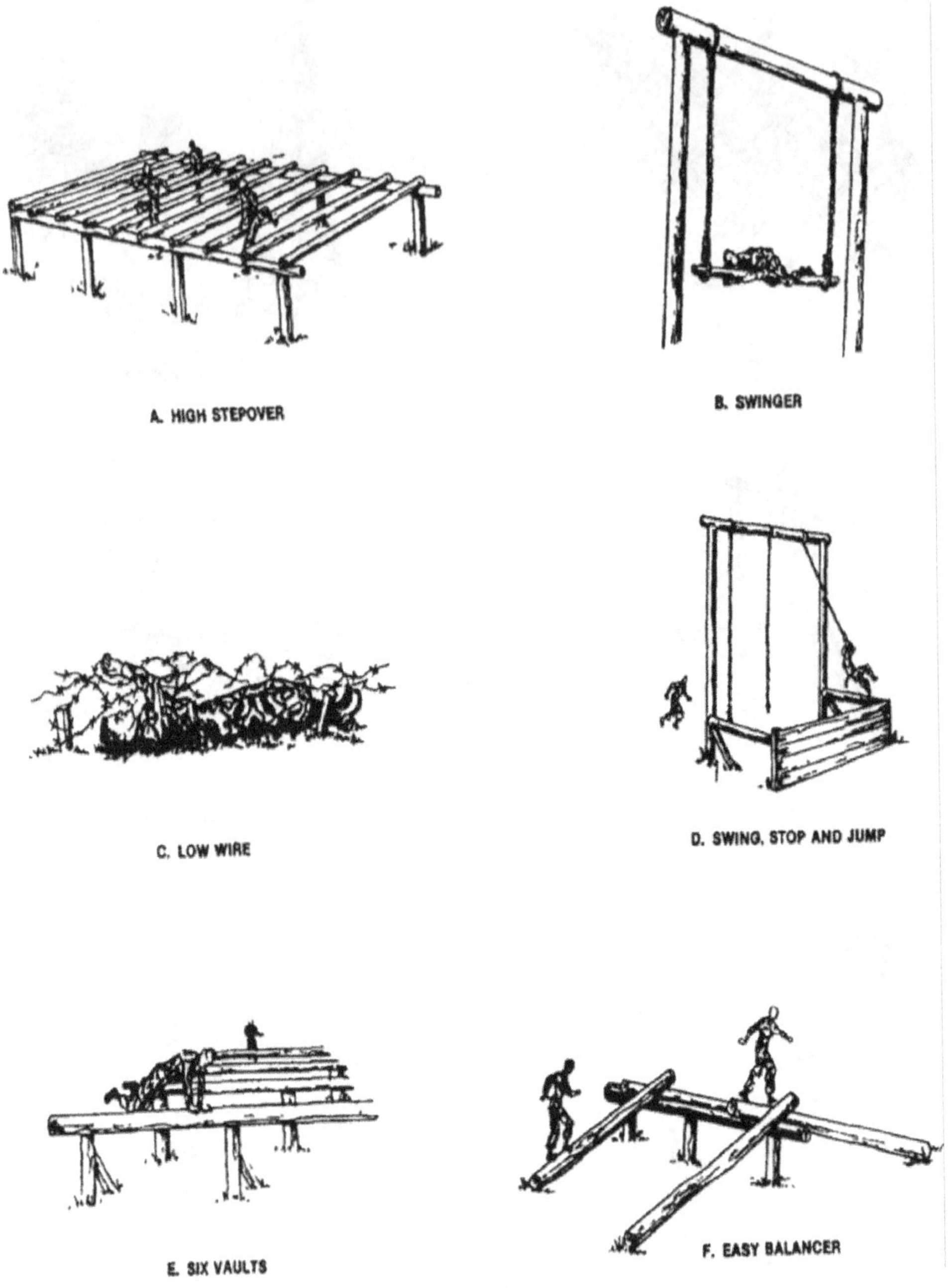

Figure 3-38.Blue Group.

A. INCLINE WALL

B. SKYSCRAPER

C. JUMP AND LAND

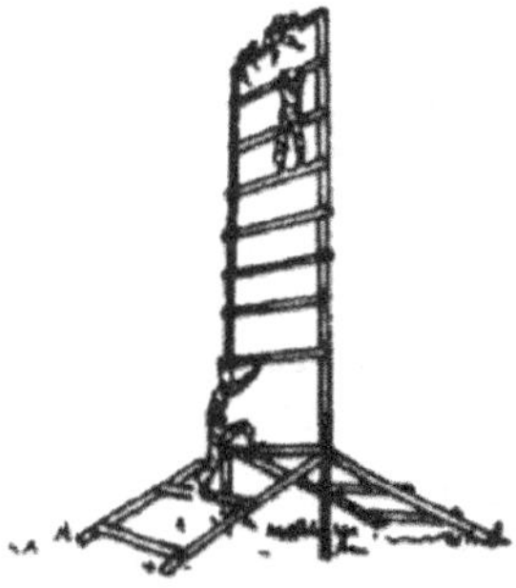

D. CONFIDENCE CLIMB

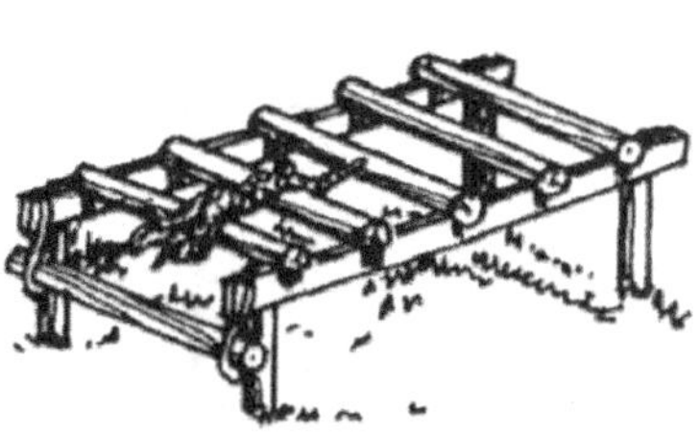

E. BELLY ROBBER

F. THE TOUGH ONE

Figure 3-39.Black Group.

Figure 3-40. Bench Conditioning, Table I.

Figure 3-41. Bench Conditioning, Table II.

PROGRESSION GUIDE

AGE GROUP	LEVEL	EXERCISES 1	2	3	4	5	6
17	A	15	18	14	15	15	250
to	B	13	16	13	13	13	235
29	C	11	14	12	11	11	[illegible]
30	A	13	14	12	13	13	200
to	B	11	13	11	11	11	185
39	C	9	12	10	9	9	165
40	A	11	11	10	11	11	150
to	B	9	10	9	9	9	135
44	C	7	9	8	7	7	120
45	A	9	8	8	9	9	100
to	B	7	7	7	7	7	90
49	C	5	6	6	5	5	80
50	A	7	6	6	7	7	75
to	B	5	5	5	5	5	70
59	C	3	4	4	3	3	60
60	A	4	5	4	4	4	50
and	B	3	4	3	3	3	40
over	C	2	3	2	2	2	30
Minutes for each exercise		2	1	1	1	2	3

Table I. Progression Guide.

Table I. Exercises.

PROGRESSION GUIDE

AGE GROUP	LEVEL	EXERCISES 1	2	3	4	5	6
17	A	17	17	17	9	19	300
to	B	15	15	15	8	17	270
29	C	13	13	13	7	15	245
30	A	15	15	15	8	17	235
to	B	13	13	13	7	15	210
39	C	11	11	11	6	13	190
40	A	13	13	13	7	15	175
to	B	11	11	11	6	13	155
44	C	9	10	9	5	11	135
45	A	11	11	11	6	13	125
to	B	9	9	9	5	11	110
49	C	7	7	7	4	9	100
50	A	9	9	9	5	11	95
to	B	7	7	7	4	9	85
59	C	5	5	5	3	7	75
60	A	6	7	7	4	9	70
and	B	5	5	5	3	7	60
over	C	4	4	4	2	5	50
Minutes for each exercise		1	1	1	1 1/2	1 1/2	6

Table II. Progression Guide.

EXERCISE 1

EXERCISE 2

EXERCISE 3

EXERCISE 4

EXERCISE 5

EXERCISE 6

Table II. Exercises.

PROGRESSION GUIDE

AGE GROUP	LEVEL	EXERCISES 1	2	3	4	5	6
17	A	10	19	19	16	10	350
to	B	9	17	17	15	9	315
29	C	8	15	15	14	8	280
30	A	9	17	17	14	9	270
to	B	8	15	15	13	8	240
39	C	7	13	13	12	7	210
40	A	8	15	15	12	8	200
to	B	7	13	13	11	7	180
44	C	6	11	11	10	6	160
45	A	7	13	13	10	7	150
to	B	6	11	11	9	6	135
49	C	5	9	9	8	5	120
50	A	6	11	11	8	6	115
to	B	5	9	9	7	5	105
59	C	4	7	7	6	4	95
60	A	5	9	9	7	5	90
and	B	4	7	7	6	4	80
over	C	3	5	5	4	3	70
Minutes for each exercise		1 1/2	1	1	1 1/2	1	6

Table III. Progression Guide.

Table III. Exercises.

PROGRESSION GUIDE

AGE GROUP	LEVEL	EXERCISES 1	2	3	4	5	6
17	A	12	9	12	24	25	400
to	B	11	8	11	22	23	380
29	C	10	7	10	21	21	360
30	A	11	8	11	23	23	305
to	B	10	7	10	21	21	290
39	C	9	6	9	20	20	275
40	A	10	7	10	20	21	225
to	B	9	6	9	18	18	215
44	C	8	5	8	16	16	205
45	A	8	6	8	16	16	175
to	B	7	5	7	14	14	165
49	C	6	4	6	12	12	155
50	A	6	5	6	13	13	135
to	B	5	4	5	11	11	130
59	C	4	3	4	10	10	120
60	A	5	4	5	10	10	100
and	B	4	3	4	9	9	95
over	C	3	2	3	8	8	90
Minutes for each exercise		1	2	1	1	1	6

Table IV. Progression Guide.

Table IV. Exercises.

PROGRESSION GUIDE

AGE GROUP	LEVEL	EXERCISES 1	2	3	4	5	6
17	A	14	13	28	14	30	450
to	B	13	12	27	13	28	430
29	C	12	11	26	12	26	410
30	A	12	12	25	12	26	350
to	B	11	11	24	11	24	330
39	C	10	10	23	10	22	310
40	A	11	11	23	11	23	250
to	B	10	10	21	10	21	240
44	C	9	9	19	9	19	230
45	A	9	9	20	9	20	200
to	B	8	8	18	8	18	190
49	C	7	7	16	7	16	180
50	A	7	7	16	7	16	170
to	B	6	6	14	6	14	155
59	C	5	5	12	5	12	140
60	A	6	6	12	6	12	115
and	B	5	5	11	5	10	110
over	C	4	4	9	4	9	105
Minutes for each exercise		2	1	1	2	1	5

Table V. Progression Guide.

Table V. Exercises.

PROGRESSION GUIDE

AGE GROUP	LEVEL	EXERCISES 1	2	3	4	5	6
17	A	17	15	32	32	35	500
to	B	16	14	30	30	33	480
29	C	15	13	28	28	31	460
30	A	15	13	30	30	31	400
to	B	14	12	28	28	29	380
39	C	13	11	26	26	27	360
40	A	13	10	27	27	27	310
to	B	12	9	25	25	25	285
44	C	11	8	23	23	23	265
45	A	11	9	23	23	23	250
to	B	10	8	21	21	21	230
49	C	9	7	19	19	19	210
50	A	9	8	19	19	19	200
to	B	8	7	17	17	17	190
59	C	7	6	15	15	15	175
60	A	8	7	15	15	17	140
and	B	7	6	13	13	15	130
over	C	5	5	10	10	12	120
Minutes for each exercise		2	1	1	1	1	6

Table VI. Progression Guide.

Table VI. Exercises.

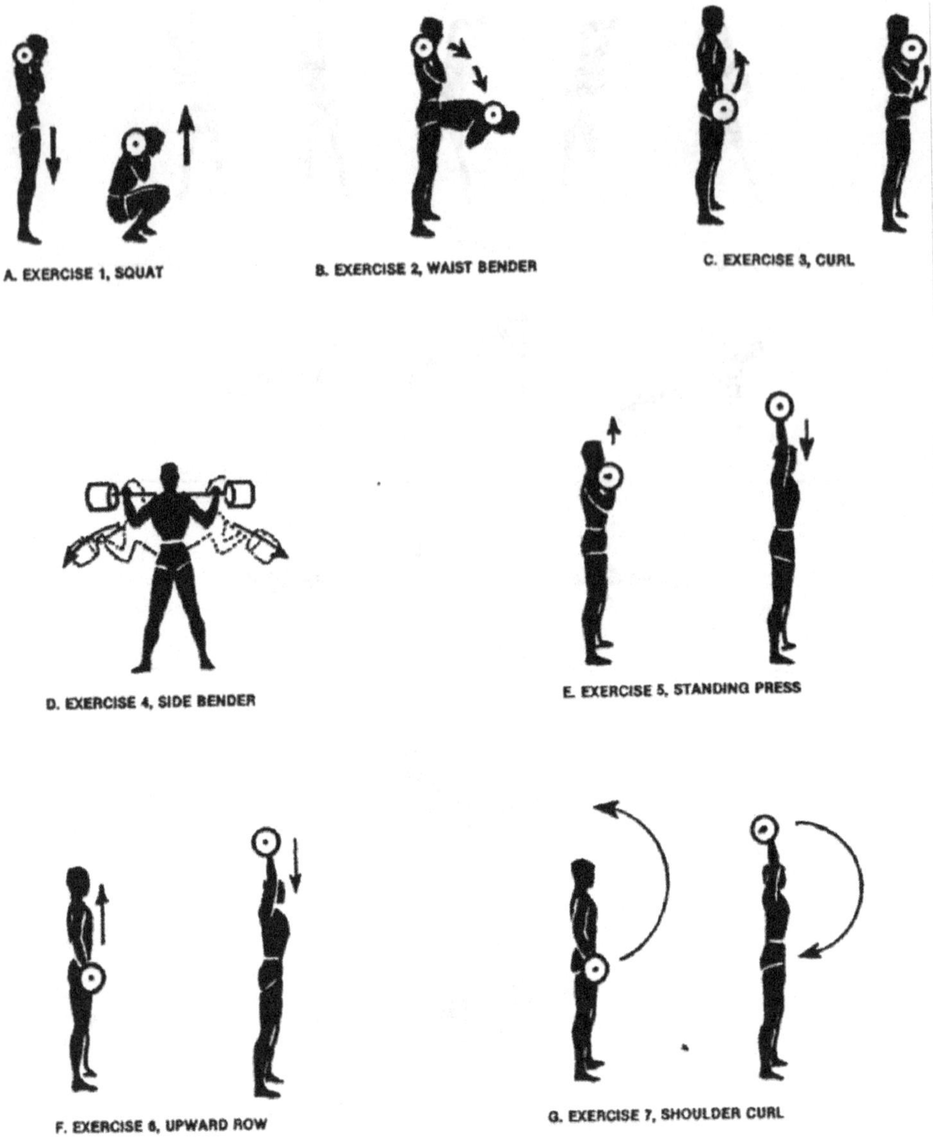

Figure 3-42. Weight Training.

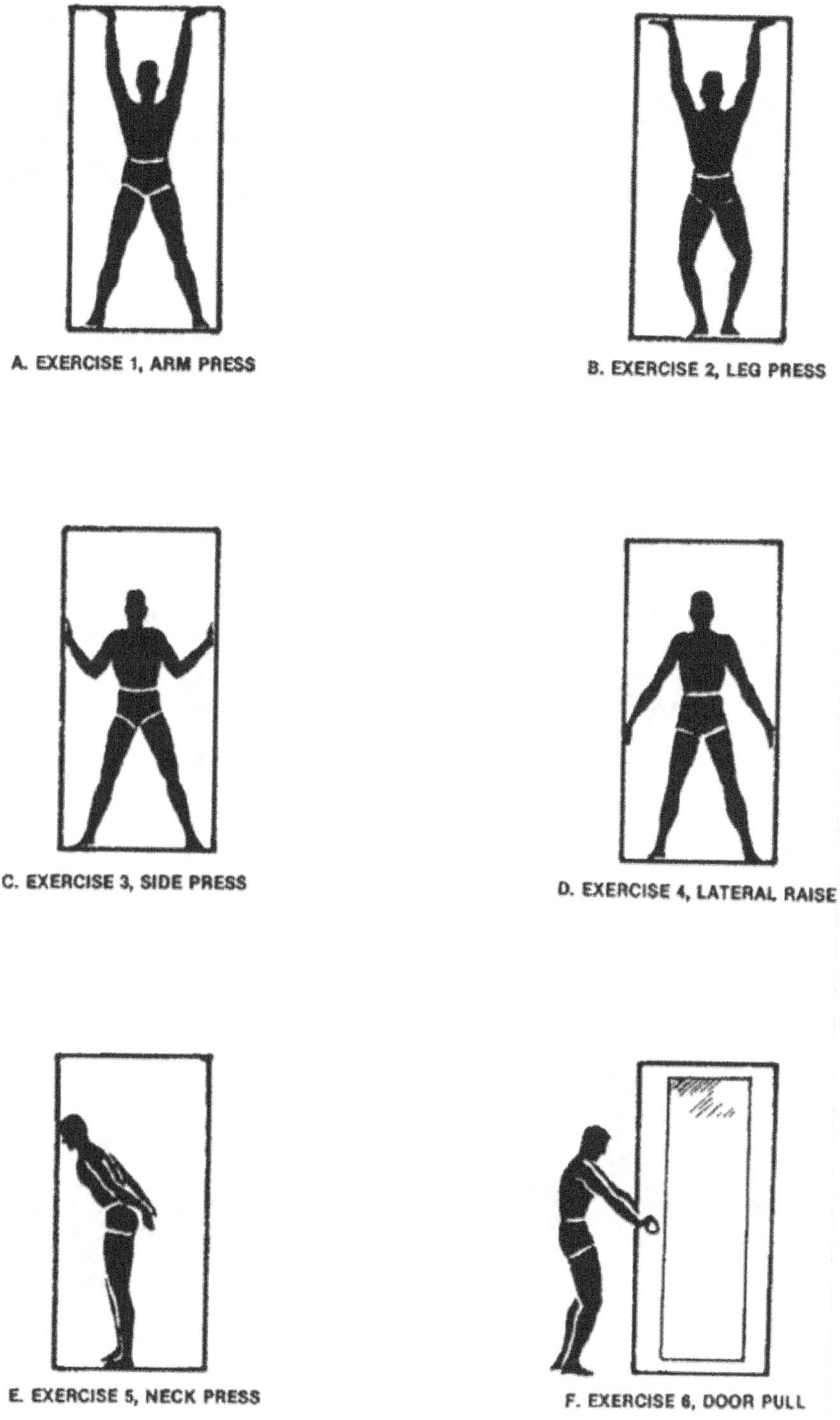

Figure 3-43. Isometric Exercises.

Figure 5-1.Combatives Table I.

Figure 5-2.Combatives Table II.

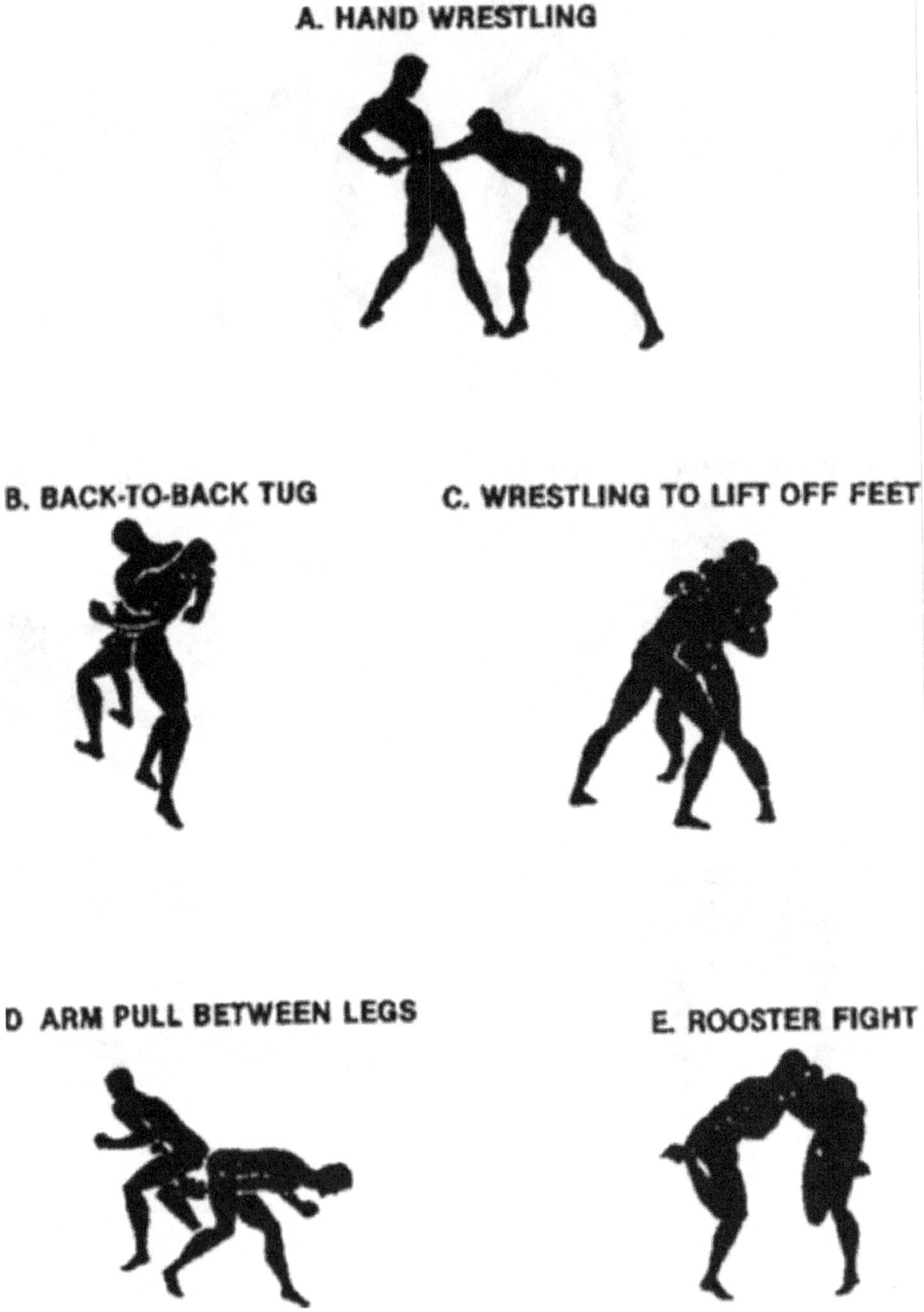

Figure. 5-3.Combatives Table III.

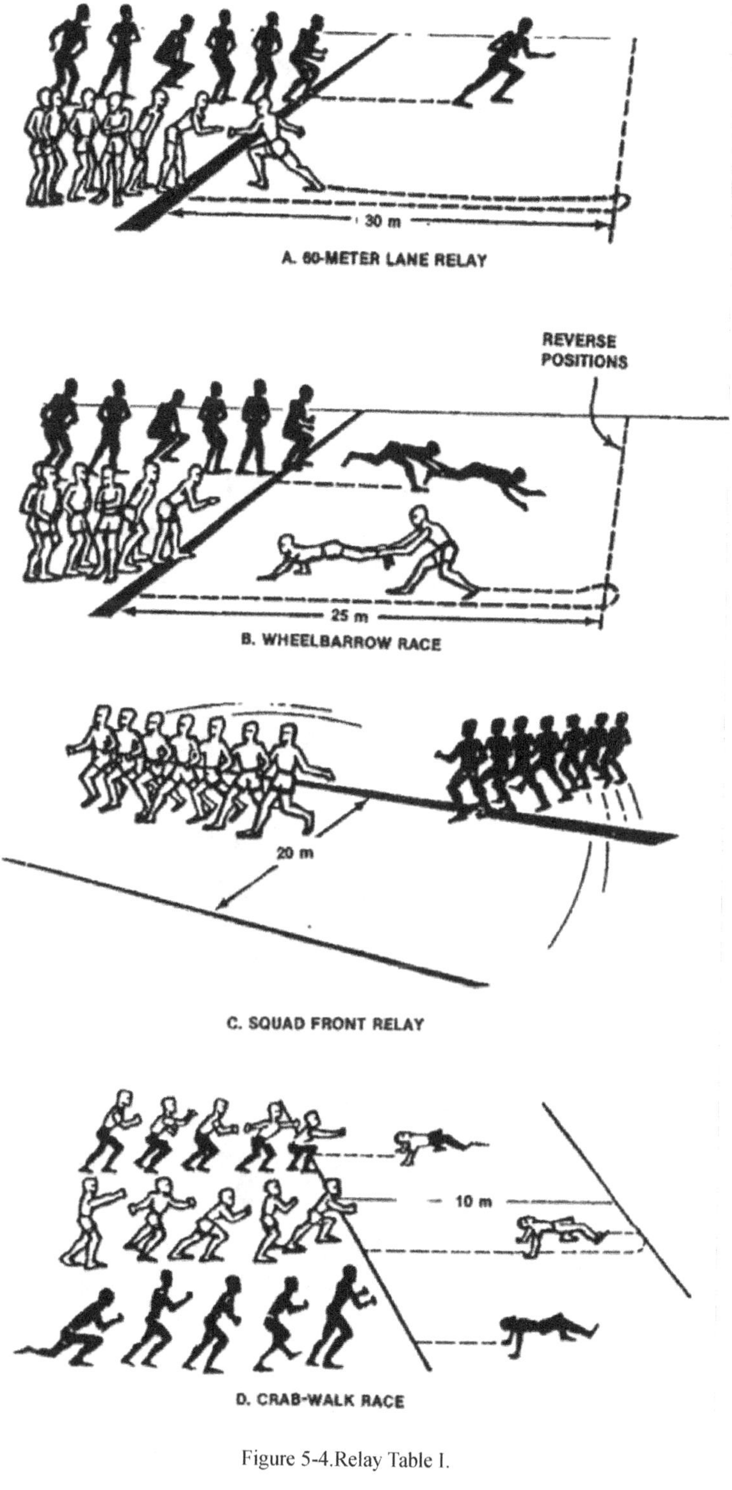

Figure 5-4.Relay Table I.

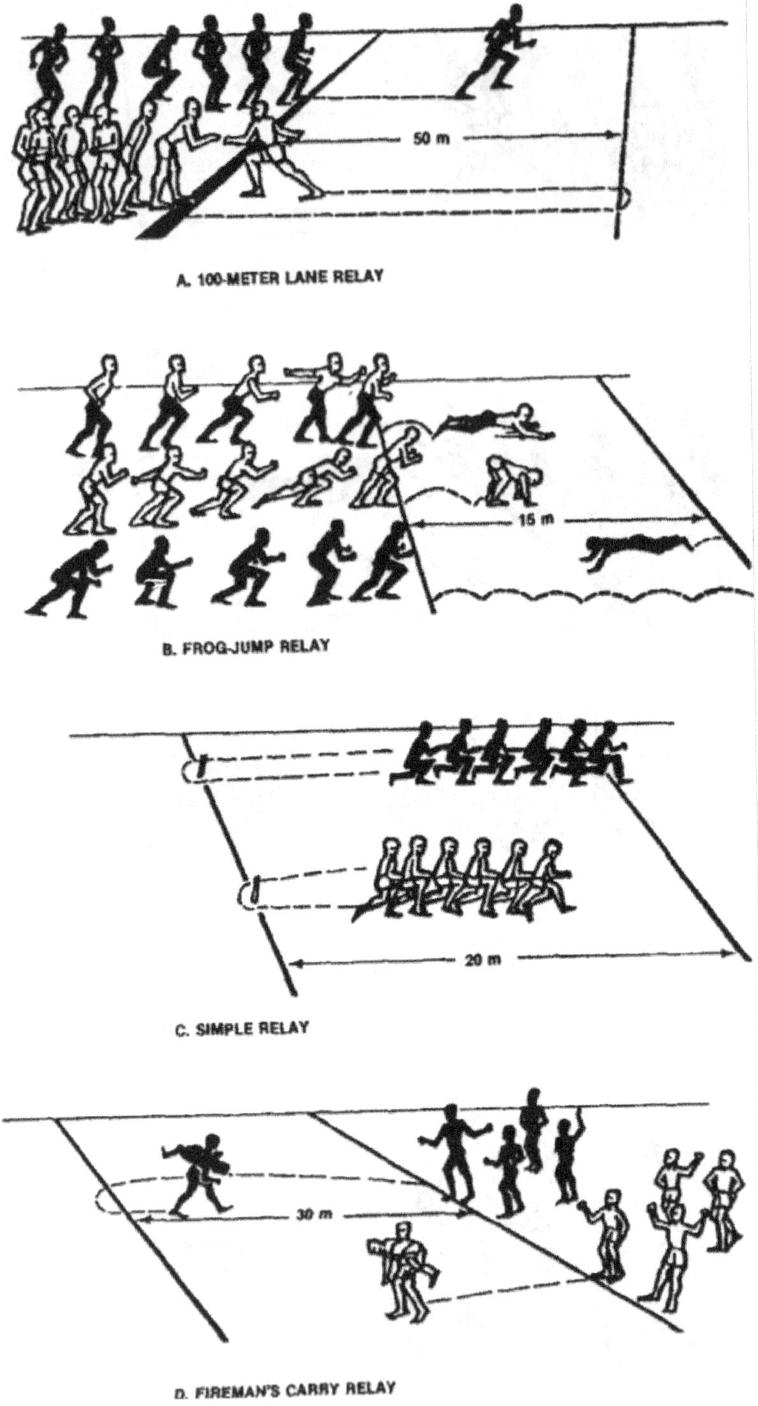

Figure 5-5.Relay Table II.

Figure 5-6.Relay Table III.

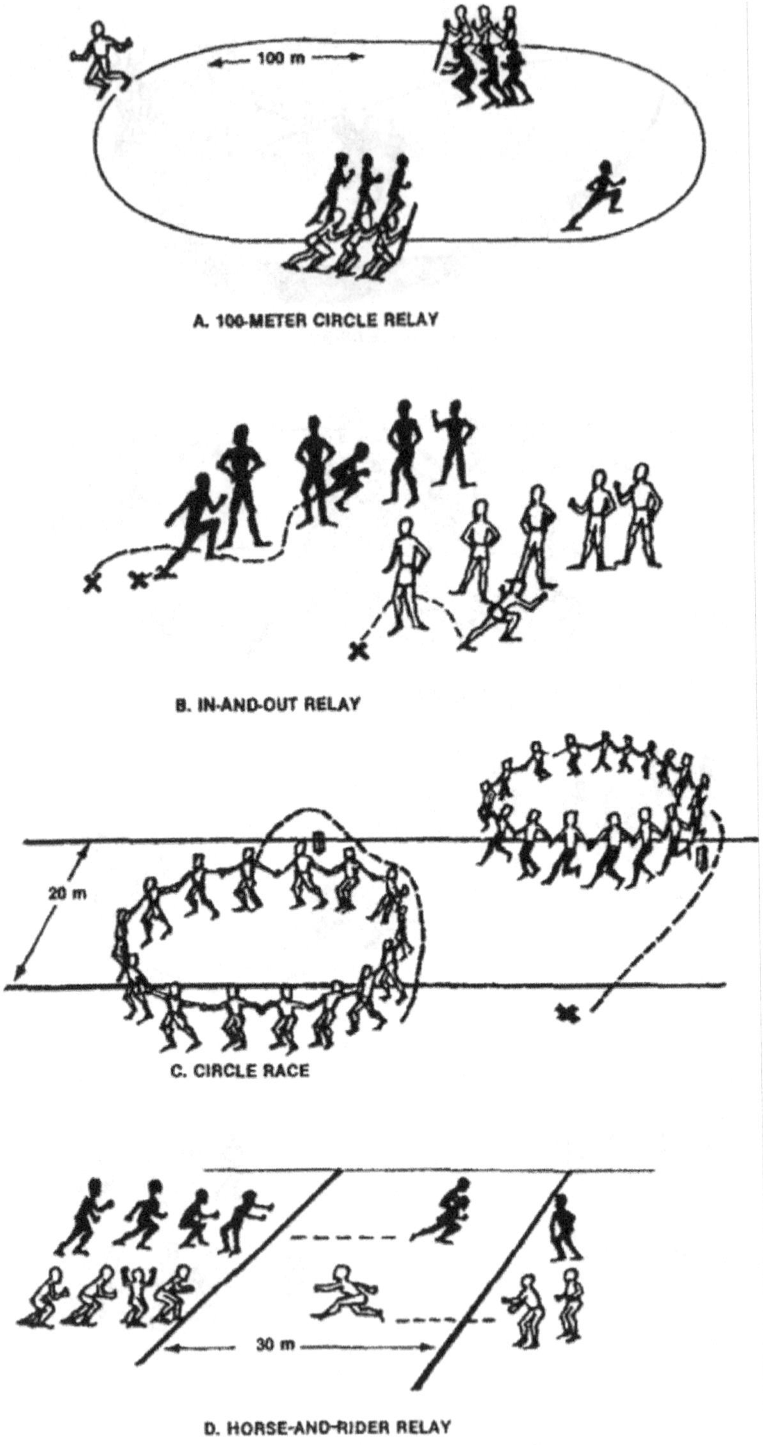

Figure 5-7.Relay Table IV.

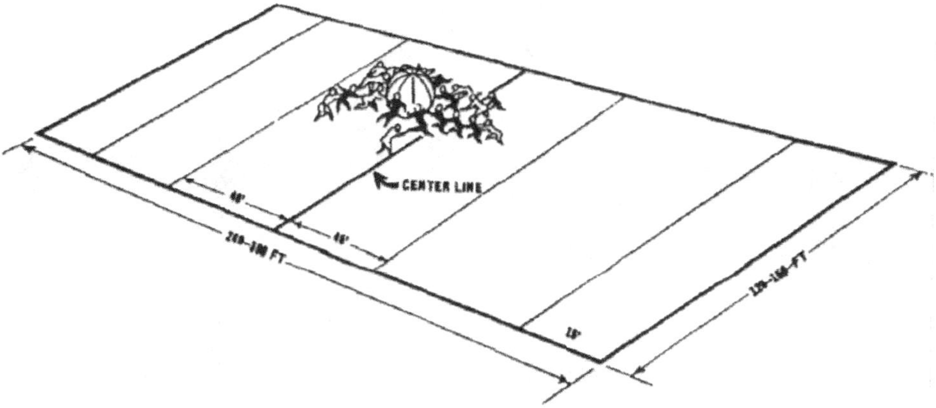

Figure 5-8.Pushball.

Figure 5-9.Line Rush.

Figure. 5-10. Human Tug-of-War.

Figure 5-11. Master of the Ring.

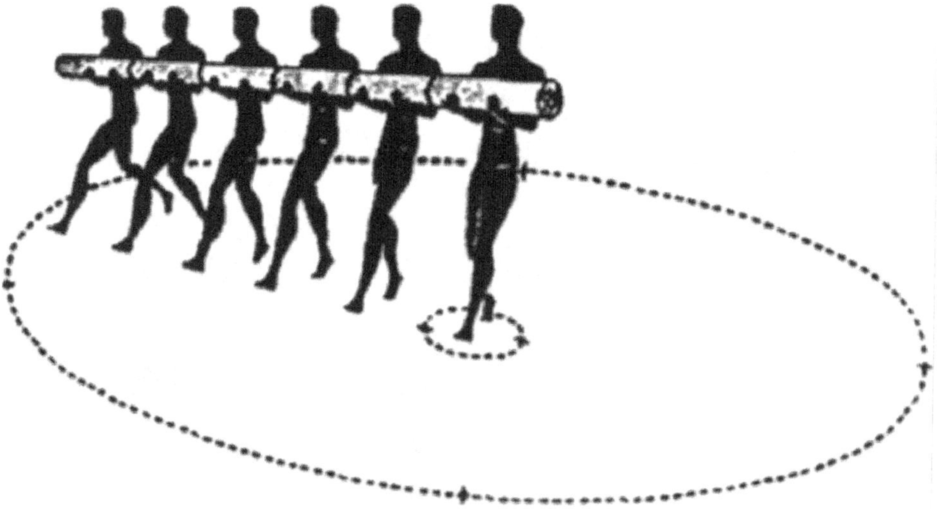

Figure 5-12.Log Pivot Circle.

Figure 5-13.Log-Rolling Race.

Figure 5-14. Prone Pushing Contest.

Figure 5-15. Shuttle Relay Race.

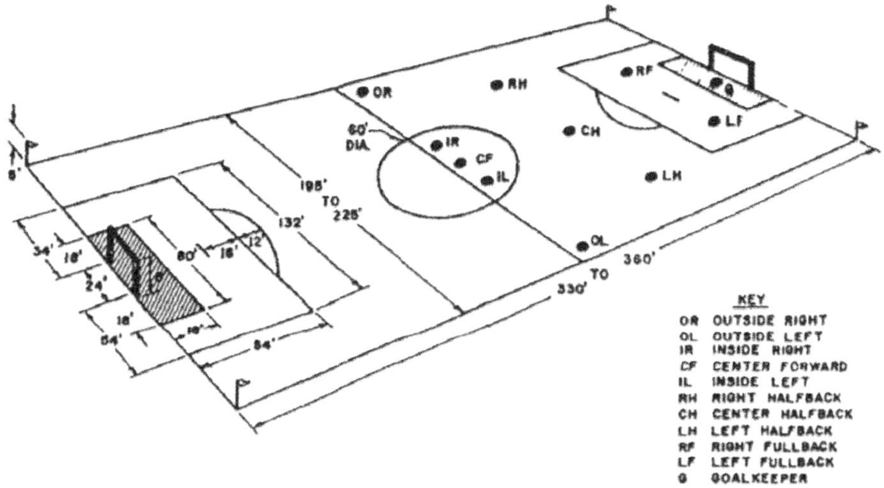

Figure 5-16. Soccer.

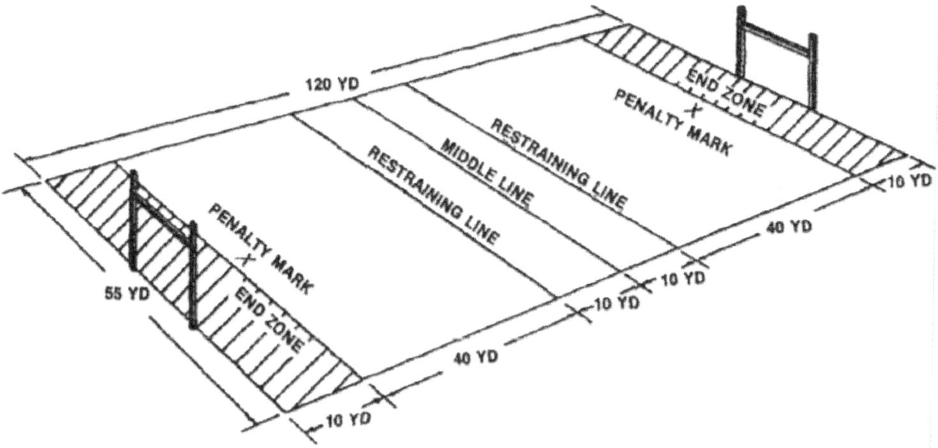

Figure 5-17.Speedball Field.

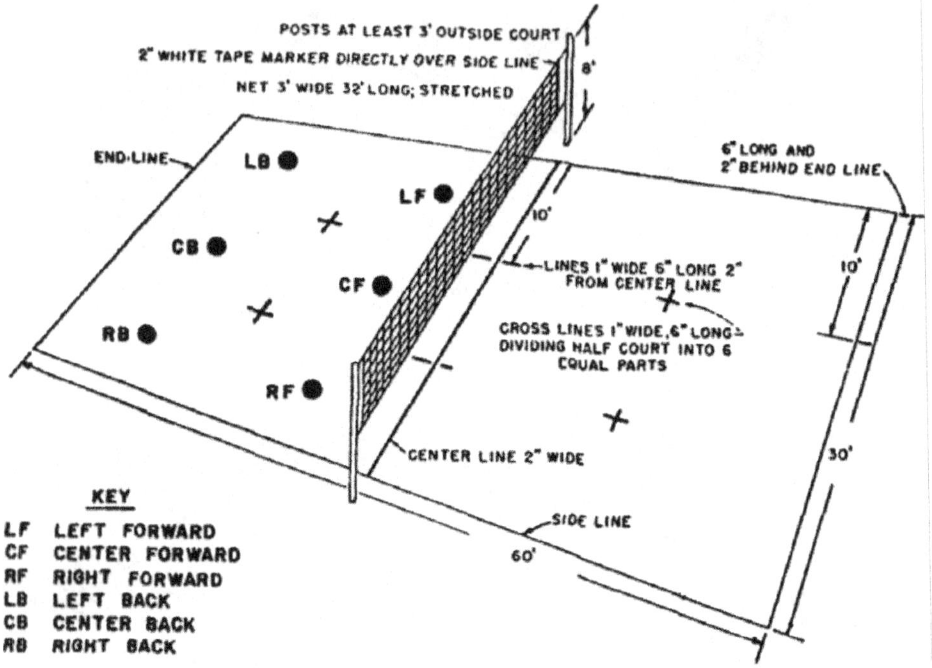

Figure 5-18. Volleyball.

PTS	UPHILL CLIMB	ROPE CLIMB	EVACUATION	FIRE AND MANEUVER	FORCED MARCH
100	1:20-BELOW	0:10-BELOW	0:25-BELOW	1:20-BELOW	26:00-BELOW
99	1:21-1:22	0:11		1 21-1:22	26:01-26:12
98	1:23-1:24	0:12	0:26	1:23-1:24	26:13-26:24
97	1:25-1:26	0:13		1:25-1:26	26:25-26:36
96	1:27-1:28	0:14	0:27	1:27-1:28	26:37-26:48
95	1:29-1:30	0:15		1:29-1:30	26:49-27:00
94	1:31-1:32	0:16	0:28	1:31-1:32	27:01-27:12
93	1:33-1:34	0:17		1 33-1.34	27:13-27:24
92	1:35-1:36	0:18	0:29	1:35-1.36	27:25-27:36
91	1:37-1:38	0:19		1:37-1:38	27:37-27:48
90	1:39-1:40	0:20	0:30	1:39-1:40	27:49-28:00
89	1:41-1:42	0:21	0:31	1:41-1:42	28:01-28:12
83	1:43-1:44	0:22	0:32	1:43-1:44	28:13-28:24
87	1:45-1:46	0:23	0:33	1:45-1:46	28:25-28:36
86	1:47-1:48	0:24	0:34	1.47-1:48	28:37-28:48
85	1:49-1:50	0:25	0:35	1.49-1:50	28:49-29:00
84	1:51-1:52	0:26	0:36	1:51-1:52	29:01-29:12
83	1:53-1:54	0:27	0.37	1:53-1:54	29:13-29:24
82	1:55-1:56	0:28	0:38	1:55-1:56	29:25-29:36
81	1:57-1:58	0:29	0:39	1:58-1:58	29:37-29:48
80	1:59-2.00	0:30	0:40	1:59-2:00	29:49-30:00
79	2.01-2:02	0:31	0:41	2:01-2:02	30:01-30:12
78	2:03-2:04	0:32	0:42	2:03-2.04	30:13-30:24
77	2:05-2:06	0.33	0:43	2:05-2:06	30:25-30:36
76	2:07-2:08	0.34	0.44	2:07-2:08	30:37-30:48
75	2:09-2:10	0:35	0:45	2:09-2:10	30:49-31:00
74	2:11-2:12	0:36	0:46	2:11-2 12	31.01-31:12
73	2:13-2:14	0:37	0.47	2.13-2:14	31:13-31:24
72	2:15-2:16	0:38	0:48	2.15-2:16	31:25-31:36
71	2:17-2:18	0:39	0:49	2:17-2:18	31:37-31:48
70	2:19-2:20	0:40	0:50	2:19-2.20	31:49-32.00
69	2.21-2:22	0:41	0:51	2:21-2:22	32:01-32:12
68	2:23-2:24	0:42	0:52	2:23-2:24	32:13-32:24
67	2:25-2.26	0:43	0:53	2:25-2:26	32:25-32:36
66	2:27-2:28	0:44	0:54	2.27-2:28	32.37-32:48
65	2:29-2:30	0.45	0.55	2.29-2 30	32.49-33:00

Figure 6-1.Physical Readiness Test Scoring Table for Male.

	EXAMPLE 1	EXAMPLE 2	EXAMPLE 3
Event #1	100 pts	100 pts	99 pts
Event #2	100 pts	99 pts	86 pts
Event #3	85 pts	75 pts (F)	80 pts
Event #4	90 pts	79 pts (F)	100 pts
Event #5	79 pts	85 pts	90 pts
Pts	454 pts	438 pts	455 pts
Final Score	79.0% (F)	69.0% (F)	91.0%
(F) = Failure			

Figure 6-2.Scoring Examples.

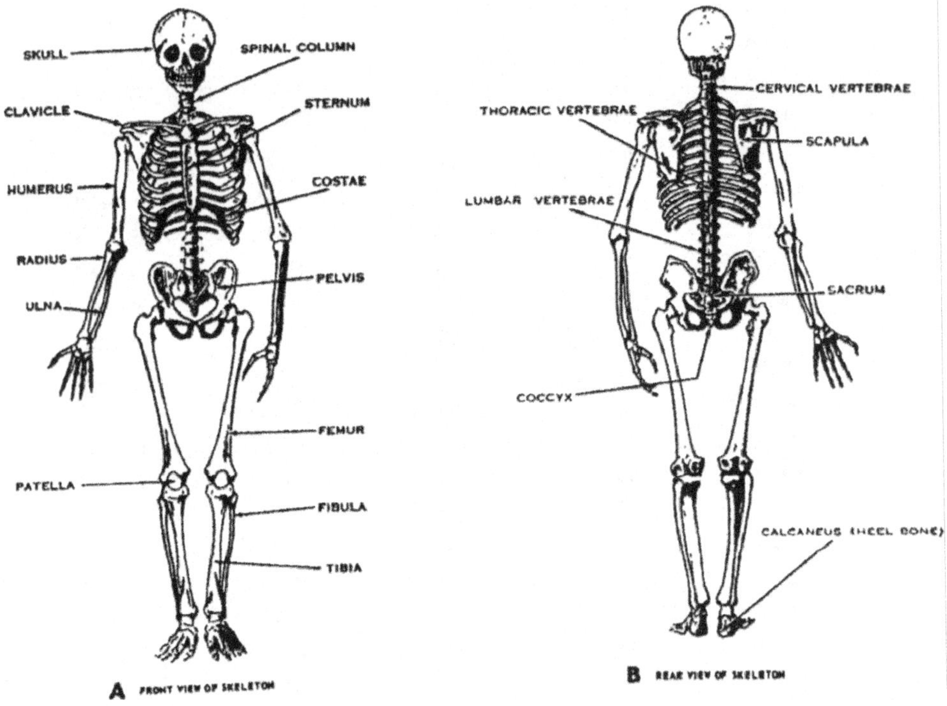

Figure 7-1.Front and Rear Views of Skeleton.

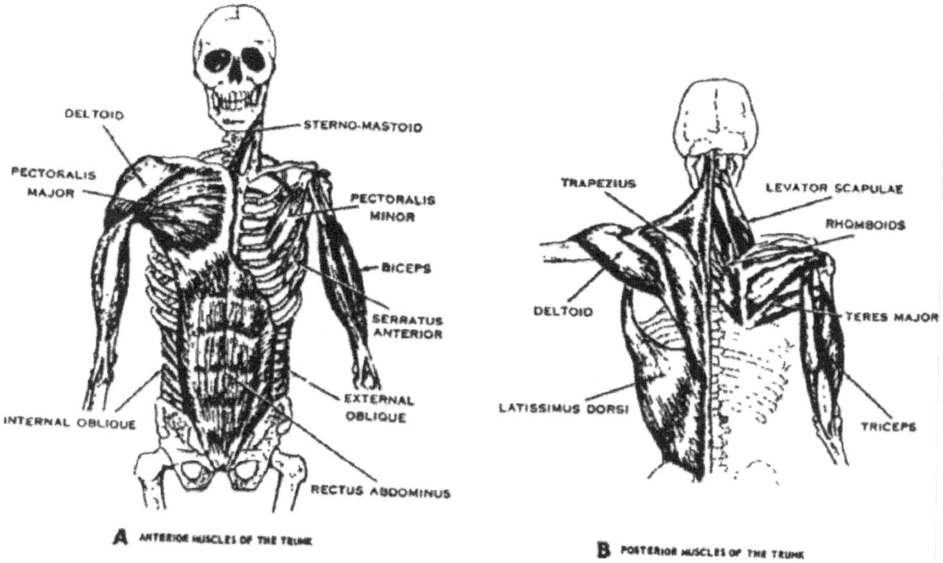

Figure 7-2.Muscles of the Trunk.

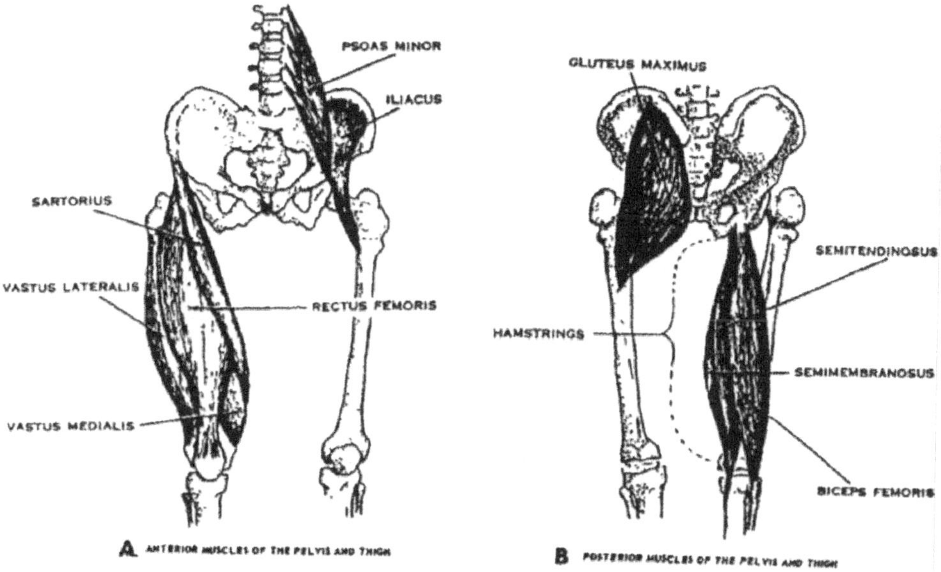

Figure 7-3.Anterior and Posterior Muscles of the Pelvisand Thigh.

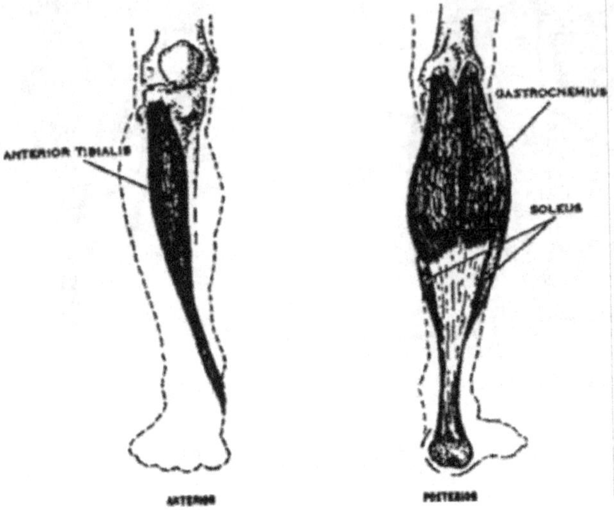

Figure 7-4. Muscles of Lower Leg.

Figure 7-5.Circulation of Blood from Heart to the Body.

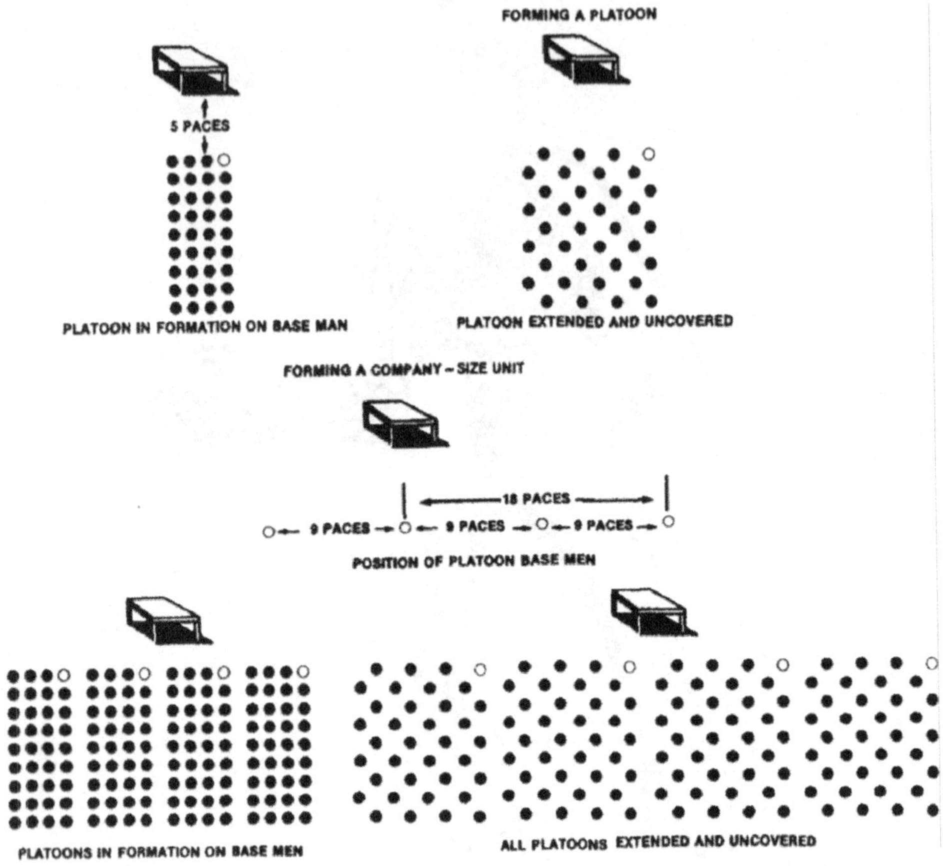

Figure A-1. Forming the Extended Rectangular Formation.

Figure A-2.Circle Formation.

Made in United States
North Haven, CT
05 November 2026

21795726R00329